MOLECULAR BIOLOGY
INTELLIGENCE
UNIT

Noncoding RNAs:
Molecular Biology and Molecular Medicine

Jan Barciszewski
Institute of Bioorganic Chemistry
of the Polish Academy of Sciences
Poznan, Poland

Volker A. Erdmann
Free University Berlin
Institute of Chemistry
Biochemistry and Pharmacy
Berlin, Germany

LANDES BIOSCIENCE / EUREKAH.COM
GEORGETOWN, TEXAS
U.S.A.

KLUWER ACADEMIC / PLENUM PUBLISHERS
NEW YORK, NEW YORK
U.S.A.

Noncoding RNAs: Molecular Biology and Molecular Medicine

Molecular Biology Intelligence Unit

Eurekah.com / Landes Bioscience
Kluwer Academic / Plenum Publishers

Softcover reprint of the hardcover 1st edition 2003

Kluwer Academic / Plenum Publishers, 233 Spring Street, New York, New York, U.S.A. 10013
http://www.wkap.nl/

Please address all inquiries to the Publishers:
Eurekah.com / Landes Bioscience, 810 South Church Street
Georgetown, Texas, U.S.A. 78626
Phone: 512/ 863 7762; FAX: 512/ 863 0081
www.Eurekah.com
www.landesbioscience.com

Noncoding RNA: Molecular Biology and Molecular Medicine edited by Jan Barciszewski and Volker A. Erdmann, Landes / Kluwer dual imprint / Landes series: Molecular Biology Intelligence Unit

ISBN: 978-1-4419-3408-6

Library of Congress Cataloging-in-Publication Data

Barciszewski, Jan.
Noncoding RNAs : molecular biology and molecular medicine / Jan Barciszewski, Volker A. Erdmann.
p. ; cm. -- (Molecular biology intelligence unit)
Includes bibliographical references and index.

1. RNA. 2. Catalytic RNA. 3. Introns.
[DNLM: 1. RNA, Untranslated--genetics. 2. RNA, Untranslated--physiology. 3. Molecular Biology. QU 58.7 B243n 2003]
I. Erdmann, V. A. (Volker A.), 1941- II. Title. III. Molecular biology intelligence unit (Unnumbered)
QP623.B37 2003
572.8'8--dc21

2003012622

CONTENTS

EDITORS

Jan Barciszewski
Institute of Bioorganic Chemistry of the Polish Academy of Sciences
Poznan, Poland
Chapter 1

Volker A. Erdmann
Free University Berlin
Institute of Chemistry
Biochemistry and Pharmacy
Berlin, Germany
Chapter 1

CONTRIBUTORS

Hubert Amrein
Department of Molecular Genetics and Microbiology
Duke University Medical Center
Durham, North Carolina, U.S.A.
Chapter 5

Werner Aufsatz
Institute of Molecular Biology
Austrian Academy of Sciences
Salzburg, Austria
Chapter 10

Jean-Pierre Bachellerie
Laboratoire de Biologie Moleculaire Eucaryote du C.N.R.S.
Université Paul-Sabatier
Toulouse, France
Chapter 12

Marisa S. Bartolomei
Howard Hughes Medical Institute and Department of Cell and Developmental Biology,
University of Pennsylvania
School of Medicine
Philadelphia, Pennsylvania, U.S.A.
Chapter 6

Alexandra Boutla
Institute of Molecular Biology and Biotechnology
Foundation for Research and Technology
Department of Biology
University of Crete
Heraklion/Crete, Greece
Chapter 8

Cristin C. Brescia
Department of Microbiology and Immunology
Medical College of Ohio
Toledo, Ohio, U.S.A.
Chapter 18

Jürgen Brosius
Institute of Experimental Pathology
ZMBE
University of Münster
Münster, Germany
Chapter 11

Anna Campalans
Institut des Sciences du Végétal
CNRS
Gif sur Yvette, France
Chapter 13

Jérôme Cavaillé
Laboratoire de Biologie Moleculaire
Eucaryote du C.N.R.S.
Université Paul-Sabatier
Toulouse, France
Chapter 12

Dana R. Crawford
Center for Immunology and Microbial
Disease
The Albany Medical College
Albany, New York, U.S.A.
Chapter 15

Martin Crespi
Institut des Sciences du Végétal
CNRS
Gif sur Yvette, France
Chapter 13

Abhaya M. Dandekar
Department of Pomology
University of California
Davis, California, U.S.A.
Chapter 9

Kelvin J.A. Davies
Ethel Percy Andrus Gerontology Center
and Division of Molecular Biology
University of Southern California
Los Angeles, California, U.S.A.
Chapter 15

John W. Day
Institute of Human Genetics
Department of Neurology
University of Minnesota
Minneapolis, Minnesota, U.S.A.
Chapter 16

Matthew A. Escobar
Department of Pomology
University of California
Davis, California, U.S.A.
Chapter 9

Alexander Hüttenhofer
Institute of Experimental Pathology
ZMBE
University of Münster
Münster, Germany
Chapter 11

Kriton Kalantidis
Institute of Molecular Biology
and Biotechnology
Foundation for Research
and Technology
University of Crete
Heraklion/Crete, Greece
Chapter 8

Tatsuo Kanno
Institute of Molecular Biology
Austrian Academy of Sciences
Salzburg, Austria
Chapter 10

Dianne Kindel
Department of Molecular Genetics
and Microbiology
Duke University Medical Center
Durham, North Carolina, U.S.A.
Chapter 5

Adam Kondorosi
Institut des Sciences du Végétal
CNRS
Gif sur Yvette, France
Chapter 13

Subhash C. Lakhotia
Cytogenetics Laboratory
Department of Zoology
Banaras Hindu University
Varanasi, India
Chapter 14

John S. Mattick
ARC Special Research Centre
for Functional and Applied Genomics
Institute for Molecular Bioscience
University of Queensland
Brisbane, Australia
Chapter 2

Antonius J. M. Matzke
Institute of Molecular Biology
Austrian Academy of Sciences
Salzburg, Austria
Chapter 10

Marjori A. Matzke
Institute of Molecular Biology
Austrian Academy of Sciences
Salzburg, Austria
Chapter 10

M. Florian Mette
Institute of Molecular Biology
Austrian Academy of Sciences
Salzburg, Austria
Chapter 10

Eric G. Moss
Cell and Developmental Biology
Fox Chase Cancer Center
Philadelphia, Pennsylvania, U.S.A.
Chapter 7

István Papp
Institute of Molecular Biology
Austrian Academy of Sciences
Salzburg, Austria
Chapter 10

Laura P.W. Ranum
Institute of Human Genetics
Department of Genetics, Cell Biology
and Development
University of Minnesota
Minneapolis, Minnesota, U.S.A.
Chapter 16

Peter Schattner
Center for Biomolecular Sciences &
Engineering
University of California
Santa Cruz, California, U.S.A.
Chapter 3

Darren D. Sledjeski
Department of Microbiology
and Immunology
Medical College of Ohio
Toledo, Ohio, U.S.A.
Chapter 18

Maciej Szymanski
Institute of Bioorganic Chemistry
Polish Academy of Sciences
Poznan, Poland
Chapter 1

Martin Tabler
Institute of Molecular Biology
and Biotechnology
Foundation for Research
and Technology
University of Crete
Heraklion/Crete, Greece
Chapter 8

Claude Thermes
Centre de Génétique Moléculaire
CNRS
Gif sur Yvette, France
Chapter 13

Henri Tiedge
Department of Physiology
and Pharmacology
SUNY Health Science Center
at Brooklyn
Brooklyn, New York, U.S.A.
Chapter 11

Mina Tsagris
Institute of Molecular Biology
and Biotechnology
Foundation for Research
and Technology
Department of Biology
University of Crete
Heraklion/Crete, Greece
Chapter 8

Raluca I. Verona
Howard Hughes Medical Institute
and Department of Cell
and Developmental Biology,
University of Pennsylvania
School of Medicine
Philadelphia, Pennsylvania, U.S.A.
Chapter 6

Jörg Vogel
Institute of Cell and Molecular Biology
Biomedical Center
Uppsala University
Uppsala, Sweden
and
Department of Molecular Genetics
and Biotechnology
The Hebrew University-Hadassah
Medical School
Jerusalem, Israel
Chapter 17

E. Gerhart H. Wagner
Institute of Cell and Molecular Biology
Biomedical Center
Uppsala University
Uppsala, Sweden
Chapter 17

Anton Wutz
Institute of Molecular Pathology
Vienna, Austria
Chapter 4

PREFACE

Noncoding RNAs—A Tale for all Seasons

Nucleic acids perform multiple tasks in living cells that range from the storage and transfer of genetic information to the catalysis of biochemical reactions. The relatively simple nucleotide composition of RNA and DNA molecules makes them easy to synthesize and to manipulate outside the confines of cells. Thus researchers may examine the ability of nucleic acids to perform various functions of molecular recognition and catalysis, and these ongoing efforts continue to add to our rapidly expanding knowledge about ligand-binding and catalytic reactions. Scientists are beginning to harness these properties to create new types of receptors and enzymes for use in basic research, biotechnology and molecular medicine (diagnostics and therapeutics).

RNA polymers are known to act as informational, structural and functional molecules in nature. RNA molecules touch many aspects of molecular biology. They are involved as (i) primers in DNA replication; (ii) messenger RNAs that carry genetic information to the ribosomes; (iii) components of the splicesome, the RNA-protein complex that removes introns from precursor mRNAs; (iv) guides which ensure that their target mRNAs specify the appropriate amino acid sequences by directing posttranscriptional nucleotide modifications, insertions and deletions; and (v) catalytic RNAs which assist in various RNA processing events and replication of viral genomes.

RNAs many also constitute drug targets, therapeutic agents and catalytic ribozymes used in a variety of chemical and biochemical applications.

The versatility of RNA function is due in part to its ability to form a variety of distinct tertiary structures. The complexity of structures can very well approach those observed for proteins.

Conformational changes and allosteric transitions in RNA structure are common in RNAs involved in biological processes such as protein synthesis, protein translocation, mRNA splicing, viral and cellular gene expression and viral RNA replication and packing. In some cases, a conformational change is an integral regulatory component of RNA function and can be modulated by interactions with other RNA and protein components. Thus nature has made extensive use of the dynamic character of RNA polymers to create molecules that respond to their local environment.

Basic scientific discoveries underscore the seminal role of RNA molecules in the utilization of genetic instructions and the versatility of these molecules in nature. Different RNA subspecies including ribosomal RNA (rRNA), transfer RNA (tRNA), messenger RNA (mRNA) and noncoding RNA (ncRNAs) are important structural, catalytic and regulatory components of living systems.

Certain RNAs fold to form catalytic centers, whereas others have structures that allow them to make specific RNA-RNA, RNA-DNA or RNA-protein complexes.

The discovery of catalytic properties of RNAs 20 years ago proved that they could catalyze reactions just as proteins do. Long before that it was shown that RNA can also store the scientific information in a similar way to DNA. Thus the realization that RNA combines features of proteins and DNA led to the "RNA World" hypothesis in which RNA molecules dominated all life activities. According to this hypothesis RNA molecules were the first forms of life on Earth, not only functioning as a reservoir of information, but also performing all the enzymatic reactions necessary to maintain life. Gradually, with the increasing complexity of living systems, they relinquished their enzymatic roles to proteins, although some RNA molecules persisted in their catalytic roles up to today as molecular fossils. With the discovery that the ribosome is one such remnant from the RNA world, the accumulating evidence for the catalytic competence of the splicesomal snRNAs makes a serious case for the splicesome to be the next in line to become a legitimate ribozyme. The modern RNA world is far richer than many had previously believed it to be. Several RNA molecules exhibit a function without being translated into proteins such as tRNAs, rRNAs, RNA in ribozymes, high affinity RNAs (aptamers) and small nuclear RNAs from splicesomes.

The newest developments in RNA technology have been elegantly summarized by Dieter Söll (Yale University): "The mind set of those in the RNA field has slowly been transformed from a somewhat pessimistic resignation to near manic optimism by events of the last twenty years."

The variety of RNA genes known today is fairly small relative to protein coding genes, although the number of members within a single RNA gene family can be substantial. Genes encoding housekeeping (constitutive) RNAs (transfer, ribosomal, small nuclear, nucleolar, telomerase etc) have also been described. These genes encode RNAs that lack open reading frames and function as their final product. Small nontranslated RNAs are engaged in a variety of molecular tasks and perform multiple functions in the cell. The genes coding for noncoding RNAs (ncRNAs) do not code proteins as most genes do, but they produce functional RNAs. Rapidly accumulating evidence indicates that such ncRNAs can play critical roles in a wide range of cellular processes from protein secretion to gene regulation. They function in diverse pathways such as dosage compensation, gene imprinting, transcriptional regulation, pre-mRNA splicing and control of mRNA translation.

Prior to genomics, one of the best approaches to understand the physiological role of unknown gene products was to develop mutant strains and analyze their new biological clever for its functions. Such analysis could sometimes take years to accomplish. In the genomic era and beyond, the best starting approach is to use *in silico* methods to compare the deduced sequence of the gene product with those of known functions, and with some luck, one might get to the results in a few hours. However, when the results of these two approaches do not appear to make any sense with each other, new experiments have to be performed.

The ultimate goal of genome projects from bacteria to human is not only to sequence their entire genomes in order to identify their complete set of genes, but also to obtain information as to when and where these genes are being expressed and whether their expression is possibly altered during disease, aging or stress.

There are two lessons that we have learned from sequencing whole genomes: first, many of the characteristics that distinguish bacterial species, such as virulence and metabolic abilities, are encoded within species-specific regions of the genome and, second, in contrast to bacterial genomes, which devote approximately 98% of their content to coding regions, only 1.5% of the human genome is coding DNA, the remaining "junk" being made up of transposons, viruses and noncoding parts.

The recent explosion of available genome sequences has initiated the need to improve an ability to organize and retrieve important sequence and structural elements in a fast, efficient and accurate manner. Currently most genomic studies focus on annotation and functional assignment of unknown proteins corresponding to open reading frames (ORFs) in the genomic sequences. In eukaryotes a small portion of the genome encodes information for amino acid sequences of proteins but the noncoding sequences, an unknown fraction, play a vital role in regulating gene expression. Noncoding RNAs make up 98% of the genome output in humans and have been suggested to act as key players in the regulation of all eukaryotic cells.

The noncoding component of genomes and the human genome in particular is receiving increased attention from biologists because of its putative role in regulation of transcription, DNA replication, chromosome pairing and chromosome condensation. It has been observed that some ncRNAs only exist in particular lineages or species, suggesting that nature must have independently created these new genes during evolution. There are some observations that give rise to questions of how new RNA genes originated and evolved. One of the possibilities of exon shuffling in protein coding genes has not been confirmed for the extant RNA genes. A direct answer might be obtained by examining an RNA gene that originated recently, because the evolutionary characteristics related to the molecular processes of origin and subsequent evolutionary dynamics of new genes may be observable in such a gene.

Emerging data indicate that a potentially important class of genes has so far escaped our detection. There is a large group of functional RNA molecules which remains hidden between and sometimes within protein coding regions (introns) and are so far unaccounted for. Yet, genome wide search of ncRNAs should not be omitted in any functional analysis. How can we find the regulatory RNAs without strong structural properties such as we observe for open reading frames? A number of computational approaches to identify genes and the cis-regulatory sequences controlling their expression have published. A promising class of methods for both gene and cis-regulatory prediction is based on comparative sequence analysis. These approaches work because functionality in evolution is conserved much more so than nonfunctional sequences. It allows us to address some questions about structure and evolution of noncoding sequences, such as: (i) what is their fraction and density, (ii) what is the length distribution of ungapped, conserved noncoding regions, (iii) what is the rate and pattern of point substitution in conserved noncoding blocks and (iv) what is the rate and pattern of index substitution in conserved noncoding strings.

This book comprises an eclectic series of 18 chapters on different aspects of noncoding RNAs from the points of view molecular biology, biotechnology, bioinformatics and molecular medicine. It begins with a general overview on structure, properties and functions of RNAs. The starting point is a discussion on the role and origin of noncoding RNAs. Most studies based on molecular data refer to the position of genomes encoding proteins. However noncoding regions constitute a notable portion of the metazoan nuclear genome. Two chapters describe different approaches to identify noncoding RNAs in genomes. Two chapters deal with chromatin associated RNAs which form a unique class of RNAs. Despite mechanistic differences the mammalian Xist and fly roXs RNAs are RNAs that have been proved to be associated with chromatin over the lengths of entire chromosomes.

There is a chapter on the H19 gene, which is one of the most extensively studied imprinted genes. It has been implicated in transcriptional control and tumorigenesis. RNA-directed methylation of DNA plays a key role in these processes. Recently very popular microRNA or interference RNA have been discussed in relation to silencing in eukaryotes. There are several chapters dealing with ncRNAs induced by different type of stress in prokaryotes and higher organisms. The functions of small nucleolar RNAs and their roles in modification of nucleic acids are also discussed in one chapter. Last but not least involvement of noncoding RNAs in pathogenesis is also summarized.

Each chapter begins with a discussion of the biological context of the topic, includes illustrations, provides current references from the primary literature and concludes with a brief summary. Most importantly, the writing is clear and to the point, allowing reading of the chapters without familiarity the original literature.

Understanding how the expression levels of thousands of genes are regulated at all times in the life of a cell and what is a role of ncRNAs in these processes remains one of the greatest challenges of molecular biology.

Linus Pauling has said that one needs to have many ideas to have a few good ones. It seems that many new, very good ones are presented in the book. In the near future it will turn out that some of them are brilliant insights that change current thinking about how the expression levels of thousands genes are regulated at all times in the life of cells by RNAs. The data presented clearly show that RNA molecules are the key players in signaling networks, and the future will certainly bring lot of surprises to the RNA researchers. Thus, there are currently no questions whether noncoding RNAs exist. The question is how ncRNAs fold, interact with other nucleic acids and proteins, and participate in remodeling of chromatin and RNA-protein particles.

We would like to thank authors of the chapters for their prompt response, which represents the first summary of our current knowledge of noncoding RNAs.

We wish to acknowledge the support of the National Foundation for Cancer Research.

We hope that the book may serve as the basis for vivid discussions between scientists in the fields of biotechnology and medicine and that it may eventually contribute to new applications of RNA molecules in these fields.

We also thank the Staff of Landes Bioscience for their cooperation.

Jan Barciszewski
Volker A. Erdmann

Chapter 1

Riboregulators: An Overview

Maciej Szymanski, Volker A. Erdmann and Jan Barciszewski

Introduction

One of the most important aspects of functioning of living organisms is the regulation of gene expression. Resolving the mechanisms regulating transcription of particular genes is crucial for understanding biological phenomena connected to changes in environmental conditions, development and diseases. Regulatory patterns of gene expression are extraordinarily diverse and complex, yet for each gene there is a proper signal that determines when it should be turned on and how much of the product is needed. The gene regulation is remarkably flexible both in response to new conditions or to accommodate evolutionary demands.

The early idea concerning molecular organization of the living systems was put forward in the simple form of the 'central dogma of molecular biology' by Francis Crick in 1958.[1] It defined a general pathway for the expression of genetic information stored in DNA, transcribed into transient messenger RNAs and translated on ribosomes with help of adapter RNAs (transfer RNAs) to produce proteins that were supposed to perform all enzymatic and structural functions in the cell. Accordingly, ribonucleic acids were regarded as molecules performing basically accessory functions and a number of protein-coding genes defined the complexity of an organism. Genes were viewed as pieces of DNA, encoding proteins following a simple 'one gene – one protein (polypeptide)' rule. It was later demonstrated that in addition to the protein-coding fragments (exons), the genes may also contain untranslated portions (introns) that are excised from primary transcripts during splicing. In many cases, though, alternative splicing provides means of production of multiple variants of a protein from the same transcription unit.[2,3] The next significant breakthrough came with findings that some cellular RNAs perform catalytic functions. Ribonuclease P RNAs and self-splicing introns were shown to be capable of mediating complex chemical reactions involved in RNA maturation pathways in the absence of proteins.[4,5] Subsequently, the technique of in vitro selection (SELEX) allowed isolation of RNA molecules with a wide range catalytic functions and ligand binding properties.[6,7] All these discoveries contributed to the realization that the chemical properties of RNA and its ability to form complex tertiary structures are the basis for many different roles which were thought to be exclusive domains of proteins. Additionally, advances in ribosome research clearly indicate that rRNAs do not merely form a structural scaffold for ribosomal proteins but play the primary role in a process of peptide bond formation and many other aspects of ribosome function.[8]

Properties of Ribonucleic Acids

RNA molecules were first identified as key components of the protein synthesis machinery. Messenger RNAs serve as intermediate molecules transferring the information on the primary structures of polypeptides from the DNA to ribosomes. About twenty years ago it became

Noncoding RNAs: Molecular Biology and Molecular Medicine, edited by Jan Barciszewski and Volker A. Erdmann.

evident that the cellular RNAs play much more diverse functions which go beyond a somewhat passive role in protein biosynthesis. The discovery of self-splicing introns in precursors of ribosomal RNA in *Tetrahymena* demonstrated the catalytic potential of RNAs.[5] The RNA of ribonuclease P, trimming the 5'-ends of some tRNA precursors, was shown to be the catalytic component capable of multiple turnover enzymatic activity.[4] Subsequently, several other naturally occurring RNA enzymes (ribozymes) involved in RNA processing were identified. They include spliceosomal U2 and U6 snRNAs which take part in the maturation of premRNAs.[9] Recently, the crystal structures of large ribosomal subunits and whole ribosomes from several organisms showed that the catalytic activity of peptidyl transferase resides in the large subunit ribosomal RNA (23/28S rRNA).[8] Although, natural ribozymes identified so far can only catalyze three types of reactions (two kinds of trans-esterification and peptide bond formation) the RNA molecules have much greater potentials.[9]

In addition to catalytically active RNAs, several new classes of RNAs were found to be key players in various cellular processes. These include, among others, small nuclear RNAs involved in pre-mRNA processing, small nucleolar RNAs responsible for rRNA maturation and modification, guide RNAs (gRNAs) directing RNA editing and RNA components of the signal recognition particles (SRP RNA) taking part in protein translocation. The outcome of more than twenty years of intensive research, and exciting discoveries of new processes where novel RNAs were implicated, clearly show that ribonucleic acids play a crucial role in all basic aspects of expression of genetic information.

RNA functions largely depend on the ability of the polynucleotide chains to form higher ordered structures. Complex folded conformations of RNA provide recognition elements for protein binding and are responsible for catalytic activities of ribozymes. RNA tertiary structures can form virtually an unlimited number of highly specific ligand binding sites. RNA can interact with chemically and structurally diverse sets of small compounds of which antibiotics are of special interest, since a majority of ribosome-targeted antibiotics bind to specific sites on ribosomal RNAs. The repertoire of four basic nucleosides (adenosine, guanosine, cytidine and uridine) that constitute building blocks of RNA polymers can be extended by numerous modifications of both bases and sugar residues. Despite many years of intensive studies, higher order structures of many RNAs remain unknown, and the precise mechanisms of RNA catalysis are poorly understood in general.

Coding vs. Noncoding Transcripts

The whole transcriptional output from the genome can be roughly divided into two major parts: protein-coding mRNAs and noncoding RNAs. For many years the majority of the noncoding portion was viewed as 'genetic junk' and regarded mostly as nonfunctional. It seems however, that it may contain additional information or functions which are equally important as the protein-coding genes. The sequence of the human genome, although still incomplete, clearly shows that its protein-coding portion accounts only for about 2% of its total length.[10] In other eukaryotic genomes, we observe that the higher the complexity of a given organism, the higher the contribution of nonprotein-coding portions in the genome. In yeast only about 30% of genomic DNA does not encode proteins, whereas in *Drosophilae* the noncoding part accounts for approximately 75% of total genome[11] (compare Table 1).

The determination of complete sequences of several eukaryotic genomes were also disappointing in another respect. The numbers of protein coding genes is much lower than initially expected. Genomic DNA of nematodes and insects contain only twice as many genes as those of yeast or some bacteria and in mammals the number is doubled relative to invertebrates.[10,11] Their variability between mammalian species is also very small. About 99% of human protein-coding genes have orthologs within the mouse genome. The variations within protein-coding genes between individuals account only for ~0.3% of all differences. It is therefore

Table 1. Contribution of protein-coding and noncoding regions in selected prokaryotic genomes

Organism	Genome Length (kbp)	Protein-Coding Part (%)	Noncoding Part (%)	Number of Genes
Eubacteria				
U. urealyticum	751	88	12	577
E. coli	4639	84	16	4000
M. leprae	3268	73	27	2584
Archaea				
P. horikoshii	1739	87	13	1636
M. jannaschii	1665	83	17	1599
S. solfataricus	2992	77	23	2610
Eukaryota				
E. cuniculi	2900	90	10	2000
S. cerevisiae	12000	71	29	5651
S. pombe	12463	57	43	4824
A. thaliana	115410	29	71	25500
C. elegans	97000	27	73	18424
D. melanogaster	180000	13	87	13600
H. sapiens	3000000	2	98	30000-40000

clear, that the protein sequences alone can not be responsible for observed variability between and within species. The diversity of life forms is therefore most probably encoded in the non-protein coding and largely unexplored portions of the genomic DNA.

In the case of mammals, at least 50% of genomic DNA contains repeated sequences mostly of transposon origin. Intervening sequences add another level of complexity in eukaryotic genomes. Introns can constitute a considerable portion of primary transcripts and are present in most of protein-coding genes in higher eukaryotes (see chapter by Mattick). Introns of certain genes contain precursors of snoRNAs.[12,13] Assessing the coding and noncoding sequences in eukaryotic genomes is more difficult than in prokaryotes. In contrast to bacteria, where usually only one strand of DNA is utilized as a template in the process of transcription, in eukaryotes, there are numerous documented cases of overlapping genes transcribed in opposite directions, often giving rise to noncoding RNAs.

The nonprotein-coding transcripts can be divided into two major groups: housekeeping and regulatory RNAs (Fig. 1). The housekeeping transcripts include constitutively expressed RNA species indispensable for correct functioning of the cell. This group includes all classes of RNAs involved in processing primary transcripts (snRNA, snoRNA, RNaseP RNA, gRNA), translation (tRNA, rRNA) and quality control of translation (tmRNA). Here also belong telomerase associated RNAs, vault RNAs and SRP RNAs. The regulatory RNAs or riboregulators constitute a much more diverse group. In a broad sense, it comprises transcripts involved in specific regulation of various aspects of gene expression both in prokaryotic and eukaryotic cells. The levels at which regulatory RNAs can influence cellular processes range from transcriptional regulation to control of translation. The mechanisms of action also vary. The most obvious way an RNA molecule can influence an expression level of a gene is by antisense interaction with its mRNA. Among regulatory RNAs we find also modulators of transcription factors and RNAs involved in modification of the chromatin structure, thus influencing the gene expression on the very early stages. Riboregulators come in different sizes. There are tiny, ~20 nt long microRNAs, ~100-200 nt long bacterial post-transcriptional regulators, and over 10 kb long transcripts in mammals.

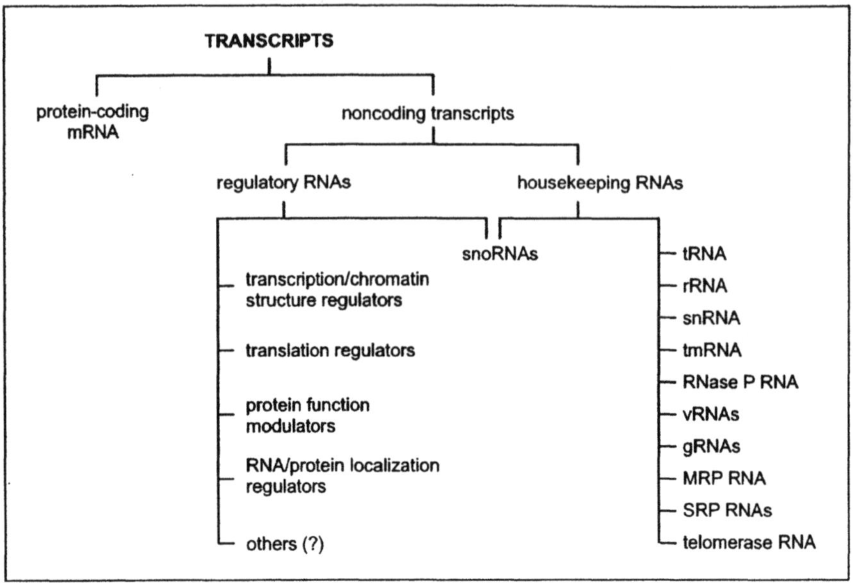

Figure 1. The transcriptome. Protein-coding and noncoding RNAs.

Most of the noncoding RNAs known to date were identified experimentally and, with the exception of most classes of housekeeping RNAs, there are no clear criteria that could be used to search for novel ncRNAs in the genomic sequences The question about the number of noncoding RNAs remains open.

Functions of RNA Regulators

It is now evident that, in all organisms the regulatory processes involving noncoding RNA molecules are very common. One of the reasons why RNA is so well suited for this purpose is an economy both on the level of single cell and the macroevolution of molecular systems. The synthesis and degradation of RNAs requires much less energy than production and destruction of proteins. Lower stability of RNA when compared to proteins is also a bonus, since the regulatory molecule used as a transient signal should be rapidly destroyed. In many cases, riboregulators may exert their action by base pairing with complementary sequences in target RNA. The protein would require a complicated RNA-binding domain.

In eukaryotes, transcriptional regulation of gene expression is often achieved on a level of chromatin structure. It has been known for some time that RNA is an abundant component of the chromatin, but its presence was attributed to nascent transcripts or RNAs involved in their processing and modification. Recent findings, suggest that some RNA molecules may play a role in the transcriptional activity of large regions of chromosomes (see chapter by Matzke et al). One genetic phenomena, often associated with noncoding RNA transcripts, is genetic imprinting which involves specific inactivation of one copy of a locus depending on its parental origin. Most mammalian imprinted genes occur in clusters that often contain noncoding RNA genes. Their expression from one parental allele often correlates with repression of linked protein coding genes, which suggests that ncRNAs are involved in silencing mechanisms (see chapter by Verona). Mammalian Xist RNA and *Drosophilae* roX-1 and roX-2 RNAs are involved in processes of X-chromosome inactivation and dosage compensation, respectively, required for equalization of levels of transcription of X-chromosome linked genes in male (XY) and female (XX) cells. In both cases, noncoding RNAs recruit protein factors responsible for specific modifications of X-chromosome chromatin (see chapters by Wutz and Kindel for details).

Some of the riboregulators function via simple antisense interactions with other RNAs. Depending on the genomic origin, endogenous antisense RNAs can be broadly divided into two categories: (a) trans-antisense RNAs transcribed from loci distinct from their putative targets and (b) cis-antisense RNAs produced from a complementary strand of the same genomic region as a target RNA.

First identified in *Caenorhabditis*, tiny lin-4 and let-7 RNAs bind to complementary regions of target mRNAs, blocking translation on the post-initiation level without affecting the stability of RNA. This mechanism provides a strict control of timing of the expression of several genes.[14,15] Systematic studies in human, *Drosophilae*, *Caenorhabditis* and *Arabidopsis* revealed that microRNAs represent a large group of trans-antisense RNAs which are common in all higher eukaryotes and probably play a crucial role in controlling the expression of developmentally regulated genes (see also chapters by Moss and Tabler et al).

Another possibility of affecting the expression level by antisense interactions is a mechanism of RNA interference (RNAi), where formation of RNA-RNA duplexes triggers an RNA degradation pathway. Since the enzymes involved in RNAi are well conserved it seems that this mechanism is commonly used in all eukaryotes[16] (see also chapter by Escobar).

Cis-antisense RNAs are also common in eukaryotes. In some instances they are associated with clusters of imprinted genes and their precise function is unknown. It is, however, well established that disruption of several of them results in severe genetic disorders. Mutations in cis-antisense RNA overlapping a Kelch-like 1 (KLHL1) gene in humans are found in patients with spinocerebellar ataxia type 8[17,18] (see also chapter by Ranum and Day). Studies on a *HFE* locus, implicated in a hereditary haemochromatosis, revealed that its expression produces an mRNA encoding an MHC class I-like protein and an antisense polyadenylated noncoding transcript. In vitro assays demonstrated that the antisense transcript represses translation but the precise mechanism in vivo is not known.[19]

In bacteria, there are several regulatory RNAs whose binding to complementary regions in target mRNAs provides a very specific and efficient regulation of gene expression on the level of translation. The effect can be either repression or stimulation depending on the location of regulatory RNA binding site. One of the best studied is an 87-nucleotide long DsrA RNA that is responsible for stimulation of translation of the stress response sigma factor (rpoS).[20] Binding of DsrA RNA to the 5'-untranslated region of rpoS mRNA competes with a secondary structure within the rpoS mRNA, that serves as a cis-acting inhibitor of translation.[21] On the other hand, DsrA interacts with both 5'- and 3'-portions of the open reading frame of H-NS mRNA blocking its translation.[22,23] DsrA has therefore a twofold influence on gene expression pattern by stimulating rpoS-stimulated and H-NS-repressed genes. There are several other genes that show sequence complementarities with DsrA RNA and seem to be very good candidates for post-transcriptional regulation. The position of DsrA binding relative to the translation start site, would determine its stimulating or repressing effect.[23] For detailed reviews on bacterial noncoding RNAs see also chapters by Wagner and Sledjeski.

Some of the noncoding RNAs may affect the activity of proteins. An RNA molecule binding to a protein can influence its structure and enzymatic or a ligand binding activity. There are three well documented cases of a regulation of gene expression by noncoding RNAs interacting with proteins.

In *E. coli* an abundant RNA species, 6S RNA, was identified over thirty years ago. For a long time its function was rather elusive, since there were no distinct phenotypes resulting from its over expression or knockout mutations. Recently, 6S RNA was found to form stable complex with σ^{70} holoenzyme of RNA polymerase thus reducing the levels of transcription of σ^{70}-dependent genes in stationary growth phase in *E. coli.*[24]

Another bacterial RNA, a product of the OxyS gene, expressed in response to oxidative stress is a negative regulator of the stress response sigma factor (rpoS). OxyS RNA binds a Hfq protein that is required for rpoS mRNA translation.[24] The same RNA can act as a negative regulator of translation of the *fhlA* mRNA using the antisense mechanism blocking the ribosome binding site.[26]

In mammals an RNA-dependent modulation of protein function was observed in the case a of steroid receptor activator SRA RNA (SRA RNA). It has been found to function as a positive regulator of several steroid hormone receptors , including receptors for androgens, estrogens, glucocorticoids and progestins. It is produced in several tissue-specific variants and mutations within the potential open reading frame did not affect its activity.[27]

Another noncoding transcript, 7SK RNA, was found to regulate the activity of RNA polymerase II. 7SK RNA forms a specific complex with the positive transcription elongation factor (P-TEFb) and inhibits its kinase and transcriptional activity.[28,29]

Noncoding RNAs were also implicated in regulation of subcellular localization of RNA and proteins. In amphibian oocytes, a prerequisite for a normal embryo development is a correct distribution of maternal mRNAs to animal and vegetal regions. At early stages of cogenesis, noncoding *Xlsirt* transcripts, containing 3 to 13 repeats of 79-81 nt long elements are localized to the vegetal cortex where they probably facilitate anchoring of other RNAs.[30,31] In *Drosophilae* a noncoding *Pgc* RNA was found to be responsible for the stability of polar granules composed of maternally produced molecules required for differentiation of pole cells into functional germ cells.[32] The nuclear form of hsr-ω RNA is involved in regulation of the distribution of heterogeneous nuclear RNA binding proteins (hnRNPs) within the nucleus[33] (see also chapter by Lakhotia).

One of the largest classes of noncoding RNAs constitute small nucleolar RNAs. A majority of snoRNAs identified to date perform housekeeping functions. They are involved in the post-transcriptional modification (2'-O-methylation and pseudouridilation) of ribosomal and small nuclear RNAs and prerRNA processing. Vertebrate telomerase RNA is also a member of the H/ACA Box class of snoRNAs.[34] There is however a growing number of developmentally regulated and tissue-specific snoRNAs with possible regulatory functions (see chapters by Bachellerie and Brosius).

In addition to noncoding transcripts which are implicated in clearly identifiable processes, there are numerous reports about RNAs for which no particular function can be ascribed. Specific noncoding RNAs have been found to be expressed in specific tissues or at certain stages of development where they may play some yet unidentified regulatory functions. The BORG RNA whose expression is induced by bone morphogenetic/osteogenic proteins (BMP/OP) was proposed to play a key role in osteoblasts differentiation, but its precise function is unknown.[35] *DD3* and *PCGEM1* RNAs are over expressed in prostate cancer cells.[36,37] Some of the activated $CD4^+$ T cells express a 17 kb long *NTT* RNA gene of unknown function which may be involved in interferon γ receptor expression regulation.[38] There is a group of noncoding transcripts whose expression is triggered by stress conditions like hydrogen peroxide treatment (adapt33/adapt15),[39,40] DNA damage (gadd7)[41] or hypoxia (aHIF)[42] or heat shock[43] (see also chapter by Crawford). There is also a growing number of noncoding RNAs involved in the development and the response to stress conditions in plants (see chapter by Crespi et al).

Perspectives

The lesson learned from the analyses of eukaryotic genomes is that the current approach to annotation of genomic sequences which basically describes only the protein-coding portion is not sufficient for understanding the genetic information as a whole. The growing number of cellular processes involving nonprotein-coding transcripts clearly shows that noncoding RNAs play a much more profound role than realized so far (Table 2). While, the prediction of

Table 2. Bacterial and eukaryotic regulatory noncoding RNAs. For sequences and references one can refer to the ncRNAs database[44]

RNA	Function
Bacteria	
6S	modulator of RNA polymerase activity
DsrA	translational repressor of *hns* and activator of *rpoS* genes
OxyS	RNA produced in response to oxidative stress and regulates expression of multiple genes
RprA	translational activator of *rpoS*
MicF	stress induced RNA, translational repressor of an outer membrane porin *ompF*
DicF	translational repressor of a cell-division gene *ftsZ*
CsrB	CsrA antagonist
UptR	RNA implicated in extracytoplasmic toxicity suppression
RyhB	regulator of discoordinate expression of genes encoded within the *sdhCDAB* operon
spot 42	regulator of discoordinate expression of genes encoded within the galETKM operon
RNAIII	virulence regulator in *Staphylococcus aureus*; influencing expression of multiple genes
BS203	*Bacillus subtilis* noncoding RNA expressed from yocI-yocJ intergenic region
Ibi	Acm1 phage-encoded RNA interfering with lipopolysaccharide synthesis
Eukaryotes	
Xist	mammalian noncoding RNA responsible for chromatin remodeling and X-chromosome inactivation
Tsix	antisense transcript from Xist locus
Enox	noncoding transcript expressed from both active and inactive X chromosomes
roX-1, rox-2s	noncoding RNAs responsiblke for dosage compensation in *Drosophila*
MHMs	noncoding transcripts from the hypermethylated regions of chicken Z chromosome
TTY2s	Y-chromosome linked testis-specific transcripts expressed in human fetal testis and adult kidney
H19	imprinted oncofetal RNA implicated in cancer progression, angiogenesis, and metastasis
Rian	maternally expressed mouse RNA, located exclusively in nucleus
KCNQ1 (KVLQT1)	a gene encoding an alpha subunit of the of the voltage gated K^+ channel, alternative splicing produces two untranslated RNAs
LIT1	imprinted antisense transcript within the KvLQT1 locus, preferentially expressed from the paternal allele in most human tissues
GNAS1as/Nespas	antisense transcripts from the GNAS1 locus
IPW	paternally expressed noncoding RNA disrupted in Prader-Willi syndrome
AIR	paternally expressed noncoding RNA required for silencing of the Igf2r Slc22a2 Slc22a3 gene cluster
MESTIT1 (PEG1-AS)	paternally expressed noncoding RNA involved in the regulation of MEST expression during development.
G8	RNA responsible for establishment of thermotolerance in *Tetrahymena thermophila*
adapt33, adapt15s	ncRNAs induced by treatment with hydrogen peroxide
gadd7	transcript associated with response to DNA damage and other stress conditions causing growth arrest

Continued on next page

Table 2. Continued

RNA	Function
AHIF	natural antisense RNA of the hypoxia inducible factor A, overexpressed in certain renal carcinomas and lymphocytes under anaerobic conditions
CR20	cucumber gene repressed by cytokinin, function unknown
TPSI1/MT4	family of plant genes induced by phosphate starvation and cytokinins; responsible for systemic phosphate starvation response
SRA	a coactivator of several steroid receptors associated with steroid receptor coactivator 1 (SRC-1)
7SK	inhibitor of kinase and transcriptional activity of the positive transcription elongation factor (P-TEFb)
BC1	RNA polymerase III transcripts expressed in the nervous tissue of rodents
Bsr	expressed in rat central nervous system
Ntab	expressed in rat central nervous system
BC200	RNA expressed in nervous cells of primates
Ks-1	RNA expressed in a honeybee central nervous system
Xlsirt	*Xenopus* oocyte-specific RNAs involved in regulation of maternal RNA distribution
hsr-ω	heat shock induced and developmentally regulated RNA in *Drosophilae*
BORG	Bone morphogenetic proteins (BMPs)/osteogenic proteins (OP) responsive RNA
CIOR	C-type natriuretic peptide-inducible osteoblastic RNA involved in stimulation of mineralization
NTT	17 kb long noncoding transcript expressed in certain activated $CD4^+$ T-cells
G90	1.5 kb-long RNA highly expressed in mouse small intestine and at lower levels in large intestine, testis and kidney
DD3, PCGEM1	human prostate specific genes overexpresed in prostate cancer cells
Hoxa11as	antisense transcript from the Hoxa11 homeobox gene involved in limbs and caudal body development
7H4	an RNA specifically expressed in the endplate zone of skeletal muscle, upregulated in early postnatal development and after denervation
CMPD associated RNA	testis specific RNA mapped to 17q breakpoint, suggested to play a role in differentiation and probably responsible for sex reversal in campomelic dysplasia (CMPD) patients.
DGCR5	a product of a gene disrupted by a balanced translocation associated with DiGeorge syndrome m— the ADU breakpoint, this gene produces several alternatively spliced RNA species during mouse and human embryogenesis.
Mei, Meus	RNA involved in control of meiosis in yeast
PAN	polyadenylated nuclear RNA encoded by human Herpesvirus 8, may be responsible for inhibition of splicing of the host genes by hybridization with a short region of U12 snRNA
NCRMS	alternatively spliced noncoding RNA in rhabdomyosarcoma (RMS), shows higher expression level in most alveolar RMS (RMS-A) than in embryonal RMS (RMS-E)
DISC2	a transcript antisense to DISC1; the two genes are implicated in molecular etiology of schizophrenia
ncR-uPAR	noncoding RNA involved in the regulation of transcription of protease activated receptor 1 (PAR-1)
G24	noncoding RNA down regulated by hsp27
G.B7	ncRNA induced by estrogen and progesterone treatment in rat mammary gland
EBER1	Epstein-Barr virus (EBV) encoded inhibitor of RNA-dependent protein kinase
GUT15	plant ncRNA with unknown function

protein-coding genes, based on the presence of open reading frames, polyadenylation signals, conserved promoter regions or splice site signals is relatively simple, although not always entirely accurate, this is not the case for most functional RNA genes. Moreover, homologous proteins can often be identified based on the similarity of amino acid sequences, while RNAs are more likely to conserve their secondary and/or tertiary structures, which does not necessarily imply extensive sequence identity. These factors make a computational approach to the identification of noncoding RNAs a difficult task (see chapter by Schattner).

The ultimate goal of genome sequencing and annotation projects is to gain knowledge not only on the DNA sequence itself, but also on the factors that influence the expression of information contained therein. We are still far away from understanding of intricate networks of gene expression that result in precise organization of living organisms. Recent discoveries clearly show that ncRNAs and the diversity of possible functions they perform may provide answers to many yet unexplained phenomena.

References

1. Crick F. Central dogma of molecular biology. Nature 1970; 227:561-563.
2. Hastings ML, Krainer AR. PremRNA splicing in the new millennium. Curr Opin Cell Biol 2001; 13:302-309.
3. Graveley BR. Alternative splicing: increasing diversity in the proteomic world. Trends Genet 2001; 17:100-107.
4. Frank DN, Pace NR. Ribonuclease P: unity and diversity in a tRNA processing ribozyme. Annu Rev Biochem 1998; 67:153-180.
5. Doherty EA, Doudna JA. Ribozyme structures and mechanisms. Annu Rev Biophys Biomol Struct 2001; 30:457-475
6. Sucheck SJ, Wong CH. RNA as a target for small molecules. Curr Opin Chem Biol 2000; 4:678-686.
7. Caprara MG, Nilsen TW. RNA: Versatility in form and function. Nature Struct Biol 2000; 7:831-833.
8. Moore PB, Steitz TA. The involvement of RNA in ribosome function. Nature 2002; 418:229-235.
9. Doudna JA, Cech TR. The chemical repertoire of natural ribozymes. Nature 2002; 418:222-228.
10. Venter JC. The sequence of the human genome. Science 2001; 291:1304-1351.
11. Adams MD. The Genome Sequence of *Drosophilae melanogaster*. Science 2000; 287:2185-2195.
12. Cavaille J, Buiting K, Kiefmann M et al. Identification of brain specific and imprinted small nucleolar RNA genes exhibiting an unusual genomic organization. Proc Natl Acad Sci USA 2000; 97:14311-14316.
13. Tycowski KT, Steitz JA. Noncoding snoRNA host genes in *Drosophilae*: expression strategies for modification guide snoRNAs. Eur J Cell Biol 2001; 80:119-125.
14. Lee RC, Feinbaum RL, Ambros V. The *C. elegans* heterochronic gene lin-4 encodes small RNAs with antisense complementarity to lin-14. Cell 1993; 75:843-854.
15. Reinhart BJ, Slack FJ, Basson M et al. The 21-nucleotide let-7 RNA regulates developmental timing in *Caenorhabditis elegans*. Nature 2000; 403:901-906.
16. Brandl S. Antisense-RNA regulation and RNA interference. Biochim Biophys Acta 2002; 1575:15-25.
17. Benzow KA, Koob MD. The KLHL1-antisense transcript (KLHL1AS) is evolutionarily conserved. Mamm Genome 2002; 13:134-141.
18. Nemes JP, Benzow KA, Moseley ML et al. The SCA8 transcript is an antisense RNA to a brain-specific transcript encoding a novel actin-binding protein (KLHL1). Hum Mol Genet 2000; 9:1543-1551.
19. Thenie AC, Gicquel IM, Hardy S et al. Identification of an endogenous RNA transcribed from the antisense strand of the *HFE* gene. Hum Mol Genet 2001; 10:1859-1866.
20. Sledjeski DD, Gupta A, Gottesman S. The small RNA, DsrA, is essential for the low temperature expression of RpoS during exponential growth in *Escherichia coli*. EMBO J 1996; 15:3993-4000.

21. Majdalani N, Cunning C, Sledjeski D et al. DsrA RNA regulates translation of RpoS message by an antisense mechanism , independent of its action as an antisilencer of transcription. Proc Natl Acad Sci USA 1998; 95:12462-12467.
22. Lease RA, Belfort M. Riboregulation by DsrA RNA: trans-actions for global economy. Mol Microbiol 2000; 38:667-672.
23. Lease RA, Cusick ME, Belfort M. Riboregulation in *Escherichia coli:* DsrA acts by RNA:RNA interactions at multiple loci. Proc Natl Acad Sci USA 1998; 95:12456-12461.
24. Wassarman KM, Storz G. 6S RNA regulates *E. coli* RNA polymerase activity. Cell 2000; 101:613-623.
25. Zhang A, Altuvia S, Tiwari A et al. The OxyS regulatory RNA represses rpoS translation and binds the Hfq (HF-I) protein. EMBO J 1998; 17:6061-6068.
26. Altuvia S, Zhang A, Argaman L et al. The *Escherichia coli* OxyS regulatory RNA represses *fhlA* translation by blocking ribosome binding. EMBO J 1998; 17:6069-6075.
27. Lanz RB, McKenna NJ, Onate SA et al. A steroid receptor co activator, SRA, functions as an RNA and is present in an SRC-1 complex. Cell 1999; 97:17-27.
28. Yang Z, Zhu Q, Luo K et al. The 7SK small nuclear RNA inhibits the CDK9/cyclin T1 kinase to control transcription. Nature 2001; 414:317-322.
29. Nguyen VT, Kiss T, Michels AA et al. 7SK small nuclear RNA binds to and inhibits the activity of CDK9/cyclin T complexes. Nature 2001; 414:322-325.
30. Kloc MG, Spohr G, Etkin LD. Translocation of repetitive RNA sequences with the germ plasm in *Xenopus* oocytes. Science 1993; 262:1712-1714.
31. Kloc M, Etkin LD. Delocalization of Vg1 mRNA from the vegetal cortex in *Xenopus* ooctes after destruction of *Xlsirt* RNA. Science 1994; 265:1101-1103.
32. Nakamura A, Amikura R, Mukai M et al. Requirement for a noncoding RNA in *Drosophilae* polar granules for germ cell establishment. Science 1996; 274:2075-2079.
33. Prasanth KV, Rajendra TK, Lal AK et al. Omega speckles - a novel class of nuclear speckles containing hnRNPs associated with noncoding hsr-omega RNA in *Drosophilae.* J Cell Sci 2000; 113:3485-3497.
34. Kiss T. Small nucleolar RNAs: an abundant group of noncoding RNAs with diverse cellular functions. Cell 2002; 109:145-148.
35. Takeda K, Ichijo H, Fujii M et al. Identification of a novel bone morphogenetic protein-responsive gene that may function as a noncoding RNA. J Biol Chem 1998; 273:17079-17085.
36. Bussemakers MJ, van Bokhoven A, Verhaegh GW et al. DD3: a new prostate-specific gene, highly over expressed in prostate cancer. Cancer Res 1999; 59:5975-5979.
37. Srikantan V, Zou Z, Petrovics G et al. *PCGEM1*, a prostate-specific gene, is over expressed in prostate cancer. Proc Natl Acad Sci USA 2000; 97:12216-12221.
38. Liu AY, Torchia BS, Migeon BR et al. The human *NTT* gene: identification of a novel 17-kb noncoding nuclear RNA expressed in activated CD4+ T cells. Genomics 1997; 39:171-184.
39. Wang Y, Crawford DR, Davies KJ. adapt33, a novel oxidant-inducible RNA from hamster HA-1 cells. Arch Biochem Biophys 1996; 332:255-260.
40. Crawford DR, Schools GP, Salmon SL et al. Hydrogen peroxide induces the expression of adapt15, a novel RNA associated with polysomes in hamster HA-1 cells. Arch Biochem Biophys 1996; 325:256-264.
41. Hollander MC, Alamo I, Fornace Jr AJ. A novel DNA damage-inducible transcript, gadd7, inhibits cell growth, but lacks a protein product. Nucleic Acids Res 1996; 24:1589-1593.
42. Thrash-Bingham CA, Tartof KD. aHIF: a natural antisense transcript over expressed in human renal cancer and during hypoxia. J Natl Cancer Inst 1999; 91:143-151.
43. Fung PA, Gaertig J, Gorovsky MA et al. Requirement of a small cytoplasmic RNA for the establishment of thermo tolerance. Science 1995; 268:1036-1039.
44. Erdmann VA, Barciszewska MZ, Szymanski M et al. The noncoding RNAs as riboregulators. Nucleic Acids Res 2001; 29:189-193.

Chapter 2

Introns and Noncoding RNAs:

The Hidden Layer of Eukaryotic Complexity

John S. Mattick

Although it is not yet widely appreciated by the molecular biological community, the vast majority of the transcriptional output of the genomes of the higher organisms is noncoding RNA, composed of introns spliced out from protein-coding transcripts, and separate noncoding RNA transcripts that are developmentally regulated and which may also be spliced. Intronic RNAs comprise around 95% of the average protein-coding transcript in humans, and have high sequence complexity with interesting patterns of conservation, suggesting that these RNAs contain information that is expressed in parallel with protein-coding sequences. In addition there are thousands of noncoding RNA genes, which appear to account for at least half of all transcripts in humans, but most have not been studied, largely because there has been no expectation that such RNAs may be common or important, although evidence is rapidly emerging that they are both. Moreover, it is now evident that there are a number of complex genetic phenomena in the higher organisms, such as RNA interference, cosuppression, transgene silencing, methylation, imprinting, and transvection, which are related through intersecting pathways, and which are mediated by or connected to RNA signaling. It has also recently been shown that intronic and other noncoding RNAs are processed into multiple smaller species (snoRNAs and microRNAs), at least some of which are capable of carrying out trans-acting regulatory functions. It also appears that chromatin architecture is influenced by RNA signals. Taken together the available evidence suggests that, far from being evolutionary hangovers or curiosities, noncoding RNAs are central to the genetic control architecture of the higher organisms, and form a higher order system for gene regulation and gene-gene communication, which enables integration of complex networks of gene activity during eukaryotic differentiation and development, via RNA-DNA/chromatin, RNA-RNA and RNA-protein interactions. In addition, this system (the cis- and trans-acting RNA-based regulatory network) would be expected to have entirely different and generally much more subtle genetic signatures compared to protein coding sequences, and probably lies at the heart of quantitative trait variation and genetic susceptibility to disease, as opposed to the more severe phenotypes associated with loss of protein function.

Introduction

It is only recently that it has started to be recognized that noncoding RNAs (ncRNAs) may have significant roles in cell and developmental biology beyond their involvement in protein synthesis, and that noncoding RNAs have begun to be studied in any systematic way. This is largely due to the fact that for the past half century it has been widely assumed, based on bacterial molecular biology and genetics, that "genes" are synonymous with proteins, i.e.,

Noncoding RNAs: Molecular Biology and Molecular Medicine, edited by Jan Barciszewski and Volker A. Erdmann. ©2003 Eurekah.com and Kluwer Academic / Plenum Publishers.

repositories of protein-coding sequences, apart from those genes which specify infrastructural RNAs (rRNAS, tRNAs, snoRNAs, snRNAs etc.) that are directly or indirectly required for mRNA processing and mRNA translation.

The notion that genes encode (just) proteins (via an mRNA intermediate) is referred to as the central dogma (DNA>RNA>protein). This is essentially true in the prokaryotes whose genomes are almost entirely comprised of closely-packed protein coding sequences with associated 5' and 3' cis-regulatory sequences that control transcriptional activity, translational efficiency and mRNA half-life, although recently it has been realized that prokaryotes do in fact contain a number of noncoding RNA genes, apart from those encoding rRNAs and tRNAs.[1-3] These may number 200 or more in *Escherichia coli*, but account for no more than about 0.5% of the total number of genes and about 0.2% of the transcriptional output. However, the situation is entirely reversed in the higher organisms, wherein the majority of the transcriptional output is noncoding RNA (see below), although this fact is not yet widely appreciated.

The presumed universality of the central dogma and the flow of genetic information was encapsulated by Jacques Monod's famous statement (reported in ref. 4) that "What was true for *E. coli* would also be true for the elephant" and ever since has strongly influenced our ideas of the nature of genetic information and the structure of genetic systems. Although Monod did suggest that RNA itself may have (other) functions, the prevailing paradigm has been that proteins are the important structural and functional components of living systems, although it has also become clear that RNA is at the core of the catalytic events underlying RNA splicing and mRNA translation.[5] The central dogma therefore not only suggests that most genes encode proteins, but also, more subtly and perhaps more insidiously, that proteins are sufficient in themselves to specify and organize the autopoeitic programming of complex biological entities, an assumption that has pervaded molecular biology for decades.

That something might be wrong with this view of the function of genetic systems in the higher organisms was first presaged by the discovery that the genes of the higher organisms are not colinear with their protein products, but in fact mosaics of mRNA sequences (exons) interspersed with intervening sequences (introns), which were separated from the pre-mRNA transcript by splicing. This was probably the biggest surprise in the history of molecular biology.[6] However it was immediately interpreted in terms of the prevailing paradigm it was assumed that the RNA products of the introns were nonfunctional (because they did not code for protein), and therefore some sort of evolutionary hangover, ostensibly dating back to the prebiotic assembly of genes. Introns were then postulated variously to have been maintained by the lack of selection pressure to streamline genomes in slower growing complex organisms (despite the fact that such organisms have only existed for a minority of the four billion year history of the eukaryotes) and/or by their evolutionary value in protein domain shuffling.[7,8] However, protein domain shuffling has clearly also occurred in prokaryotes as evidenced by the modular domain structure of many bacterial signaling proteins.[9] There is no doubt that the exon-intron structure of eukaryotic genes is exploited to produce different protein isoforms in real time by alternative splicing, but this does not mean that this is the essential raison d'être for introns, especially as the length of introns has no bearing on splicing per se. Following from the general assumption that introns are nonfunctional have been a number of subsidiary assumptions, including that the distribution of introns in modern organisms reflects the ancient structure of genes,[8] and that introns are rapidly degraded after excision from premRNA,[10] both of which are probably incorrect.[11,12]

In addition, despite the traditional emphasis on protein-coding genes, protein-coding sequences represent only a tiny minority of the genomic complement of the higher organisms. The majority of the genomic output of the higher organisms is in fact ncRNA. This fact and its implications are just beginning to be considered.

Noncoding RNAs Represent the Majority of Genomic Output in Mammals and Probably All Complex Organisms

Around 97-98% of the transcriptional output from the human genome is non-protein-coding RNA.[13] This estimate is based firstly upon the fact that intronic RNA comprises 95%, on average, of primary protein-coding transcripts (pre-mRNAs),[14,15] and secondly on a range of evidence that suggests that there are a very large number of noncoding RNA transcripts (ncRNAs) whose spliced products do not contain substantial open reading frames (although some of these may encode small bioactive peptides) and that these represent at least half and possibly three quarters of all transcripts. Many of these primary ncRNA transcripts are polyadenylated and also spliced, some alternatively (see below).

The conclusion that ncRNAs account for between half and three quarters of all transcripts in humans and other animals is drawn from a number of independent lines of evidence, including: (i) early hybridization kinetic studies on heterogenous nuclear RNA versus mRNA which indicated a large difference in complexity which can not, in retrospect, be accounted for by introns alone;[12,13,16] (ii) the large numbers of antisense and "intergenic" transcripts that have been reported ad hoc in the literature (see e.g., refs. 17-27), although with some exceptions these have rarely been given names nor studied further; (iii) detailed analyses of particular loci, such as the Gnas imprinted locus in mice and the Bithorax region in *Drosophila*, which show that noncoding RNAs comprise the majority of transcription from these well studied regions, which may well be representative of the genome as a whole;[28-31] (iv) the finding that many of the recently discovered microRNAs are derived from "intergenic" regions that were not previously recognized as being transcribed;[32] (v) the recent study using Affymetrix chips which showed that the numbers of detectable transcripts from human chromosomes 21 and 22 was at least an order of magnitude higher than could be accounted for by the known or predicted exons;[33] and (vi) the realization that there are thousands of clones in well constructed cDNA libraries which appear to be derived from noncoding RNAs (see below). At least some of these RNAs (e.g., from the Gnas locus), which were previously unknown, have been experimentally validated and shown to be imprinted (C. Wells, personal communication).

While some have expressed the view that eukaryotic transcription is inherently sloppy and that there is a large background of nonspecific transcription which could account for some of these observations,[34] in all cases that have been examined these noncoding RNAs have been found to be developmentally regulated, i.e., expressed in a gender, tissue- or cell-specific manner (see below).

Thus it may be that the true number of "genes", as defined as segments of the genome that are transcribed to produce functional information, may be much higher than anticipated, with a significant proportion, and probably the majority in complex organisms, producing noncoding RNAs. This would also resolve the discrepancy between the estimates of human gene numbers based on genome sequence analysis (30-40,000)[14,15] and EST cluster analysis (65-70,000)[35] as the latter appear to contain substantial numbers of sequences that have no open reading frame (see below).

In addition, while protein coding sequences comprise only a tiny fraction (1.5-2%) of the human genome, a very much larger proportion of the genome is in fact transcribed. Since the average human protein coding transcript contains 95% intronic sequences,[14,15] then at least 30-40% of the genome is actually transcribed, a rather astounding fact that is generally unappreciated. If one allows an equal number of nonprotein-coding transcripts of similar average size, then at least 60-80% of the genome is transcribed. (Many noncoding RNAs are of the order of 20kb in length, similar to the average length of protein-coding transcripts. Examples include *Xist*, *NTT* and *bxd*, which range from around 17-26kb in length.[28,36,37]) It may be that over 100% of the human genome is transcribed, if one takes into account the substantial numbers of loci that express overlapping sense and antisense transcripts (see below). Noncoding RNAs also appear to comprise the majority of the transcriptional output in other animals, such as *Drosophila* and *C. elegans*, where the intron:exon ratio is approximately 1:1, and where, at least in *Drosophila*, molecular genetic analyses of well studied loci have revealed a significant proportion of ncRNAs. For example, only three of the seven major transcripts identified in the bithorax-abdominalAB region encode proteins, but all seven are developmentally regulated and the interruption or deletion of the DNA that encodes them has known phenotypic consequences.[28,29] In the puffer fish *fugu rubripes*, which has a minimal genome with little repetitive DNA, the average intron:exon ratio in protein-coding genes is around 2:1, i.e., there is still much more unique sequence information in introns than in protein-coding sequences, despite the minimization of the average intron size in this organism.[38] In addition, conventional protein coding genes occupy only one third of the Fugu genome, despite the conclusion that there appears to be "rapid deletion of nonfunctional sequences" from this genome.[38] It seems highly likely that much of the remainder, wherein protein-coding genes are not present and which has been referred to as "gene bare",[38] or in the case of the human genome as "gene deserts",[15] may in fact be transcribed into ncRNAs.

At the minimum it appears that at least 50%, and probably the large majority, of the human genome, and of the genomes of other complex organisms, is transcribed, primarily into non-protein-coding RNAs, which means that either these genomes are replete with useless transcription, or that the expressed introns and other ncRNAs represent a hitherto overlooked genetic output that has unexpected but important function(s).

Do Introns and ncRNAs Represent a Second Tier of Gene Expression

The clue to understanding the possible significance of the massive amount of non-protein-coding transcriptional output in the higher organisms lies considering the situation from a different perspective, starting with introns. As noted above, when introns were first discovered, it was assumed that because they did not code for protein, they must be nonfunctional. This view was reinforced by the observation that introns are generally less conserved than exons, and that many introns, including those in human genes, are very small. However, promoter sequences of genes, which are well established to be functional, are also less conserved than their associated protein coding sequences, presumably because the constraints on protein structure are much tighter than those of regulatory sequences, be they acting in cis or trans (see below), and because regulatory sequences can (and do) evolve much faster. In addition, it is now reasonably well established that modern nuclear introns are the evolutionary descendants of group II self-splicing introns which invaded eukaryotic genes late in their evolution, and which have since evolved in situ independently of one another, but in parallel with their associated exons.[11,39-47] Thus it is entirely plausible that some introns, but not all, may have evolved to produce functional RNAs, post splicing and further processing (see below), that contribute additional information into the system.[11]

If one entertains the possibility that introns may have evolved to produce transacting RNA signals, the only possible general function of the signals would be networking, since these sequences are expressed in parallel with protein-coding sequences (and with exonic ncRNA sequences—see below). That is, intronic RNAs may convey secondary information into the system about gene expression status, via efference signals that modulate other parts of the molecular genetic network, through RNA-DNA, RNA-RNA and RNA-protein interactions that regulate chromatin architecture (and epigenetic memory), transcription, splicing, mRNA translation and RNA stability/turnover, among others, for most of which there is experimental evidence (see below). This hypothesis also predicts that introns will be processed post-splicing to produce multiple smaller signals (rather than simply degraded to ribonucleotides), which will intersect with many different parts of the regulatory circuitry of the cell, and that some (many) genes will have evolved to only produce (noncoding) RNAs, as higher order regulators in the network, thereby generally explaining the expansion of both intronic sequences and ncRNA genes in the higher organisms. Moreover, this network may have been a fundamental advance in the genetic operating system of these organisms, and may lie at the heart of the programmed responses which direct eukaryotic differentiation and development.[11-13]

Thus, there may be two (interacting) tiers of genetic information that are produced in the higher organisms, and three types of "genes" – those that encode only proteins (such as those encoding some histones and regulatory switches like Sry, but which are rare), those that encode both proteins and associated intronic RNAs, or "efference RNAs" (eRNA), and those that encode ncRNAs, which may or may not also produce associated intronic RNAs, and which may or may not be further processed into smaller signals, like microRNAs (see below).

At face value such a system has enormous advantages over a regulatory circuitry that relies simply on protein feedback loops, especially when attempting to integrate large sets and different levels of gene activity. If this is so, it further suggests that the evolution of a more advanced genetic operating system based on a highly parallel RNA-based communication network may have been the fundamental prerequisite for the emergence of complex organisms. It also implies that the basis of species diversity and quantitative trait variation in complex organisms is primarily embedded in the control architecture of the system, rather than structural variation in the protein components themselves (although this will also contribute). This in turn has considerable implications for understanding the genetics of the higher organisms and the genetic factors underpinning complex traits, which are discussed further later in this article. The following sections will discuss the evidence for a genetic regulatory system based on introns and ncRNAs.

Introns

There are several circumstantial lines of evidence that suggest that introns may have evolved function. Firstly, the relative percentage of the genome that is occupied by introns broadly correlates with developmental complexity.[40,48] Secondly, despite varying loads of repetitive DNAs and transposon-derived sequences, introns are largely comprised of unique sequence of high complexity. This is true in a minimal vertebrate genome (*Fugu rubripes*) as well as in much larger mammalian genomes, such as human.[14,15,38] In the latter case, it is interesting to note that at least some of the repetitive sequences are also transcribed, and may also be evolving and contributing to the regulatory circuitry of the cell.[14] As noted above, not all introns will have evolved function, and many are small, probably vestigial remnants of past insertions. However, even in very compact genomes like *Fugu*, many introns are very large,[38] which is presumably indicative of those that have acquired information.

Thirdly, although they are usually less conserved than their associated protein coding sequences (although there are notable exceptions), intron sequences show intriguing patterns of conservation, often in large blocks, in many types of genes, including those encoding homeotic proteins, T-cell receptors, cytoskeletal proteins and enzymes involved in RNA editing.[49-63] This was not apparent until the onset of large-scale genome sequencing, but comparison of syntenic regions of mammalian genomes shows extensive regions of conservation within introns, separated by regions of low homology, a pattern which is also observed in "intergenic" regions, many of which probably encode ncRNAs.[49] Presumably these areas of conservation reflect conserved sequences whose functions are important in mammalian biology, whereas those that vary may contain genus- or species-specific signals that have evolved in different lineages.

Fourthly, there are known examples of introns that produce functional information, most notably those that encode small nucleolar RNAs (snoRNAs), a group of more than 100 stable RNA molecules which are involved in a variety of functions ranging from 2'-O-methylation and pseudouridylation of various classes of RNAs, through nucleolytic processing of rRNAs to the synthesis of telomeric DNA.[64] All snoRNAs are derived from introns of other genes, about half of which encode proteins related to ribosome or nucleolar biology, including various ribosomal proteins (L1, L5, L7, L13, S1, S3, S7, S8, S13 and others), ribosome-associated proteins (e.g.,eIF-4A), nucleolar proteins (nucleolin, laminin, fibrillarin), the heat shock protein hsc70 and the cell-cycle regulated protein RCC1, among others.[65-71] The remainder, some of which are developmentally regulated, are derived from genes which have no open reading frame, and whose exons may or may not have other functions.[72-75] These observations are perhaps the visible tip of a much bigger iceberg, and may be exemplary in several respects: (i) they confirm that at least some genes have a dual output, in which both exons and introns contain separate functional information, as predicted; (ii) they suggest that some ncRNA genes, especially those that have an intron-exon structure, may be derived from genes (possibly after gene duplication) that have lost protein coding capacity, but have retained or acquired functions associated with their RNA products, derived either from their introns (like snoRNAs) or exons, or both; (iii) snoRNAs are a relatively stable and ubiquitous and stable class of noncoding RNAs, whereas the RNA products that may be processed from other introns, of which there may be tens of thousands, may be more transient and present in much lower individual abundance in the cell, and therefore much more difficult to detect; (iv) snRNAs are processed from introns by specific pathways involving double stranded RNAse III-related enzymes, exonucleases and helicases,[76-79] similar to those involved in RNA interference and the production of microRNAs from other RNA substrates (see below). At least some microRNAs have been mapped to introns of known genes,[32] although whether they are derived from these introns, or from some other transcript across the same region remains to be determined. However, the possibility remains that introns are generally processed after splicing excision from their primary transcripts by these types of pathways, to produce smaller functional species.

Lastly, at least some intronic RNAs are relatively stable[80-82] and not degraded rapidly as has often been assumed.[10,12] While few studies have been done on the fate of intronic RNAs, the available evidence suggests that intronic RNAs are easily detectable in the cell after their excision from the primary transcript, with some being trafficked to particular subcellular locations.[81] Given the example of snoRNAs, it seems more likely that introns, in general, may be processed post-splicing to produce smaller RNAs, many of which may be functional, but which are have not been readily detected in biochemical analyses, because of the small size and complexity of this population. In this context it is also worth noting that adenosine deaminases that act on RNA (ADARs), which are a family of RNA-editing enzymes that convert adenosine to inosine within double-stranded regions of RNA and which play a role in RNA interference,[83] have been shown to edit intronic and other noncoding RNA sequences in both *C. elegans* and human.[84]

Noncoding RNAs

It is not the purpose of this article to provide a catalog of the currently known ncRNAs, as these almost certainly comprise only a tiny subset of the actual number that are expressed in different species. Nonetheless, there are many ncRNAs that have been well documented[85-125] and which give some insight into this poorly explored facet of the genomic output of the higher organisms. These include, as exemplars:

- *Pgc*, which is required for germ cell formation in *Drosophila*;[92]
- *Hsr-ω*, which is heat shock-inducible and appears to carry out house-keeping functions in *Drosophila*;[93]

- *CR20*, whose expression is repressed by cytokinin repressed by gene in cucumber cotyledons;[94]
- *BC200*, which is limited to the anthropoid lineage of the primates and is expressed specifically in neuronal cells;[95]
- *7H4*, *Ntab* and *BC1*, which are specifically expressed in the rat nervous system;[96-99]
- *adapt33*, which is induced by oxidants in hamster HA-1 cells;[100]
- *Bic*, which is expressed as two spliced and alternatively polyadenylated transcripts in lymphoid/hematopoietic tissues, and which is strongly upregulated in certain B-cell lymphomas;[101,102]
- *Hp53int1*, which is encoded within the first intron of the human p53 tumor suppressor gene and is induced during differentiation of myeloid leukemia cells;[103]
- *DISC2*, implicated in the molecular etiology of schizophrenia;[27]
- The Y-chromosome-specific *TTY2* family, which is expressed in human testis and kidney,[104] and other noncoding RNAs present in spermatozoa;[105]
- *Xist/Tsix* and *roX1/2*, which are involved in X chromosome dosage compensation in mammals and insects respectively;[36,106-108]
- a variety of noncoding / antisense RNAs which are expressed from or near imprinted loci;[106,109-112]
- *ncR-uPAR*, which regulates human protease-activated receptor-1 gene during embryogenesis;[113]
- *FGF-AS*, which is differentially spliced to produce coding and noncoding RNAs in different tissues and which regulates the expression of fibroblast growth factor-2;[114]
- *Bsr*, which is preferentially expressed subsets of differentiating cells but not in proliferating cells in the rat central nervous system, especially the pareo- and archicortex, amygdala, thalamus and hypothalamus;[115]
- *NTT*, which is expressed in activated CD4+ T cells;[37]
- *BORG*, which is induced in mouse myoblast cell lines during trans-differentiation into osteoblastic cells following exposure to bone morphogenetic proteins;[116]
- *H19*, which is an abundant hepatic fetal-specific noncoding RNA implicated as a tumor suppressor;[117,118]
- *SCA8*, within which an expansion of a *CTG* trinucleotide repeat underlies the neurodegenerative disorder spinocerebellar ataxia type 8;[25]
- *RMRP*, mutations of which underlie the recessively inherited developmental disorder, cartilage-hair hypoplasia *(CHH)*, with manifestations including short stature, defective cellular immunity, and predisposition to several cancers;[119]
- *Mei* RNA and five *Meu RNAs* induced during meiosis in *Schizosaccharomyces pombe*;[120,121]
- *MHM* RNAs from the chicken Z chromosome which accumulate in the nucleus of female cells adjacent to the *DMRT1* locus;[122]
- *Tts-1*, which is associated with telomeres in the developmentally arrested, long-lived dauer stage of *C. elegans*;[123]

Other examples include the *bxd* locus of *Drosophila* which produces a 27kb transcript early in embryogenesis. *bxd* has a number of large introns, and is subject to differential splicing to give various small (~1.2kb) polyA+RNAs which do not contain any significant open reading frame.[28,124,125] The expression of this transcript is highly regulated during embryogenesis, in a pattern that is partially reflexive of *Ubx* transcript, which encodes the homeiotic protein bithorax.[124,126] A number of *bxd* insertional mutations have no effect on the amount or the size of the *bxd* polyA+RNA, suggesting that this species is irrelevant to the observed phenotypes, and that the real import of the transcription and processing of this gene is to produce intronic RNAs.[125] In addition virtually the entire human and mouse β-globin gene clusters which span over 70kb is transcribed, including the "locus control region" and the intergenic regions between the ε-Gγ-Aγ-δ-β genes, into transcripts that are separate from the globin transcripts themselves.[17] There are also a wide variety of ncRNAs involved in viral infection (see e.g., ref. 127).

The numbers of noncoding RNA genes in mammals may be in fact in the order of tens of thousands, and in all likelihood will outnumber those encoding proteins, perhaps by a factor of two or three. This has been suggested by whole chromosome transcript analyses of chromosomes 21 and 22[33] and others (T. Gingeras, personal communication), and is starting to become evident in the more comprehensive and carefully constructed cDNA libraries, such as those developed by RIKEN.

Most cDNA collections are comprised of incomplete reverse transcripts ("expressed sequence tags" or ESTs) in which it is impossible to conclude that an open reading frame is absent, but RIKEN have taken a systematic approach to characterizing the mouse transcriptome, by cloning and sequencing full-length cDNAs, physical mapping of these sequences to their corresponding genes in the mouse genome, and determining the expression of these genes in different tissues (see https://genomec.gsc.riken.go.jp/matrics/ef/fantom2/). This approach has many benefits including definition of 5' promoter regions and the proper analysis of the sequences and coding potential of genomic transcripts, at least of those that have polyA tails. Of the 33,409 transcription units in the RIKEN Fantom2 database, almost half do not contain a substantial open reading frame (ORF) (greater than 100 codons), and which therefore are candidate ncRNA transcripts. Almost 4,500 of these are strong candidates for ncRNAs by stringent criteria,[128] which is clearly an underestimate of the total, as some known ncRNAs fail these criteria.[128] A large number of these transcripts are represented by more than one independent clone and many have been shown to be differentially expressed in different tissues, and thus are real transcripts, not genomic contamination of the cDNA library. Some of these transcripts may or may not encode small peptides, but in either case they represent new genes expressed from the mouse genome. In addition there are over 2,400 other clones in the Fantom2 database that are antisense to known or predicted protein coding sequences (and hence have been clustered with these sequences as "sense-antisense" pairs rather than with the "unknownESTs).[128] Some of these transcripts, such as previously unknown antisense transcripts from Gnas locus have been experimentally verified and shown to be imprinted (C. Wells, personal communication). The database contains homologs of rat *NTAB, 7H4*, human *NTT* and hamster *adapt33A*.

In addition, it has recently been discovered that both plants and animals contain large numbers of microRNAs (miRNAs),[32,129-136] short RNAs (21-22 nucleotides) that are produced by post-transcriptional processing of ncRNAs by enzymes, including Dicer and RDE homologs, which also participate in the phenomenon of RNA interference, and which involves double stranded RNA binding and cleavage. These microRNAs are derived from precursor RNAs that are sourced from "intergenic regions" of the genome and from sequences that have been annotated as introns,[32,134] although it is not yet known whether (some of) the precursors are intronic RNAs per se or other transcripts that are derived from the region in question. The latter is clearly the case where the microRNA sequence is antisense to the intron, and on the limited evidence available there seems to be no strong bias in the orientation of miRNAs relative to that of the intron. In only one case has the primary transcript which is the likely precursor of a miRNA been identified, the BIC RNA referred to above, and where the miRNA sequence is located in a conserved region of the exons (rather than the introns) of this gene.[102,131] At least some transcripts give rise to multiple miRNAs,[32] and analysis of their sequence strongly suggests that the specificity of miRNA processing is guided by the secondary (stem-loop) structure of the precursor RNA.[32,132] Taken together, this suggests firstly that many genomic regions that have been annotated as "intergenic" (i.e., sequences between protein-coding genes) do in fact contain (ncRNA) genes, and secondly that many RNAs may be processed by pathways involving double stranded RNAses, helicases and exonucleases. This may be the tip of a much bigger iceberg of large numbers of small trans-acting RNAs (such as snoRNAs and miRNAs) that are produced from ncRNA and intronic precursors, and whose incidence has until recently gone unnoticed, because of the complexity of this population and its biochemical characteristics.

miRNAs may have a role in tissue specification or cell lineage decisions.[136] In plants, most miRNAs appear to target protein-coding sequences, frequently of transcription factors, leading to the suggestion that in plants miRNAs have a major function in targeting mRNAs for destruction, perhaps as part of a differentiation program,[135] although others small dsRNAs clearly have a role in directing DNA methylation.[137,138] In animals, many microRNAs are tissue-specific and many are conserved among different species, in some cases over large evolutionary distances (e.g., nematodes, insects, fishes and mammals).[32,131] Some of these RNAs, such as the archetypal *lin-4* and *let-7* in *C. elegans*, have known developmental functions, and act by binding to the 3' end of target mRNAs.[136] On this basis it is believed that animal miRNAs are modulators of target mRNA translation and stability, which is well established in some cases, but since most target RNAs remain to be identified, it is also possible that such RNAs actually transmit a variety of signals and fulfil a variety of functions, including the modulation of chromatin architecture, transcription and alternative splicing (see below).

Origin and Nomenclature of Noncoding RNAs

The classification of ncRNAs is a problem. ncRNAs may be derived from unspliced transcripts, such as NTT,[37] but more often than not are assembled from the exons of spliced genes, such as those encoding *H19* and *Xist*, which may also be alternatively spliced or polyadenylated.[139,140] Various ncRNAs may also be derived from the further processing of such exons, such as the miRNAs that are apparently derived from the *BIC* gene,[102,131] and from the introns of both protein-coding and ncRNA genes, as exemplified by the snoRNAs.

The ad hoc development of this burgeoning field of ncRNAs has introduced nomenclature difficulties, similar in some respects to those that afflict the naming of proteins and protein families. Apart from the traditional classes of noncoding RNAs (rRNAs, tRNAs, snoRNAs, small nuclear RNAs, and spliceosomal RNAs), the naming of which refers either to their function or their subcellular location, recent papers have described new functional ncRNAs (such as *NTT* and *Xist*) which reflect their biology in one respect or another, as well as a variety of new classes of ncRNAs including small interfering RNA (siRNA), small temporal RNA (stRNA), microRNA (miRNA), and efference RNA (eRNA), among others. Some of these classes are related—stRNAs are a subclass of miRNAs,[32] and both are produced by a pathway which overlaps with that which produces siRNAs, which are double stranded small RNAs which mediate RNA interference (the targeted destruction of homologous RNAs). In reality ncRNAs are likely to fulfill a very wide range of functions, by a wide variety of mechanisms, in different cells and organisms, the range of which is only now beginning to be explored, and thus any descriptors or classification scheme should be taken with a grain of salt, especially at this stage. However, if there is an output of functional RNA signals from the introns of genes whose exons encode proteins or other ncRNAs, the term efference RNA (to borrow a term from neurobiology, wherein such secondary signals are required for motor coordination and cognition) may be apt to describe such RNA signals in general, irrespective of their particular function, targets or mechanism of action.[12,13]

A New Definition of a Gene—A Transcription Cluster with cis-Acting Regulatory Elements

Such considerations also require a reassessment of the definition of a gene. Traditionally, the genetic definition of a gene has been based on an inheritable phenotype irrespective of

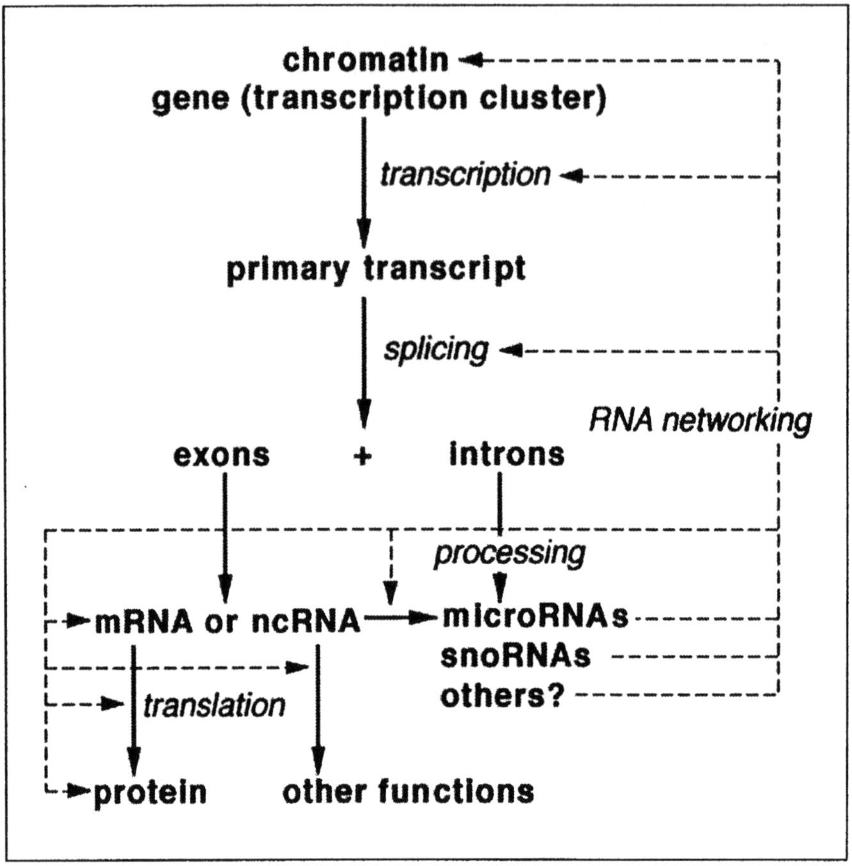

Figure 1. Revised definition of a gene as a transcription cluster, and outline of the flow of expressed genetic information in eukaryotes, showing the potential networks mediated by intronic and exonic ncRNAs.

where the relevant information and variation is encoded, whereas the biochemical definition has tended to emphasize the definition of a gene as a protein-coding region with associated regulatory signals. Given the expanding universe of ncRNAs with functional (i.e., genetic) activity and the fact that many transcripts are initiated from alternate promoters with alternate splicing and alternate polyadenylation signals, the biochemical definition of a gene should perhaps be expanded to encompass a "transcription cluster" (defined as a separate locus producing one or more related transcripts from the same strand) and its associated cis-acting regulatory elements, which may encode one or more related protein isoforms and/or one or more different ncRNAs (such as the miRNA precursor that specifies seven different miRNAs), with different functions in different cells and tissues (Fig. 1).

Complex Genetic Phenomena Involving RNA

The importance of RNA signaling to eukaryotic cell and developmental biology is also indicated by the range of complex genetic phenomena in the higher eukaryotes (including cosuppression, gene silencing, RNA interference, genomic imprinting, DNA methylation, and possibly transvection and position effect variegation) that have been shown to involve RNA, in most cases via intersecting pathways.[137,141-150] Cosuppression, post-transcriptional gene silencing, and RNA interference are all related phenomena involving dsRNA-targeted destruction of mRNAs, and at least in plants, is also linked to gene silencing via methylation of DNA.[148-154] Methylation is RNA-directed in plants and probably also in animals.[137,144] Imprinting in animals involves antisense RNAs and DNA methylation.[106,109,111] The link between DNA methylation, specific antisense RNAs, cosuppression, transcriptional and post-transcriptional gene silencing, and RNAi suggests that RNA-directed DNA methylation is involved in epigenetic gene regulation throughout the eukaryotes.[137] Cosuppression has also been reported in animals[154-160] and, at least in *Drosophila* and *C. elegans*, is dependent on Polycomb-group proteins,[150,156,157] as is transgene silencing,[161] which implicates not only RNA but also the structure of chromatin complexes in cosuppression and gene silencing.[149,150,162,163]

This suggests that trans-acting RNA signals can influence chromatin structure (and hence gene activity)[164] via Polycomb-group proteins, which also influence an apparently unrelated and poorly understood genetic phenomenon, transvection, which itself has been implicated in genomic imprinting and X chromosome inactivation in mammals,[165,166] and which is probably linked to RNA signaling.[12] The Polycomb-group proteins are believed to control the correct spatial expression of key developmental regulators by changing the structure of chromatin,[167-169] although they do not appear to bind DNA per se and their target specificity is not known. They are thought to be responsible for the maintenance of transcriptional regulation by providing a "cellular memory mechanism throughout development",[169-173] but what determines their activity in different lineages is unknown. However, as noted already, Polycomb-group proteins have been shown to influence cosuppression and gene silencing, both of which are RNA-dependent,[145,149,174] leading to the suggestion that trans-acting RNAs may direct the gene-specific binding of Polycomb complexes.[150] Indeed, it has been recently suggested that the chromodomain, which occurs in a variety of chromatin-associated regulatory proteins including Polycomb-group proteins, is an RNA binding module (see below) and in the case of MOF and MSL-3 in *Drosophila* binds the ncRNA *roX2* involved in dosage compensation.[175] In addition, it has been shown that the locus control region and intergenic regions of the human β-globin locus is transcribed in erythroid but not in nonerythroid cells, but can be induced to do so in the latter following transient transfection with a β-globin gene via a process called "transinduction" that depends on the transcription but not translation of the plasmid-encoded transgene.[17] Interestingly the transgene colocalises to the β-globin locus in these cells, which implicates homologous sequence pairing in this process, and is reminiscent of some of the features of transvection.[12,17]

RNA Binding and Signaling Proteins

Proteins which are involved in RNA processing and signaling, and those that may recognize higher order structures involving RNA, probably comprise the largest class of proteins in the higher eukaryotes. A wide variety of "transcription factors" have been shown to have affinity for RNA, including zinc finger proteins such as Sp1 that have a comparable or greater affinity for RNA-DNA hybrids than for double-stranded DNA, which interestingly is strand-specific,[176] and WT1 whose splice variants can apparently bind both to chromatin and to spliceosomes. The Y-box (cold shock) proteins which are also considered to be transcription factors are known to be able to bind RNA.[177-179] The adenosine deaminases that act on dsRNAs (ADARS) and play a role in RNA interference[83] have been shown to contain domains related to winged helix-turn-helix domains and the globular domain of histone H5[180] which binds Z-DNA[180-182] and/or catalyzes its formation.[183] As noted above the chromodomain, which appears to be an RNA binding module[175] and which determines the specificity of binding to particular chromatin locations,[184] is found in a wide variety of chromatin-associated regulatory proteins with a wide variety of other functional domains, including the chromodomain-helicase-DNA binding (CHD) family, the retinoblastoma-binding protein (RBP) family, the histone acetyltransferase (HAT) and histone methyltransferase families, the heterochromatin protein 1 (HP1) family, and the SWI3 family,[162] all of which implicates RNA signaling in the targeting of chromatin complexes and in the positive and negative regulation of gene expression during differentiation and development, as well as in chromatin remodeling, histone methylation and acetylation.

Possible Functions of ncRNAs

ncRNAs will undoubtedly fulfil a wide range of functions within cells, as it would be expected that nature will exploit any interactions and chemistries that have functional value. Many different ncRNAs with different functions in eukaryotic cell and developmental biology have already been described. However, if intronic and exonic ncRNAs are not simply idiosyncratic genetic outputs that have particular functions, but have evolved to form a higher-order regulatory architecture within the higher organisms which conveys a range of efference signals as part of a cascade of programmed response networks during differentiation and development, then one can make a range of specific predictions about the range of functions that these RNAs might fulfill at various levels of the control of molecular genetic activity via RNA-DNA, RNA-RNA and RNA-protein interactions.

Firstly, as discussed above, there is already a strong case that all epigenetic phenomenon, i.e., the cellular memory of developmental history and its effect on current gene activation status, may be directed by trans-acting RNAs, as RNA signaling appears to be involved in DNA methylation, imprinting, transvection, position effect variegation, chromatin remodeling and domain activation and repression, and in histone methylation and acetylation (J. S. Mattick, in preparation).

Secondly, it is possible that trans-acting RNAs are routinely, if not obligatorily, required for the initiation of transcription itself, i.e., that sequence-specific RNA signals are involved in the targeting and recruitment of transcription factors and transcription complexes at appropriate places in the genome, whose general availability has already been determined by longer-term epigenetic mechanisms that have also been controlled by RNA signaling. In this context it is of interest to note that the β-globin LCR which is considered to be the archetypal long distance transcriptional "enhancer" is itself specifically transcribed in erythroid cells.[17] It has also recently been shown that steroid receptor transactivation requires a ncRNA called SRA,[185] and it appears that this may be just one example of a very large family of such RNAs. It has also been shown that a noncoding RNA 7SK is involved in the transcriptional activation of the proto-oncogene *c-myc*.[186] The idea that a trans-acting ncRNA may act to recruit transcription factors to sequence-specific locations in the chromosome has a lot of merit, as it would potentially solve the problem that most "transcription factors" have relatively loose binding consensus sites that occur in very many places throughout the genome, as well as provide an explanation as to why many transcription factors appear to have an equal if not higher affinity for nucleic acid structures involving RNA than for duplex DNA. If this is the case, it would also provide a mechanism for endogenously programmed transcriptional responses during differentiation and development, and suggests that cellular receptor and signaling cascades that intersect with nuclear transcription factors serve mainly to integrate endogenous programming with exogenous information in relation to cell position, morphogenic gradients and physiological status.

Thirdly, it seems highly likely that alternate splicing is regulated by trans-acting RNAs. This is another programmed response that is of fundamental importance in differentiation and development, but its mechanistic basis is at present not understood. Very few protein factors which specifically regulate alternate splicing have been described, one exception being Sex-lethal in *Drosophila* (for review see ref. 187), and mutational studies have shown that only very few bases, usually located at or near the intron/exon boundary, are required for splice site selection.[188,189] Splicing is, in essence, an RNA catalyzed process.[5,43] A parsimonious mechanism for alternative splicing, which could occur as part of programmed responses in particular cell lineages, is that alternate splice sites are activated or (more likely) inhibited by trans-acting RNAs (produced from other loci) which recognize the site in question. Indeed, it has been shown that synthetic oligoribonucleotides can modulate splicing patterns in cell culture,[190-193] in which case this is also likely to occur naturally in vivo.

Fourthly it is well established that trans-acting RNAs (such as *lin-4* and *let-7*) can affect mRNA stability and translation.[136] In addition, it would be expected that many ncRNAs will interact with proteins and participate in cell signaling pathways. There are many proteins that bind RNA through Zn finger domains, RRMs and other motifs, and that include signaling domains such as SH3-domain binding domains (see e.g., refs. 177, 194, 195).

A New Genetics

The idea that introns and ncRNAs may comprise a second tier of gene expression in the higher organisms that enables the integration and coordination of complex suites of gene expression and programmed responses during differentiation and development has been discussed in detail elsewhere,[11-13,196] but in general it brings a entirely new perspective to the consideration of how information may be stored and communicated in complex autopoeitic systems. The principles that emerge have much in common with other complex information processing systems that underlie neural and computer function, which include component multitasking and endogenous signals (referred to as the "hidden layer" in neural systems),[12] and bring into consideration important emerging concepts in terms of the structure of networks (scale-free networks, small world networks and recurrent dynamical neural networks) that may underpin metabolic, signaling, regulatory and epigenetic pathways and interactions.[12,197-199]

Little thought has thus far been given to the nature of the control architecture which underpins the programs that are enacted during differentiation and development, and the fact that the genetic and phenotypic signatures of point mutations in the control architecture (cis-acting regulatory sequences and trans-acting regulatory RNAs) are likely to be fundamentally different from mutations in the protein components. Complex organisms require two levels of genetic programming for their autopoeitic development from a fertilized embryo. The genomes of these organisms must specify the functional components of the system, mainly proteins, which have been the primary focus of genetic and genomic research to date. Damage to these components (by mutation) that makes them nonfunctional or mis-functional is very obvious and usually gives a severe phenotype, for example in the case of monogenic diseases such as cystic fibrosis and thalassemia, just as damaging the functional components of any structure is obvious. However, these genomes of complex organisms must also specify the control architecture which deploys these components in sophisticated suites of differentiation and development. Damage to this architecture is much more subtle, because of the nature and complexity of this information, which primarily affects quantitative trait variation.

Traditionally it has been assumed that this control architecture is embedded in the cis-acting control sequences (promoters and enhancers) which regulate gene expression in conjunction with combinations of trans-acting proteins, but this may not be sufficient. In any case, like introns, these cis-acting promoter sequences are much less conserved than protein-coding regions, and are presumably much more plastic. Indeed known (single base) mutations in promoters that have strong effects on phenotype are very rare, although such effects are more pronounced with deletions (see e.g., ref. 125). Only about 1% of all known human mutations/ variations that have known phenotypic consequences are in flanking or intergenic nonprotein-coding sequences that are classified (or assumed to be) promoters or enhancers. The same is true for introns and ncRNAs in general, for the simple reason that a single base change may have profound effects on the amino-acid codon and hence structure-activity of a protein, but may have only subtle effects on cis- and trans-interactions that involve nucleic acid sequences. However, while it has been assumed that (benign) protein structural variation will form the main basis of quantitative trait variation between individuals and species, the likelihood is that the majority of this variation, which will include susceptibility to disease, will in fact occur in non-protein-coding sequences that affect the patterns and levels of protein expression, and operate both in cis (as conventional promoter/enhancer elements) and in trans (as ncRNAs communicating signals within the network). This is beginning to come to light as more advanced genome mapping projects tracking quantitative trait variation are completed. For example, a significant proportion of the variation between modern maize and its wild ancestor teosinte occurs in noncoding regulatory regions near the *tb1 (teosinte branched1)* gene which affects the pattern of expression of *tb1* and results in the production of a short branches tipped by ears (corn) rather than long branches with tassels at their tips, whereas the structure (coding sequence) of the *tb1* protein itself is unaltered.[200] In addition a significant proportion of the quantitative trait variation in muscle mass between wild and domestic pigs can be ascribed to a point mutation (variation) within a noncoding region of the *Igf2* gene (L. Andersson, personal communication).[201]

The fact that introns and noncoding RNAs dominate the transcriptional output of the genomes of humans and other complex organisms suggests that a second tier of gene expression has emerged in the higher eukaryotes. It seems highly likely that the introns have evolved to transmit secondary signals in parallel with protein-coding sequences (and at least in some cases, in parallel with exonic ncRNAs), which enable networking of gene activity. If this is the case it would not be surprising if some (and indeed many) genes had evolved to express only ncRNAs, as higher order regulators in the network. This also indicates that the major facet of the evolution of the higher organisms was not simply an expanded repertoire of proteins and protein isoforms, but a larger set of genomic instructions embedded in introns and ncRNAs. I suggest that these RNAs constitute a critically important endogenous genetic communication system within (and perhaps between) cells, which conveys a range of primary and secondary (efference) signals as part of a cascade of programmed response networks capable of implementing stored sequences of dynamical activities in response to internal and external stimuli during differentiation and development. That is, although only 1.5% of the human genome codes for proteins (the functional catalytic, structural and signaling components of the system), the majority of the genome encodes a larger set of secondary information in cis-acting promoters and trans-acting RNA sequences which determines how these components are deployed. In these instructions, which dominate the genomes of complex organisms, is embedded the majority of information encoding the phenotypic variation that distinguishes different individuals and different species. An elephant is not like *E. coli*.

References

1. Rivas E, Klein RJ, Jones TA et al. Computational identification of noncoding RNAs in *E. coli* by comparative genomics. Curr Biol 2001; 11:1369-1373.
2. Argaman L, Hershberg R, Vogel J et al. Novel small RNA-encoding genes in the intergenic regions of *Escherichia coli*. Curr Biol 2001; 11:941-950.
3. Klein RJ, Misulovin Z, Eddy SR. Noncoding RNA genes identified in AT-rich hyperthermophiles. Proc Natl Acad Sci USA 2002; 99:7542-7547.
4. Judson HF. In: Simon, Schuster, eds. The Eighth Day of Creation: The Makers of the Revolution in Biology. New York: 1979.
5. Schroeder R. Dissecting RNA function. Nature 1994; 370:597-598.

6. Williamson B. DNA insertions and gene structure. Nature 1977; 270:295-297.
7. Gilbert W. Why genes in pieces? Nature 1978; 271:501.
8. Gilbert W, Marchionni M, McKnight G. On the antiquity of introns. Cell 1986; 46:151-154.
9. Croft L, Beatson SA, Whitchurch CB et al. An interactive web-based *Pseudomonas aeruginosa* genome database: discovery of new genes, pathways and structures. Microbiol 2000; 146:2351-2364.
10. Sharp PA, Konarksa MM, Grabowski PJ et al. Splicing of messenger RNA precursors. Cold Spring Harb Symp Quant Biol 1987; 52:277-285.
11. Mattick JS. Introns: evolution and function. Curr Opin Genet Dev 1994; 4:823-831.
12. Mattick JS, Gagen MJ. The evolution of controlled multitasked gene networks: the role of introns and other noncoding RNAs in the development of complex organisms. Mol Biol Evol 2001; 18:1611-1630.
13. Mattick JS. Noncoding RNAs: the architects of eukaryotic complexity. EMBO Rep 2001; 2:986-991.
14. Lander ES, Linton LM, Birren B et al. Initial sequencing and analysis of the human genome. Nature 2001; 409:860-921.
15. Venter JC, Adams MD, Myers EW et al. The sequence of the human genome. Science 2001; 291:1304-1351.
16. In: Davidson EH, ed. Gene activity in early development. New York: Academic Press, 1976.
17. Ashe HL, Monks J, Wijgerde M et al. Intergenic transcription and transinduction of the human beta-globin locus. Genes Dev 1997; 11:2494-2509.
18. Weil D, Power MA, Webb GC et al. Antisense transcription of a murine FGFR-3 psuedogene during fetal developement. Gene 1997; 187:115-122.
19. Vanhee-Brossollet C, Vaquero C. Do natural antisense transcripts make sense in eukaryotes? Gene 1998; 211:1-9.
20. Munroe SH, Lazar MA. Inhibition of c-erbA mRNA splicing by a naturally occurring antisense RNA. J Biol Chem 1991; 266:22083-22086.
21. Islam N, Poitras L, Gagnon F et al. Antisense and sense poly(A)-RNAs from the Xenopus laevis pyruvate dehydrogenase gene loci are regulated with message production during embryogenesis. Gene 1996; 176:9-16.
22. Lehner B, Williams G, Campbell RD et al. Antisense transcripts in the human genome. Trends Genet 2002; 18:63-65.
23. Knee R, Murphy PR. Regulation of gene expression by natural antisense RNA transcripts. Neurochem Int 1997; 31:379-392.
24. Lipman DJ. Making (anti)sense of noncoding sequence conservation. Nucl Acids Res 1997; 25:3580-3583.
25. Nemes JP, Benzow KA, Koob MD. The SCA8 transcript is an antisense RNA to a brain-specific transcript encoding a novel actin-binding protein (KLHL1). Hum Mol Genet 2000; 9:1543-1551.
26. Potter SS, Branford WW. Evolutionary conservation and tissue-specific processing of Hoxa 11 antisense transcripts. Mamm Genome 1998; 9:799-806.
27. Millar JK, Wilson-Annan JC, Anderson S et al. Disruption of two novel genes by a translocation cosegregating with schizophrenia. Hum Mol Genet 2000; 9:1415-1423.
28. Lipshitz HD, Peattie DA, Hogness DS. Novel transcripts from the Ultrabithorax domain of the bithorax complex. Genes Dev 1987; 1:307-322.
29. Sanchez-Herrero E, Akam M. Spatially ordered transcription of regulatory DNA in the bithorax complex of *Drosophila*. Development 1989; 107:321-329.
30. Wroe SF, Kelsey G, Skinner JA et al. An imprinted transcript, antisense to Nesp, adds complexity to the cluster of imprinted genes at the mouse Gnas locus. Proc Natl Acad Sci USA 2000; 97:3342-3346.
31. Peters J, Wroe SF, Wells CA et al. A cluster of oppositely imprinted transcripts at the Gnas locus in the distal imprinting region of mouse chromosome 2. Proc Natl Acad Sci USA 1999; 96:3830-3835.
32. Lau NC, Lim LP, Weinstein EG et al. An abundant class of tiny RNAs with probable regulatory roles in *Caenorhabditis elegans*. Science 2001; 294:858-862.
33. Kapranov P, Cawley SE, Drenkow J et al. Large-scale transcriptional activity in chromosomes 21 and 22. Science 2002; 296:916-919.
34. Dennis C. The brave new world of RNA. Nature 2002; 418:122-124.

35. Zhuo D, Zhao WD, Wright FA et al. Assembly, annotation, and integration of UNIGENE clusters into the human genome draft. Genome Res 2001; 11:904-918.
36. Hong YK, Ontiveros SD, Strauss WM. A revision of the human XIST gene organization and structural comparison with mouse Xist. Mamm Genome 2000; 11:220-224.
37. Liu AY, Torchia BS, Migeon BR et al. The human NTT gene: identification of a novel 17-kb noncoding nuclear RNA expressed in activated CD4+ T cells. Genomics 1997; 39:171-184.
38. Aparicio S, Chapman J, Stupka E et al. Whole-genome shotgun assembly and analysis of the genome of *Fugu rubripes*. Science 2002; 297:1301-1310.
39. Cavalier-Smith T. Intron phylogeny: a new hypothesis. Trends Genet 1991; 7:145-148.
40. Palmer JD, Logsdon Jr JM. The recent origins of introns. Curr Opin Genet Dev 1991; 1:470-477.
41. Lambowitz AM, Belfort M. Introns as mobile genetic elements. Annu Rev Biochem 1993; 62:587-622.
42. Stoltzfus A, Spencer DF, Zuker M et al. Testing the exon theory of genes: the evidence from protein structure. Science 1994; 265:202-207.
43. Jacquier A. Group II introns: elaborate ribozymes. Biochimie 1996; 78:474-487.
44. Cho G, Doolittle RF. Intron distribution in ancient paralogs supports random insertion and not random loss. J Mol Evol 1997; 44:573-584.
45. Logsdon JMJ. The recent origins of spliceosomal introns revisited. Curr Opin Genet Dev 1998; 8:637-648.
46. Wolf YI, Kondrashov FA, Koonin EV. No footprints of primordial introns in a eukaryotic genome. Trends Genet 2000; 16:333-334.
47. Eickbush TH. Molecular biology: Introns gain ground. Nature 2000; 404:940-941.
48. Deutsch M, Long M. Intron-exon structures of eukaryotic model organisms. Nucl Acids Res 1999; 27:3219-3228.
49. Mayor C, Brudno M, Schwartz JR et al. VISTA : visualizing global DNA sequence alignments of arbitrary length. Bioinformatics 2000; 16:1046-1047.
50. Garbe JC, Pardue ML. Heat shock locus 93D of *Drosophila* melanogaster: a spliced RNA most strongly conserved in the intron sequence. Proc Natl Acad Sci USA 1986; 83:1812-1816.
51. Rieger M, Franke WW. Identification of an orthologous mammalian cytokeratin gene. High degree of intron sequence conservation during evolution of human cytokeratin 10. J Mol Biol 1988; 204:841-856.
52. Tournier-Lasserve E, Odenwald WF, Garbern J et al. Remarkable intron and exon sequence conservation in human and mouse homeobox Hox 1.3 genes. Mol Cell Biol 1989; 9:2273-2278.
53. Lloyd C, Gunning P. Noncoding regions of the gamma-actin gene influence the impact of the gene on myoblast morphology. J Cell Biol 1993; 121:73-82.
54. Starke T, Gogarten JP. A conserved intron in the V-ATPase A subunit genes of plants and algae. FEBS Lett 1993; 315:252-258.
55. Koop BF, Hood L. Striking sequence similarity over almost 100 kilobases of human and mouse T-cell receptor DNA. Nature Genet 1994; 7:48-53.
56. Bagavathi S, Malathi R. Introns and protein revolution - an analysis of the exon/intron organisation of actin genes. FEBS Lett 1996; 392:63-65.
57. John TR, Smith JJ, Kaiser II. A phospholipase A2-like pseudogene retaining the highly conserved introns of Mojave toxin and other snake venom group II PLA2s, but having different exons. DNA Cell Biol 1996; 15:661-668.
58. Rosby O, Alestrom P, Berg K. High-degree sequence conservation in LPA kringle IV-type 2 exons and introns. Clin Genet 1997; 52:293-302.
59. Kazmierczak B, Bullerdiek J, Pham KH et al. Intron 3 of HMGIC is the most frequent target of chromosomal aberrations in human tumors and has been conserved basically for at least 30 million years. Cancer Genet Cytogenet 1998; 103:175-177.
60. Aruscavage PJ, Bass BL. A phylogenetic analysis reveals an unusual sequence conservation within introns involved in RNA editing. RNA 2000; 6:257-269.
61. Sun L, Li Y, McCullough AK et al. Intron conservation in a UV-specific DNA repair gene encoded by chlorella viruses. J Mol Evol 2000; 50:82-92.
62. Yatsuki H, Watanabe H, Hattori M et al. Sequence-based structural features between Kvlqt1 and Tapa1 on mouse chromosome 7F4/F5 corresponding to the Beckwith-Wiedemann syndrome re-

gion on human 11p15.5: long-stretches of unusually well conserved intronic sequences of Kvlqt1 between mouse and human. DNA Res 2000; 7:195-206.
63. Jareborg N, Birney E, Durbin R. Comparative analysis of noncoding regions of 77 orthologous mouse and human gene pairs. Genome Res 1999; 9:815-824.
64. Kiss T. Small nucleolar RNAs: an abundant group of noncoding RNAs with diverse cellular functions. Cell 2002; 109:145-148.
65. Prislei S, Michienzi A, Presutti C et al. Two different snoRNAs are encoded in introns of amphibian and human L1 ribosomal protein genes. Nucl Acids Res 1993; 21:5824-5830.
66. Sollner-Webb B. Novel intron-encoded small nucleolar RNAs. Cell 1993; 75:403-405.
67. Bachellerie JP, Nicoloso M, Qu LH et al. Novel intron-encoded small nucleolar RNAs with long sequence complementarities to mature rRNAs involved in ribosome biogenesis. Biochem Cell Biol 1995; 73:835-843.
68. Maxwell ES, Fournier MJ. The small nucleolar RNAs. Annu Rev Biochem 1995; 64:897-934.
69. Nicoloso M, Qu LH, Michot B et al. Intron-encoded, antisense small nucleolar RNAs: the characterization of nine novel species points to their direct role as guides for the 2'-O- ribose methylation of rRNAs. J Mol Biol 1996; 260:178-195.
70. Rebane A, Tamme R, Laan M et al. A novel snoRNA (U73) is encoded within the introns of the human and mouse ribosomal protein S3a genes. Gene 1998; 210:255-263.
71. Filipowicz W. Imprinted expression of small nucleolar RNAs in brain: Time for RNomics. Proc Natl Acad Sci USA 2000; 97:14035-14037.
72. Pelczar P, Filipowicz W. The host gene for intronic U17 small nucleolar RNAs in mammals has no protein-coding potential and is a member of the 5'-terminal oligopyrimidine gene family. Mol Cell Biol 1998; 18:4509-4518.
73. Smith CM, Steitz JA. Classification of gas5 as a multi-small-nucleolar-RNA (snoRNA) host gene and a member of the 5'-terminal oligopyrimidine gene family reveals common features of snoRNA host genes. Mol Cell Biol 1998; 18:6897-6909.
74. Tycowski KT, Shu MD, Steitz JA. A mammalian gene with introns instead of exons generating stable RNA products. Nature 1996; 379:464-466.
75. Tanaka R, Satoh H, Moriyama M et al. Intronic U50 small-nucleolar-RNA (snoRNA) host gene of no protein-coding potential is mapped at the chromosome breakpoint t(3;6)(q27;q15) of human B-cell lymphoma. Genes Cells 2000; 5:277-287.
76. van Hoof A, Parker R. The exosome: a proteasome for RNA? Cell 1999; 99:347-350.
77. van Hoof A, Lennertz P, Parker R. Three conserved members of the RNase D family have unique and overlapping functions in the processing of 5S, 5.8S, U4, U5, RNase MRP and RNase P RNAs in yeast. EMBO J 2000; 19:1357-1365.
78. Allmang C, Petfalski E, Podtelejnikov A et al. The yeast exosome and human PM-Scl are related complexes of 3' —> 5' exonucleases. Genes Dev 1999; 13:2148-2158.
79. Allmang C, Kufel J, Chanfreau G et al. Functions of the exosome in rRNA, snoRNA and snRNA synthesis. EMBO J 1999; 18:5399-5410.
80. Clement JQ, Qian L, Kaplinsky N et al. The stability and fate of a spliced intron from vertebrate cells. RNA 1999; 5:206-220.
81. Clement JQ, Maiti S, Wilkinson MF. Localization and stability of introns spliced from the Pem homeobox gene. J Biol Chem 2001; 276:16919-16930.
82. Qian L, Vu MN, Carter M et al. A spliced intron accumulates as a lariat in the nucleus of T cells. Nucl Acids Res 1992; 20:5345-5350.
83. Bass BL. Double-stranded RNA as a template for gene silencing. Cell 2000; 101:235-238.
84. Morse DP, Aruscavage PJ, Bass BL. RNA hairpins in noncoding regions of human brain and Caenorhabditis elegans mRNA are edited by adenosine deaminases that act on RNA. Proc Natl Acad Sci USA 2002; 99:7906-7911.
85. Erdmann VA, Barciszewska MZ, Szymanski M et al. The noncoding RNAs as riboregulators. Nucl Acids Res 2001; 29:189-193.
86. Erdmann VA, Szymanski M, Hochberg A et al. Collection of mRNA-like noncoding RNAs. Nucl Acids Res 1999; 27:192-195.
87. Erdmann VA, Barciszewska MZ, Hochberg A et al. Regulatory RNAs. Cell Mol Life Sci 2001; 58:960-977.

88. Erdmann VA, Szymanski M, Hochberg A et al. Noncoding, mRNA-like RNAs database Y2K. Nucl Acids Res 2000; 28:197-200.
89. Askew DS, Xu F. New insights into the function of noncoding RNA and its potential role in disease pathogenesis. Histol Histopathol 1999; 14:235-241.
90. Eddy SR. Noncoding RNA genes. Curr Opin Genet Dev 1999; 9:695-699.
91. Eddy SR. Noncoding RNA genes and the modern RNA world. Nature Rev Genet 2001; 2:919-929.
92. Nakamura A, Amikura R, Mukai M et al. Requirement for a noncoding RNA in *Drosophila* polar granules for germ cell establishment. Science 1996; 274:2075-2079.
93. Lakhotia SC, Sharma A. The 93D (hsr-omega) locus of *Drosophila*: noncoding gene with house-keeping functions. Genetica 1996; 97:339-348.
94. Teramoto H, Toyama T, Takeba G et al. Noncoding RNA for CR20, a cytokinin-repressed gene of cucumber. Plant Mol Biol 1996; 32:797-808.
95. Kuryshev VY, Skryabin BV, Kremerskothen J et al. Birth of a gene: locus of neuronal BC200 snmRNA in three prosimians and human BC200 pseudogenes as archives of change in the Anthropoidea lineage. J Mol Biol 2001; 309:1049-1066.
96. Velleca MA, Wallace MC, Merlie JP. A novel synapse-associated noncoding RNA. Mol Cell Biol 1994; 14:7095-7104.
97. French PJ, Bliss TV, O'Connor V. Ntab, a novel noncoding RNA abundantly expressed in rat brain. Neurosci 2001; 108:207-215.
98. Lin Y, Brosius J, Tiedge H. Neuronal BC1 RNA: coexpression with growth-associated protein-43 messenger RNA. Neurosci 2001; 103:465-479.
99. Muslimov IA, Titmus M, Koenig E et al. Transport of neuronal BC1 RNA in Mauthner axons. J Neurosci 2002; 22:4293-4301.
100. Wang Y, Crawford DR, Davies KJ. adapt33, a novel oxidant-inducible RNA from hamster HA-1 cells. Arch Biochem Biophys 1996; 332:255-260.
101. Tam W, Ben-Yehuda D, Hayward WS. bic, a novel gene activated by proviral insertions in avian leukosis virus-induced lymphomas, is likely to function through its noncoding RNA. Mol Cell Biol 1997; 17:1490-1502.
102. Tam W. Identification and characterization of human BIC, a gene on chromosome 21 that encodes a noncoding RNA. Gene 2001; 274:157-167.
103. Reisman D, Balint E, Loging WT et al. A novel transcript encoded within the 10-kb first intron of the human p53 tumor suppressor gene (D17S2179E) is induced during differentiation of myeloid leukemia cells. Genomics 1996; 38:364-370.
104. Makrinou E, Fox M, Lovett M et al. Tty2: a multicopy y-linked gene family. Genome Res 2001; 11:935-945.
105. Miller D, Briggs D, Snowden H et al. A complex population of RNAs exists in human ejaculate spermatozoa: implications for understanding molecular aspects of spermiogenesis. Gene 1999; 237:385-392.
106. Kelley RL, Kuroda MI. Noncoding RNA genes in dosage compensation and imprinting. Cell 2000; 103:9-12.
107. Lee JT, Davidow LS, Warshawsky D. Tsix, a gene antisense to Xist at the X-inactivation centre. Nature Genet 1999; 21:400-404.
108. Luikenhuis S, Wutz A, Jaenisch R. Antisense transcription through the Xist locus mediates Tsix function in embryonic stem cells. Mol Cell Biol 2001; 21:8512-8520.
109. Sleutels F, Barlow DP, Lyle R. The uniqueness of the imprinting mechanism. Curr Opin Genet Dev 2000; 10:229-233.
110. Sleutels F, Barlow DP. Investigation of elements sufficient to imprint the mouse Air promoter. Mol Cell Biol 2001; 21:5008-5017.
111. Takada S, Tevendale M, Baker J et al. Delta-like and gtl2 are reciprocally expressed, differentially methylated linked imprinted genes on mouse chromosome 12. Curr Biol 2000; 10:1135-1138.
112. Johnston C, Newall A, Brockdorff N et al. Enox, a novel gene that maps 10 kb upstream of Xist and partially escapes X inactivation. Genomics 2002; 80:236.
113. Madamanchi NR, Hu ZY, Li F et al. A noncoding RNA regulates human protease-activated receptor-1 gene during embryogenesis. Biochim Biophys Acta 2002; 1576:237-245.

114. Li AW, Murphy PR. Expression of alternatively spliced FGF-2 antisense RNA transcripts in the central nervous system: regulation of FGF-2 mRNA translation. Mol Cell Endocrinol 2000; 162:69-78.
115. Komine Y, Tanaka NK, Yano R et al. A novel type of noncoding RNA expressed in the rat brain. Brain Res Mol Brain Res 1999; 66:1-13.
116. Takeda K, Ichijo H, Fujii M et al. Identification of a novel bone morphogenetic protein-responsive gene that may function as a noncoding RNA. J Biol Chem 1998; 273:17079-17085.
117. Wrana JL. H19, a tumour suppressing RNA? Bioessays 1994; 16:89-90.
118. Hurst LD, Smith NG. Molecular evolutionary evidence that H19 mRNA is functional. Trends Genet 1999; 15:134-135.
119. Ridanpaa M, van Eenennaam H, Pelin K et al. Mutations in the RNA component of RNase MRP cause a pleiotropic human disease, cartilage-hair hypoplasia. Cell 2001; 104:195-203.
120. Sato M, Shinozaki-Yabana S, Yamashita A et al. The fission yeast meiotic regulator Mei2p undergoes nucleocytoplasmic shuttling. FEBS Lett 2001; 499:251-255.
121. Watanabe T, Miyashita K, Saito TT et al. Comprehensive isolation of meiosis-specific genes identifies novel proteins and unusual noncoding transcripts in *Schizosaccharomyces pombe*. Nucl Acids Res 2001; 29:2327-2337.
122. Teranishi M, Shimada Y, Hori T et al. Transcripts of the MHM region on the chicken Z chromosome accumulate as noncoding RNA in the nucleus of female cells adjacent to the DMRT1 locus. Chromosome Res 2001; 9:147-165.
123. Jones SJ, Riddle DL, Pouzyrev AT et al. Changes in gene expression associated with developmental arrest and longevity in *Caenorhabditis elegans*. Genome Res 2001; 11:1346-1352.
124. Akam ME, Martinez-Arias A, Weinzierl R et al. Function and expression of ultrabithorax in the *Drosophila* embryo. Cold Spring Harb Symp Quant Biol 1985; 50:195-200.
125. Hogness DS, Lipshitz HD, Beachy PA et al. Regulation and products of the Ubx domain of the bithorax complex. Cold Spring Harb Symp Quant Biol 1985; 50:181-194.
126. Irish VF, Martinez-Arias A, Akam M. Spatial regulation of the Antennapedia and Ultrabithorax homeotic genes during *Drosophila* early development. EMBO J 1989; 8:1527-1537.
127. Chao YC, Lee ST, Chang MC et al. A 2.9-kilobase noncoding nuclear RNA functions in the establishment of persistent Hz-1 viral infection. J Virol 1998; 72:2233-2245.
128. Okazaki Y, Furuno M, Kasukawa T et al. Analysis of the mouse transcriptome based on functional annotation of 60,770 full-length cDNAs. Nature 2002; 420:563-573.
129. Lee RC, Ambros V. An extensive class of small RNAs in *Caenorhabditis elegans*. Science 2001; 294:862-864.
130. Lagos-Quintana M, Rauhut R, Lendeckel W et al. Identification of novel genes coding for small expressed RNAs. Science 2001; 294:853-858.
131. Lagos-Quintana M, Rauhut R, Yalcin A et al. Identification of tissue-specific microRNAs from mouse. Curr Biol 2002; 12:735-739.
132. Mourelatos Z, Dostie J, Paushkin S et al. miRNPs: a novel class of ribonucleoproteins containing numerous microRNAs. Genes Dev 2002; 16:720-728.
133. Reinhart BJ, Weinstein EG, Rhoades MW et al. MicroRNAs in plants. Genes Dev 2002; 16:1616-1626.
134. Llave C, Kasschau KD, Rector MA et al. Endogenous and silencing-associated small RNAs in plants. Plant Cell 2002; 14:1605-1619.
135. Rhoades M, Reinhart B, Lim L et al. Prediction of plant microRNA targets. Cell 2002; 110:513.
136. Pasquinelli AE, Ruvkun G. Control and developmental timing by microRNAs and their targets. Annu Rev Cell Dev Biol 2002; 28:28.
137. Wassenegger M. RNA-directed DNA methylation. Plant Mol Biol 2000; 43:203-220.
138. Matzke MA, Matzke AJ, Pruss GJ et al. RNA-based silencing strategies in plants. Curr Opin Genet Dev 2001; 11:221-227.
139. Lin WL, He XB, Svensson K et al. The genotype and epigenotype synergize to diversify the spatial pattern of expression of the imprinted H19 gene. Mech Dev 1999; 82:195-197.
140. Memili E, Hong YK, Kim DH et al. Murine Xist RNA isoforms are different at their 3' ends: a role for differential polyadenylation. Gene 2001; 266:131-137.

141. Judd BH. Mutations of zeste that mediate transvection are recessive enhancers of position-effect variegation in *Drosophila melanogaster*. Genetics 1995; 141:245-253.
142. Brenton JD, Ainscough JF, Lyko F et al. Imprinting and gene silencing in mice and *Drosophila*. Novartis Found Symp 1998; 214:233-244.
143. Broday L, Lee YW, Costa M. 5-azacytidine induces transgene silencing by DNA methylation in Chinese hamster cells. Mol Cell Biol 1999; 19:3198-3204.
144. Fire A. RNA-triggered gene silencing. Trends Genet 1999; 15:358-363.
145. Jones L, Hamilton AJ, Voinnet O et al. RNA-DNA interactions and DNA methylation in post-transcriptional gene silencing. Plant Cell 1999; 11:2291-2302.
146. Wu CT, Morris JR. Transvection and other homology effects. Curr Opin Genet Dev 1999; 9:237-246.
147. Bosher JM, Labouesse M. RNA interference: genetic wand and genetic watchdog. Nature Cell Biol 2000; 2:E31-36.
148. Mette MF, Aufsatz W, van Der Winden J et al. Transcriptional silencing and promoter methylation triggered by double-stranded RNA. EMBO J 2000; 19:5194-5201.
149. Morel J, Mourrain P, Beclin C et al. DNA methylation and chromatin structure affect transcriptional and post-transcriptional transgene silencing in *Arabidopsis*. Curr Biol 2000; 10:1591-1594.
150. Sharp PA. RNA interference-2001. Genes Dev 2001; 15:485-490.
151. Montgomery MK, Fire A. Double-stranded RNA as a mediator in sequence-specific genetic silencing and cosuppression. Trends Genet 1998; 14:255-258.
152. Dernburg AF, Zalevsky J, Colaiacovo MP et al. Transgene-mediated cosuppression in the *C. elegans* germ line. Genes Dev 2000; 14:1578-1583.
153. Fagard M, Boutet S, Morel JB et al. AGO1, QDE-2, and RDE-1 are related proteins required for post- transcriptional gene silencing in plants, quelling in fungi, and RNA interference in animals. Proc Natl Acad Sci USA 2000; 97:11650-11654.
154. Ketting RF, Plasterk RH. A genetic link between cosuppression and RNA interference in *C. elegans*. Nature 2000; 404:296-298.
155. Cameron FH, Jennings PA. Inhibition of gene expression by a short sense fragment. Nucl Acids Res 1991; 19:469-475.
156. Pal-Bhadra M, Bhadra U, Birchler JA. Cosuppression in *Drosophila*: gene silencing of Alcohol dehydrogenase by white-Adh transgenes is Polycomb dependent. Cell 1997; 90:479-490.
157. Pal-Bhadra M, Bhadra U, Birchler JA. Cosuppression of nonhomologous transgenes in *Drosophila* involves mutually related endogenous sequences. Cell 1999; 99:35-46.
158. Bingham PM. Cosuppression comes to the animals. Cell 1997; 90:385-387.
159. Bahramian MB, Zarbl H. Transcriptional and posttranscriptional silencing of rodent alpha1(I) collagen by a homologous transcriptionally self-silenced transgene. Mol Cell Biol 1999; 19:274-283.
160. Plasterk RH, Ketting RF. The silence of the genes: Curr Opin Genet Dev 2000; 10:562-567.
161. Birchler JA, Bhadra MP, Bhadra U. Making noise about silence: repression of repeated genes in animals. Curr Opin Genet Dev 2000; 10:211-216.
162. Jones DO, Cowell IG, Singh PB. Mammalian chromodomain proteins: their role in genome organisation and expression. Bioessays 2000; 22:124-137.
163. Jedrusik MA, Schulze E. A single histone H1 isoform (H1.1) is essential for chromatin silencing and germline development in *Caenorhabditis elegans*. Development 2001; 128:1069-1080.
164. Struckenholz C, Kageyama Y, Kuroda MI. Guilt by association: noncoding RNAs, chromosome-specific proteins and dosage compensation in *Drosophila*. Trends Genet 1999; 15:454-458.
165. Tsai JY, Silver LM. Escape from genomic imprinting at the mouse T-associated maternal effect (Tme) locus. Genetics 1991; 129:1159-1166.
166. Marahrens Y. X-inactivation by chromosomal pairing events. Genes Dev 1999; 13:2624-2632.
167. Gould A. Functions of mammalian Polycomb group and trithorax group related genes. Curr Opin Genet Dev 1997; 7:488-494.
168. Campbell RB, Sinclair DA, Couling M et al. Genetic interactions and dosage effects of Polycomb group genes of *Drosophila*. Mol Gen Genet 1995; 246:291-300.
169. Gebuhr TC, Bultman SJ, Magnuson T. Pc-G/trx-G and the SWI/SNF connection: developmental gene regulation through chromatin remodeling. Genesis 2000; 26:189-197.

170. Kennison JA. The Polycomb and trithorax group proteins of *Drosophila*: trans-regulators of homeotic gene function. Annu Rev Genet 1995; 29:289-303.
171. Hanson RD, Hess JL, Yu BD et al. Mammalian trithorax and polycomb-group homologues are antagonistic regulators of homeotic development. Proc Natl Acad Sci USA 1999; 96:14372-14377.
172. Jacobs JJ, van Lohuizen M. Cellular memory of transcriptional states by Polycomb-group proteins. Semin Cell Dev Biol 1999; 10:227-235.
173. Caldas C, Aparicio S. Cell memory and cancer—the story of the trithorax and Polycomb group genes. Cancer Metastasis Rev 1999; 18:313-329.
174. Jones AL, Thomas CL, Maule AJ. De novo methylation and cosuppression induced by a cytoplasmically replicating plant RNA virus. EMBO J\ 1998; 17:6385-6393.
175. Akhtar A, Zink D, Becker PB. Chromodomains are protein-RNA interaction modules. Nature 2000; 407:405-409.
176. Shi Y, Berg JM. Specific DNA-RNA hybrid binding by zinc finger proteins. Science 1995; 268:282-284.
177. Ladomery M. Multifunctional proteins suggest connections between transcriptional and post-transcriptional processes. Bioessays 1997; 19:903-909.
178. Matsumoto K, Wolffe AP. Gene regulation by Y-box proteins: coupling control of transcription and translation. Trends Cell Biol 1998; 8:318-323.
179. Shnyreva M, Schullery DS, Suzuki H et al. Interaction of two multifunctional proteins. Heterogeneous nuclear ribonucleoprotein K and Y-box-binding protein J Biol Chem 2000; 275:15498-15503.
180. Herbert A, Schade M, Lowenhaupt K et al. The Zalpha domain from human ADAR1 binds to the Z-DNA conformer of many different sequences. Nucl Acids Res 1998; 26:3486-3493.
181. Herbert A, Alfken J, Kim YG et al. A Z-DNA binding domain present in the human editing enzyme, double-stranded RNA adenosine deaminase. Proc Natl Acad Sci USA 1997; 94:8421-8426.
182. Herbert A, Lowenhaupt K, Spitzner J et al. Chicken double-stranded RNA adenosine deaminase has apparent specificity for Z-DNA. Proc Natl Acad Sci USA 1995; 92:7550-7554.
183. Kim YG, Lowenhaupt K, Maas S et al. The Zab domain of the human RNA editing enzyme ADAR1 recognizes Z-DNA when surrounded by B-DNA. J Biol Chem 2000; 275:26828-26833.
184. Platero JS, Hartnett T, Eissenberg JC. Functional analysis of the chromo domain of HP1. EMBO J 1995; 14:3977-3986.
185. Lanz RB, McKenna NJ, Onate SA et al. A steroid receptor coactivator, SRA, functions as an RNA and is present in an SRC-1 complex. Cell 1999; 97:17-27.
186. Krause MO. Chromatin structure and function: the heretical path to an RNA transcription factor. Biochem Cell Biol 1996; 74:623-632.
187. Singh R. RNA-protein interactions that regulate premRNA splicing. Gene Expr 2002; 10:79-92.
188. McKeown M. Alternative mRNA splicing. Annu Rev Cell Biol 1992; 8:133-155.
189. Watakabe A, Tanaka K, Shimura Y. The role of exon sequences in splice site selection. Genes Dev 1993; 7:407-418.
190. Sierakowska H, Gorman L, Kang SH et al. Antisense oligonucleotides and RNAs as modulators of premRNA splicing. Methods Enzymol 2000; 313:506-521.
191. Kole R, Sazani P. Antisense effects in the cell nucleus: modification of splicing. Curr Opin Mol Ther 2001; 3:229-234.
192. Mercatante DR, Kole R. Control of alternative splicing by antisense oligonucleotides as a potential chemotherapy: effects on gene expression. Biochim Biophys Acta 2002; 1587:126-132.
193. Mercatante DR, Sazani P, Kole R. Modification of alternative splicing by antisense oligonucleotides as a potential chemotherapy for cancer and other diseases. Curr Cancer Drug Targets 2001; 1:211-230.
194. Shamoo Y, Abdul-Manan N, Williams KR. Multiple RNA binding domains (RBDs) just don't add up. Nucl Acids Res 1995; 23:725-728.
195. Kennedy D, Wood SA, Ramsdale T et al. Identification of a mouse orthologue of the human ras-GAP-SH3-domain binding protein and structural confirmation that these proteins contain an RNA recognition motif. Biomed Pept Proteins Nucleic Acids 1996; 2:93-99.
196. Mattick JS. Noncoding RNAs: a regulatory role? Nature Encyclopaedia of the Human Genome 2002; in press.

197. Albert R, Jeong H, Barabasi AL. Error and attack tolerance of complex networks. Nature 2000; 406:378-382.
198. Amaral LA, Scala A, Barthelemy M et al. Classes of small-world networks. Proc Natl Acad Sci USA 2000; 97:11149-11152.
199. Bhalla US, Iyengar R. Emergent properties of networks of biological signaling pathways. Science 1999; 283:381-387.
200. Wang R-L, Stec A, Hey J et al. The limits of selection during maize domestication. Nature 1999; 398:236-239.
201. Amarger V, Nguyen M, Laere AS et al. Comparative sequence analysis of the INS-IGF2-H19 gene cluster in pigs. Mamm Genome 2002; 13:388-398.

CHAPTER 3

Computational Gene-Finding for Noncoding RNAs

Peter Schattner

Abstract

Computer gene-finding programs have been quite successful at locating protein-coding genes in both prokaryotic and eukaryotic genomes. However these programs—which use genomic features such as long open-reading-frames and codon signatures—are not designed to identify noncoding RNA (ncRNA) genes. As a result ncRNA-specific gene-finders have been required.

The first successful attempts at computational ncRNA gene-finding focussed on ncRNAs with well-characterized primary sequences and/or secondary structures, such as tRNAs or methylation-guide snoRNAs. In addition user-configurable RNA-motif search programs were developed. These programs search for RNAs by looking for user-specified primary-sequence motifs and stable secondary-structures as indicated by increased Watson-Crick base-pairing or low calculated free energies. However, to date, these RNA-motif searching programs have had only modest success at finding ncRNAs.

Recently, computational ncRNA gene-finders have been developed which show promise of locating a much larger number of previously undetected ncRNAs. Some of the most successful are based on comparative sequence analysis between genomes of related species. Others exploit base-composition signatures of ncRNAs or use new methods for RNA sequence alignment and secondary-structure prediction. With these approaches, numerous previously undetected ncRNAs have been predicted and subsequently experimentally confirmed in species including *Escherichia coli* and the hyperthermophiles *Methanococcus jannaschii* and *Pyrococcus furiosus*.

This chapter will review the strategies employed in the principal computational ncRNA gene-finders. We will compare the successes of the different approaches as well as their limitations. Finally, we will consider the impact that these new computational methods are having on our picture of the world of ncRNAs.

Introduction

With the sequencing of the human genome, the scientific community has completed the first stage in compiling a complete "parts list" for the human body. The next phases—identifying the genomic location of the "parts" (i.e., the genes), discovering their functions, and determining how they are regulated are at a much less advanced stage. For the more familiar protein-coding genes there has been considerable progress in at least finding their genomic locations—by experimental means, such as by the building of cDNA libraries, as well as by

Noncoding RNAs: Molecular Biology and Molecular Medicine, edited by Jan Barciszewski and Volker A. Erdmann.

computational methods. This task is not yet complete as is evidenced by the continuing debates as to the total number of human genes. But there at least appears to be a growing consensus of the approximate number of genes (at least to within a factor of two or three) as well as their genomic locations.

In the world of ncRNA genes there is nothing resembling such a consensus yet. Few authors speculate as to the total number of ncRNA genes, even in small genomes—and when they do, the estimates vary widely. For example, when two groups predicted and experimentally confirmed several novel ncRNAs in *Escherichia coli* in 2001, one group wrote[1] "we think it unlikely that there are many more than 50 sRNAs (i.e., small ncRNAs) encoded by the *E. coli* chromosome" while the other group[2] predicted that "a significant number of our 275 candidate loci do indeed correspond to independent ncRNA genes". And this is for the relatively compact *E. coli* genome whose complete genome sequence had been already known for four years.

One does not need to look far for reasons for this lack of consensus. Until recently, genome-wide screening for ncRNAs was quite limited. Traditional EST-based methods for RNA-screening were primarily designed to look for RNAs with lengths greater than 200 base pairs (bp) and with poly-A tails—i.e., for protein-coding mRNAs. To some extent the technology was deliberately skewed away from detecting ncRNAs because of the (somewhat self-fulfilling) beliefs that ncRNAs were few in number and not of much biological interest.

The last two years, however, have seen a significant increase of activity in identifying and characterizing ncRNAs. Experimental efforts[3,4] have yielded dramatic results—showing that ncRNAs are far more numerous and presumably have greater biological significance than had been anticipated. These results, though clearly exciting, are outside the scope of the present review; the reader is referred to the chapter of this volume by J.P. Bachellerie and J. Cavaille or the original papers.[3,4] Until recently, computational ncRNA searches have also had only limited success. In this case the reason has been less lack of interest, than the difficulty of developing effective gene-finders. However, the last two years has also seen increasing success in ncRNA-gene discovery by new kinds of computational approaches.

Because of these developments, it seems timely to review the status of computational ncRNA gene-finding. I will summarize the different strategies which have been developed for this purpose—indicating the principal strengths and limitations of each . Particular attention will be paid to algorithms which have been developed within the last two years. In the next section we begin with a brief review of the basic strategies for ncRNA computational gene-finding and a comparison with the simpler case of gene-finding for protein-coding genes. The following two sections consider gene-finders targeting ncRNAs whose primary sequence and secondary structure are at least partially known and conserved. First we focus on customized programs that search for a single RNA class. Then we examine more general search programs that can be reconfigured by the user to target a variety of ncRNAs. Next we look at the more difficult task of searching for ncRNAs when we have little or no idea of their consensus primary sequence or secondary structure. In the final section we summarize the accomplishments and limitations of these programs and speculate on their future development.

RNA sequence alignment, RNA secondary-structure prediction and the identification of RNA motifs in mRNA sequences will only be discussed to the extent that they have impacted ncRNA gene-finding. For further information on these topics the reader is encouraged to consult recent articles and reviews[5-9] and references therein.

Gene-Finding for Protein-Coding Genes and ncRNAs

We begin with a brief review of the methods used for finding protein-coding genes. In the recent article "Gene-finding approaches in eukaryotes,"[10] Stormo notes that there are two main components to a gene-finder: the type of information—or what to look for, and the algorithm—or how to look for that information.

Sequence Information—Signals, Content Statistics and Similarity

Stormo groups sequence information into three basic classes: "signals", "content statistics" and similarity to known genes. Sequence signals may include promoters, terminators, poly-A-addition and transcription-factor binding sites, splice sites, start and stop codons and CpG islands. In addition, if characteristic distances are known between these individual signals, the distances themselves can serve as sequence signals.

Already here we begin to see the challenge of ncRNA gene-finding. Except for splice sites in the occasional multiple-exon ncRNA, only the usually weakly-conserved promoter and terminator signals (and possibly other poorly known transcription binding sites) will be present in ncRNA genes.

Content statistics—i.e., nonrandom variations in base sequence—are also useful clues for finding protein-coding genes. Especially in prokaryotes, open-reading frame (ORF) length alone can serve as a statistically significant gene marker. In addition, codon statistics can be exploited. Species-specific variations in the selection among synonymous codons can be utilized. Since specific pairs of amino acids are often adjacent in proteins, constraints on more probable sequences of bases in a gene can be found. Moreover, since the third codon base is often degenerate, nonrandom relationships in mutations in homologous genes can be exploited. None of these codon-specific statistical variations are available in ncRNA gene-finding.

Finally, the rapid growth in the number of known protein sequences has made it increasingly likely that a new protein-coding gene will have at least some homology with an already-known protein. As a result, sequence-similarity methods (e.g., BLAST searches[11]) are often effective in gene hunting for protein-coding genes.

Again the situation with ncRNAs is more challenging. Far fewer ncRNA sequences are available in the databases. ncRNA sequences can be compared only at the nucleotide level—not as translated amino acids—and, except for ribosomal RNAs (rRNAs), ncRNA sequences are generally short. Consequently, distinguishing weakly conserved genes from random "hits" is more difficult when searching for ncRNAs than for protein-coding genes. Moreover, even in cases where there are large RNA families, sequence conservation is often at the secondary-structure level, i.e., what is conserved are base pairings rather than the individual base sequence. As a result, except for rRNAs or RNAs with well-conserved homologs in closely-related species, conventional sequence-similarity methods have had limited success in ncRNA gene identification. For other RNAs, different methods have been required.

The one class of sequence signals that ncRNAs do have, that are not present in protein-coding genes, are those that result from secondary-structure constraints. For example, many ncRNAs have specific base-pairings, computable low-free-energy folding patterns, unusual base-composition variations or characteristic cross-species patterns of mutations occurring in complementary pairs. As we will see, these secondary-structure signatures have been central in the design of many ncRNA gene-finders.

Algorithms

Nearly all of the current gene-finders for protein-coding-genes, such as Genscan,[12] Genie[13] and Glimmer,[14] use probabilistic algorithms—typically implementing some form of a Hidden Markov Model (HMM).[15] A central feature of these algorithms is a dynamic-programming protocol which is used to "train" the model—i.e., to maximize the selectivity between a training-set of verified genes and a negative training-set of similar sequences which do not represent genes.

However, HMMs are not able to model sequences with secondary structure and consequently are ill-suited for ncRNA gene-modeling. Moreover, as we shall see, fully probabilistic algorithms that can model secondary structure tend to have long CPU execution times and to require a larger training set than may be available. Consequently we will see a variety of deterministic and partially probabilistic algorithms—such as weight-matrix approaches[15,16]—in addition to the fully probabilistic gene-finders. In addition, with ncRNAs there is often only limited training data. Consequently, choices commonly must be made between training with only a small number of sequences for the specific ncRNA gene being sought, or using a larger number of training sequences, that however include a wider assortment of ncRNA genes.

A final issue in ncRNA-gene searches is determining precisely where to search. Should the gene-finding program work equally well on all genomes or only on those with specific properties (e.g., ones with high AT content)? Should the program scan the entire genome or only specific genomic regions? These considerations are generally irrelevant in conventional gene-finders. Gene-finders for protein-coding-genes usually search the entire genome, except for regions of multiply-repeated subsequences. And, aside from the fact that gene-finders are typically designed for either prokaryotic genomes or eukaryotic ones, protein-coding gene-finders generally work comparably well on most genomes. On the other hand, because of the often subtler signals in ncRNA gene searches, it is sometimes advantageous to restrict such searches either to only a portion of the genome (typically to the regions that do not overlap any known protein-coding gene) or to only a limited range of genomes with specific characteristics.

Custom-Designed ncRNA Gene-Finders

We begin with a discussion of programs that are custom-designed to find a single type of ncRNA. Among these the most successful have been gene-finders for the tRNAs and the methylation guide snoRNAs.

tRNA Gene-Finders

The earliest ncRNA gene-finders were custom programs designed to search for tRNAs. During the 1980's and early 1990's, increasingly detailed models of tRNA sequence and secondary structure were developed,[17-19] culminating in tRNAscan by Fichant and Burks[20] and the Pol3Scan linear search algorithm of Pavesi et al.[21] tRNAscan had a sensitivity (i.e., a "true positive" rate) of 95.1% with false positive rate of 0.37 / megabasepair(Mbp).[22] Pol3scan had a sensitivity of 98.6% with false positive rate of 0.23 / Mbp.[22] Both programs used weight matrices taken from the sequences of the known tRNA genes, though the motifs each program looked for were somewhat different. Pol3scan searched only for primary sequence signals—tRNA sequence motifs, transcriptional control elements and terminator sequences from eukaryotic tRNAs—while tRNAscan used tRNA sequence motifs combined with secondary-structure patterns.

Although the sensitivities and specificities of tRNAscan and Pol3scan were impressive, they were not yet fully satisfactory for genomic scanning. For example, a false positive rate of 0.37 / Mbp would imply approximately 1100 false positives in the 3,000 MB human genome. Interestingly, the 7 known tRNAs missed by Pol3scan were (with a single exception) different from the 19 tRNAs missed by tRNAscan.[21] Considering that the two programs looked for somewhat different motifs, this result is not so surprising. But it did suggest that a more powerful gene-finder might be possible if the two programs were combined.

The next improvement in tRNA searching came with the introduction of the COVE program[23] in 1994. COVE is a reconfigurable gene-finder and will therefore be covered in the following section. However, COVE's principal success to date has been in searching for tRNAs, where it achieved a sensitivity of 99.8% with an estimated false positive rate less than 2 / gigabase.[22] Unfortunately, this sensitivity came at a cost; COVE's algorithm could only scan at approximately 20 base pairs / sec[22]—far too slow for routine genomic screening.

Consequently the next efforts at tRNA searching involved developing faster algorithms. FASTRNAscan[24] ran faster than tRNAscan or Pol3scan and far faster than COVE. However, its sensitivity and specificity were worse than that of COVE. tRNAScan-SE[25] on the other hand managed to match the sensitivity and specificity of COVE while decreasing COVE's running time by a factor of 15,000. tRNAScan-SE accomplished this by first running tRNAscan and Pol3scan with more permissive "cutoff values"—in order to rapidly scan a genome for candidate hits. Subsequently the candidate-gene list was passed to COVE where the potential hits were subjected to COVE's stringent testing. Since COVE only had to test a relatively small proportion of the original candidate sequence (1 – 10%), its slow execution speed was not a problem. tRNAScan-SE is now used routinely to rapidly screen newly sequenced genomes for tRNA genes, achieving sensitivities of 99.5% with an expected false positive rate of only 0.07 / gigabase.[22]

Unfortunately, it has not generally been possible to match the success of tRNAScan-SE in searches for other classes of ncRNAs. Probabilistic gene-finders like COVE generally work best when the RNA being sought is highly conserved in primary and secondary structure, and when sequences in multiple species are known and available for use in training. Consequently, tRNAs were ideal targets. Over 1000 tRNA sequences from multiple species have been in the databases[26] for many years. However, for many other types of ncRNAs only a few examples are known and there is little data available for model training.

Searches for Methylation-Guide snoRNAs

One group of RNAs for which sufficient data has become available to develop successful custom gene-finders are the methylation-guide snoRNAs. An early attempt at snoRNA gene-finding used a combination of standard sequence pattern –recognition programs.[27] This approach however generates a large number of false-positive "hits". Consequently, the snoRNA gene search-space was limited to vertebrate introns, since previously identified snoRNAs had been found in these sequences. This approach resulted in the identification of 9 previously unknown methylation-guide snoRNAs.[27]

However, performing a genome-wide search for snoRNAs required a custom-designed, probabilistic search program. Such a program[28]—known as snoscan—was able to predict 22 previously undetected *S. cerevisiae* methylation-guide snoRNAs which were subsequently experimentally confirmed. For 12 of the 22 snoRNAs, the associated methylation site had previously not been known. Snoscan also facilitated identification of snoRNAs throughout the domain of Archaea, after a seed training set was biochemically isolated from a single species.[29]

Despite its successes, snoscan has its limitations. If the snoRNA methylation site is unknown—and hence can not be used as part of the snoRNA signature—the S/N ratio of the program decreases (i.e., the number of false positive increase.) Consequently snoscan has been difficult to apply to larger genomes with unknown methylation sites. In addition, snoscan is only designed to detect methylation-guide snoRNAs, Locating the less-well-conserved pseudouridylation-guide snoRNAs by computational means has not been accomplished to date.

Customized Gene-Finders for Other Classes of ncRNAs

Beyond tRNAs and methylation guide snoRNAs, there are few examples of custom ncRNA gene-finders which have successfully identified new RNA genes. Dandekar and Sibbald predicted several trans-splicing RNAs in a search of the EMBL database of which some candidates were later confirmed experimentally.[30] Lisacek et al[31] developed a weight-matrix-based program which successfully found 132 of 143 known group I catalytic RNAs; however, no new RNAs were predicted with this program. In addition, very recently, custom programs for locating microRNAs [C. Burge, personal communication] and tmRNAs[31a] have been developed, with promising initial results.

Reconfigurable ncRNA Gene-Finders

In addition to the customized RNA gene-finders, user-configurable programs to search for RNA motifs have been developed since the late 1980's.[32,33] With these programs, the user specifies either a "descriptor file", a set of "production rules" or else a multiple-sequence alignment to describe the class of RNAs being searched for.

In programs using the descriptor-file approach, the descriptors typically include primary-sequence motifs, secondary-structure patterns, and gap-lengths between motifs . In most programs of this class, the user can also set additional search parameters such as the allowed number of mismatches in a motif or whether G-U base pairs should be accepted as matches in secondary-structure stems. A typical descriptor file is shown in Figure 1.

In addition to the programs of refs. 32-33, descriptor-file RNA search programs include RNAMOT,[34,35] RNABOB,[36] Overseer,[37] Patscan/Patsearch,[38] Palingol,[39] and RNAMOTIF.[40] With the exception of the recently introduced RNAMOTIF, these programs are all deterministic and consequently have limitations when searching for sequence motifs that are weakly conserved. RNAMOTIF, which is based on the earlier RNAMOT program, is an attempt to introduce the elements of probabilistic searching while maintaining the user-programmable descriptor-file interface of the earlier programs. Specifically, RNAMOTIF introduces user-supplied "score functions" that can incorporate statistical, thermodynamic or other information into the motif-evaluation procedure. Recently, RNAMOTIF has successfully searched for signal recognition protein (SRP) RNAs using an empirical scoring function based on observed biases in nucleotide and base-pair frequencies and loop lengths. Using this scoring function, RNAMOTIF was able to locate SRP RNAs in seven previously unannotated prokaryotic genomes.[40]

A different approach to user-configurability was taken by Searls and collaborators who introduced the concepts of "context-free-grammars"(CFGs) from the field of computational linguistics to ncRNA gene-finding. CFGs are elegant models of ncRNA sequence and secondary-structure in which the descriptor file is replaced with a set of production rules (see Fig. 1). The production rules describe how to generate all the allowed structures of the model (e.g., the class of RNA structures). In this sense they are very similar to the production rules of "regular expressions" from computer science—or their probabilistic counterpart, Hidden Markov Models. However , CFGs can model more complex structures than regular expressions and as a result are able to model RNA secondary structure in addition to primary sequence. For additional details on CFGs in RNA structure modeling, the reader is referred to the original articles[41,42] and earlier reviews.[15] In practice, the CFG models did not predict any new ncRNAs; their importance has been more in laying the foundation for the stochastic-context free-grammar (SCFG) models that followed.

The third class of user-configurable RNA-motif and RNA-gene-finders rely on the input of sequence alignments of known RNAs to train the gene-finder, rather than using either descriptor-files or grammar production rules. The idea is that rather than have the user manually extract the critical features in a family of RNA sequences, the program will do it automatically.

The first examples of this class of programs were the stochastic context free grammars COVE by Eddy and Durbin[23] and the SCFG program of Sakakibara et al[43] These programs are fully probabilistic extensions of the CFG models of RNAs. The Sakakibara program requires a structurally-annotated, multiple-sequence alignment for training. COVE, on the other hand, can—with relatively "ideal" training data—be trained in three different ways: using a multiple-sequence alignment—with or without structural annotation—or just with a set of unaligned sequences.[44] However, in more realistic situations, a structurally-annotated, multiple-sequence alignment for training is also important for COVE to perform well.[44]

The SCFGs have demonstrated advantages over their deterministic predecessors in cases where training data with many aligned sequences were available—e.g., for tRNAs. However, in

A:

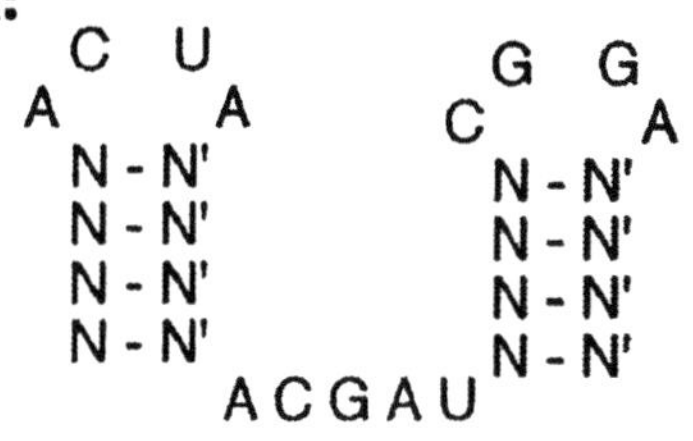

B:

H1 s1 H1 s2 H2 s3 H2

H1 4:4 0
s1 A C U A
s2 A C G A U
H2 4:4 0
s3 C G G A

C:

$$B \rightarrow H_A\ S\ H_B$$
$$H_A \rightarrow a H_{A1} u \mid c H_{A1} g \mid g H_{A1} c \mid u H_{A1} a$$
$$H_{A1} \rightarrow a H_{A2} u \mid c H_{A2} g \mid g H_{A2} c \mid u H_{A2} a$$
$$H_{A2} \rightarrow a H_{A3} u \mid c H_{A3} g \mid g H_{A3} c \mid u H_{A3} a$$
$$H_{A3} \rightarrow a L u \mid c L g \mid g L c \mid u L a$$
$$L \rightarrow a L_1 \ ; L_1 \rightarrow c L_2 \ ; L_2 \rightarrow u L_3 \ ; L_3 \rightarrow a$$
$$S \rightarrow a S_1 \ ; S_1 \rightarrow c S_2 \ ; S_2 \rightarrow g S_3 \ ; S_3 \rightarrow a S_4 \ ; S_4 \rightarrow u$$

Figure 1. A simple hypothetical double stem loop as described by a descriptor file and a Context Free Grammar. A) graphical representation of secondary structure. N indicates any one of the four bases. N' is the complement of N. B) Description of structure using the RNAMOT file descriptor language. C) Structure description using production rules of a Context-Free Grammar. Capital letters indicate CFG "Non-terminals". Note: The production rules for the second stem loop are omitted for brevity. For more details on the file descriptor languages see refs. 34-39. For more details on CFGs see ref. 42 and chapter 10 of ref. 15.

most cases, such extensive training data is not available. In addition, SCFG's cannot describe nonplanar RNA structures such as pseudoknots nor—at least in their current implementations—can they model nonexponential gap-length distributions. In addition, they are also complex; users of SCFG approaches typically face steep learning curves. Finally, and perhaps most importantly, the use of stochastic context free grammars has been limited by their being computationally expensive. Typical SCFG memory costs are of $O(N^3)$ and time costs are of $O(N^4)$ for a sequence of length N.[45] The speed limitation can sometimes be addressed by the use of a fast preprocessor, as was done by tRNAscan-SE. And recent work indicates that the memory demands of SCFGs may also be decreased.[45] However, to date, these technical limitations have limited the widespread application of SCFGs in ncRNA gene-finding.

To address some of the limitations of the SCFG's, a probabilistic model called ERPIN[46] was recently introduced. ERPIN uses weight matrices rather than a SCFG to probabilistically model an RNA sequence alignment. ERPIN requires a secondary-structure annotation along with a trusted multiple-sequence alignment for training. When such an annotated multiple-sequence alignment is available—e.g., with tRNAs—ERPIN performs well. In contrast to the SCFG's, ERPIN can handle RNA pseudoknots and its "reasonable run times"[46] are listed among its advantages compared to SCFGs. On the other hand, ERPIN has only limited capability for handling complex helix-stem "indels"—i.e., insertion and deletions of large structured regions within a base-paired stem. Consequently, ERPIN would be expected to have more difficulty than the SCFG's in handling RNAs with highly variable secondary structures such as Rnase-P RNA.

In any case, ERPIN as well as the SCFGs require multiple-sequence alignments—generally with structural annotations—for training. As a result, the future success of these methods will depend on improvements in the accuracy of RNA sequence alignment and RNA structure prediction. Within the last year, programs such as Dynalign[7] and Foldalign[8] have shown that when RNA sequence alignment and RNA structure prediction are performed simultaneously, the results can be significantly improved compared to when the two operations are performed separately. Foldalign has been used to create sequence alignments for use by COVE in mRNA motif-finding, with encouraging results.[47] In principal, this strategy of supplying more accurate training-alignments to probabilistic RNA motif-finders should improve the results of ncRNA gene-finding as well.

Although the idea of a user-configurable ncRNA gene-finder is appealing, these programs have had only limited success at actually finding new ncRNAs to date. Apart from tRNA searching and the application of RNAMOTIF to SRP RNA searches, there have been few confirmed ncRNA-gene predictions by programs of this class. Gaspin et al identified and experimentally confirmed 46 previously unknown *Pytococcus* methylation-guide snoRNAs using the program Palingol along with genomic sequence comparisons.[47a] Another example was the identification of hammerhead ribozyme RNAs in schistosome satellite DNA by Cedergren and collaborators using RNAMOT.[48] In addition, Cedergren's group predicted other ncRNAs by running RNAMOT against the genomic databases—however, their papers do not indicate whether any of these putative ncRNAs were subsequently experimentally confirmed.[49,50] In fairness, it should be noted that most of the user-configurable RNA-motif finders were not designed primarily to be ncRNA gene-finders. Instead, their main objective has been to detect sequence-motifs and secondary-structure motifs in mRNA, at which they have had more success.[51,52]

De novo ncRNA Gene-Finding—Searching for Genes without a priori Knowledge of Sequence or Structure

As we have seen, when sufficient training data is available, computational searches for well-characterized ncRNAs can be quite successful. But what if one wants to search for RNAs for which there are few examples or none at all? At first this sounds impossible. How can one search for RNAs without knowledge of their primary sequence motifs or secondary structure? Yet it is possible to design such searches and, remarkably, in the last year such methods have succeeded in finding many new ncRNAs.

Algorithms to find completely unknown RNAs have fallen into three classes. The first group are based on finding stable secondary structures. Others exploit variations in ncRNA base-composition relative to the genomic background . Finally, some use genomic sequence comparisons among related species.

Structure-Based de novo Gene-Finding

The first de novo methods were based on secondary-structure computations. These methods exploited the observation that calculated thermodynamic free energies of ncRNAs are generally lower than those of random sequences with the same base composition. Hence the gene-finding program would segment the genome into fragments of the typical ncRNA-length (e.g., 100 or 200 base pairs) and compare the computed minimum free energies of the sequence fragments with randomized versions of the same sequences. If the calculated free energy of the sequence was significantly less than that of the randomly shuffled sequences, then one would predict the presence of an ncRNA gene.[53,54] One example of this approach was the program of Chen et al that searched for RNA pseudoknots.[55] Alternately one could look for genomic sequence fragments capable of being folded into specific types of RNA secondary structures in the spirit of RNA-folding programs such as Mfold[56] and ViennaRNA.[57] Looking for folding patterns—rather than computing free energies—had the advantage of generally being faster, while producing similar predictions.[58] Unfortunately, these methods have not led to the discovery of new ncRNAs. Moreover, computer experiments with known targets and randomized sequences suggested that secondary-structure computations by themselves would never be successful for de novo ncRNA gene-finding.[58]

Gene-Finders Using Base-Composition Variations

The same paper[58] which demonstrated the limitations of using secondary-structure alone to search for ncRNAs, also suggested that (G+C)%, i.e., the percent of G and C bases in a sequence, might serve as a signature for the presence of an ncRNA gene. Subsequently, three groups[59-61] have successfully applied this idea to de novo gene-finding. Klein et al[59] and Schattner[60] searched for ncRNAs in thermophillic archaebacteria with high (A+T)% genomic backgrounds. Klein et al combined (G+C)% with the QRNA comparative genomics method[59] described below to search for ncRNAs in *M. jannaschii* and *P. furiosus*. Schattner examined variations in multiple base-composition statistics including (G+C)% , (G-C)% difference and dinucleotide frequency variations. Among these statistics (G+C)% and the frequency in the 'CpG' dinucleotide were observed to vary significantly between ncRNAs and the genome in the thermophile *M. jannaschii*. (Although the increased occurrence of CpG dinucleotides in *M. jannaschii* ncRNAs is reminiscent of the CpG islands of mammalian protein-coding-gene regions, there is currently no evidence indicating that they are in any way related.) Predictions from the two investigations in *M. jannaschii* were similar, though not identical. Northern blots performed by Klein[59] showed that 4 of the 6 *M. jannaschii* ncRNAs predicted by both approaches are in fact expressed. In addition, Klein et al. predicted and experimentally verified 7 new ncRNAs in *P. furiosus*.

One of the conclusions of these two groups—that base-composition oriented gene-finding is primarily useful only with thermophiles—is somewhat discouraging. Nevertheless, the authors did suggest ways that the method may have wider applicability. Klein et al noted that ncRNAs may be found in nonthermophiles by first finding their homologs in a thermophillic species. Schattner observed that even in some nonthermophiles, such as *Caenorhabditis elegans*, significant base-composition variations between ncRNAs and the background exist. Although in *C. elegans* these variations are not sufficient to serve as a de novo gene-finder by themselves (as seen in Fig. 2), they may still be useful as a supplementary component of a gene-finder that also includes secondary-structure or primary-sequence motifs.

RNAGenie, developed by Carter et al,[61] incorporates base-composition variations—along with primary sequence motifs and free-energy calculations—in a "neural network" ncRNA gene-finder. Their work is particularly interesting since it was applied to *E. coli* and other nonthermophiles that might not be expected to be good candidates for a (G+C)% based

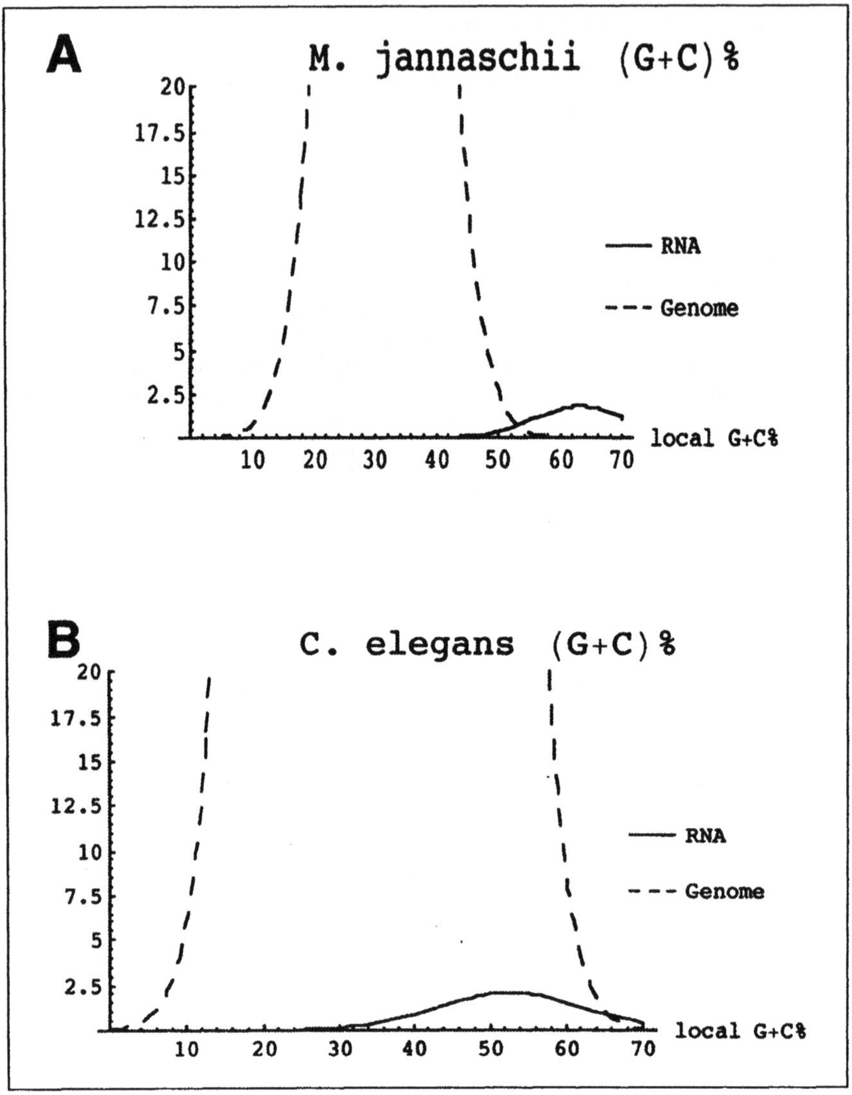

Figure 2. Separation of RNA's and genomic background using G+C%. Vertical axes indicate estimated relative number of 100 bp subsequences. Note that peak of curve for number of genomic sequences is truncated. RNA estimate assumes ratio of protein coding genes to ncRNA genes is approximately equal to that in *S. cerevisae*. Graphs are shown as normally distributed for purpose of illustration—actual distribution of G+C% may vary. A) *M. jannaschii* RNA and genome G+C% distributions are separated enough to enable discrimination between RNA and background populations. B) *C. elegans* chromosome X. Although *C. elegans* ncRNA and genomic G+C% population means are significantly different, ncRNA distribution can not be distinguished from that of the background by G+C% alone. (modified from ref. 60 by permission, copyright 2002 Oxford University Press.)

gene-finder. Using their method, Carter et al find 370 putative novel ncRNAs in *E. coli*. Although no experimental testing of these candidate RNAs was performed in their original work, 13 of their predictions have been subsequently confirmed with Northern analysis.[61] In addition, Carter et al listed 10 previously known ncRNAs which had not been in their training set, seven of which were successfully found by RNAGenie. However, until a systematic verification of their predictions is performed, it will be unclear how many of their remaining 350 candidates are true ncRNAs and how many are simply false positives.

Gene-Finding Using Comparative Genomics

Perhaps the most exciting development in the area of de novo ncRNA gene-finding has come from three recent studies based on comparative genomics.[1,2,64] Each of these methods looks for regions of homology among two or more related genomes. The idea is that regions of biological importance—e.g., loci of ncRNAs—will be more conserved than regions that do not have any genes. So far these algorithms have been applied only to intergenic regions to avoid the large number of false positives likely to arise from homologous protein-coding genes.

In the method of Wasserman et al,[1] local cross-species sequence conservation was the only bioinformatic signature used. Since this resulted in a large number of putative "hits", Wasserman et al complemented their computational search with an experimental screen using micro-arrays. When applied to the *E. coli* genome (with comparisons to 5 related bacterial genomes), their method predicted 60 new ncRNAs of which 18 have been experimentally confirmed.[63]

In contrast, Argaman et al[64] combined the search for conserved sequences among bacterial genomes with a computational screen for nearby promoter and terminator sequence motifs. Since Argaman et al were searching in *E. coli*—where promoter and terminator sequence motifs are known and relatively well conserved—the method worked well. They made 24 predictions of novel ncRNAs of which 14 have been experimentally verified.[63] However, their method is difficult to apply to nonbacterial genomes for which promoter and terminator sequences are much less well conserved, or even to other bacterial genomes that have different promoter signatures.

Probably the most promising of the comparative approaches is the QRNA program of Rivas and Eddy.[2,62] Their method not only searches for regions of cross-species homology, but also examines the nature of the mismatches that occur among the aligned sequences. The key idea, illustrated in Figure 3, is that if a region contains a protein-coding gene, then mismatches between homologous sequences should frequently correspond to a synonymous codon or a codon for a closely related amino acid. In contrast, if the region contains an ncRNA, then a higher percentage of substitutions should occur in complementary pairs such that the underlying ncRNA secondary structure is preserved despite the substitutions of the individual bases. Finally, if the region does not contain any gene, then the distribution of interspecies mismatches should correspond to their background base frequencies. The appeal of this approach is that it should apply to any genome for which related sequenced genomes are available. Knowledge of promoter and terminator consensus sequences is not required. And since QRNA uses comparative information specific to RNA secondary-structure (in contrast to the methods of ref. 1 or ref. 64), it may be able to find ncRNAs while searching an entire genome—and not just the intergenic regions. QRNA has already been applied successfully to the *E. coli*,[2] *M. jannaschii* and *P. furiosus*[59] genomes and shows promise to being applicable to a wide range of additional genomes. Of course, since QRNA relies on secondary-structure signatures, it will have difficulty finding ncRNAs that have little or no secondary structure.

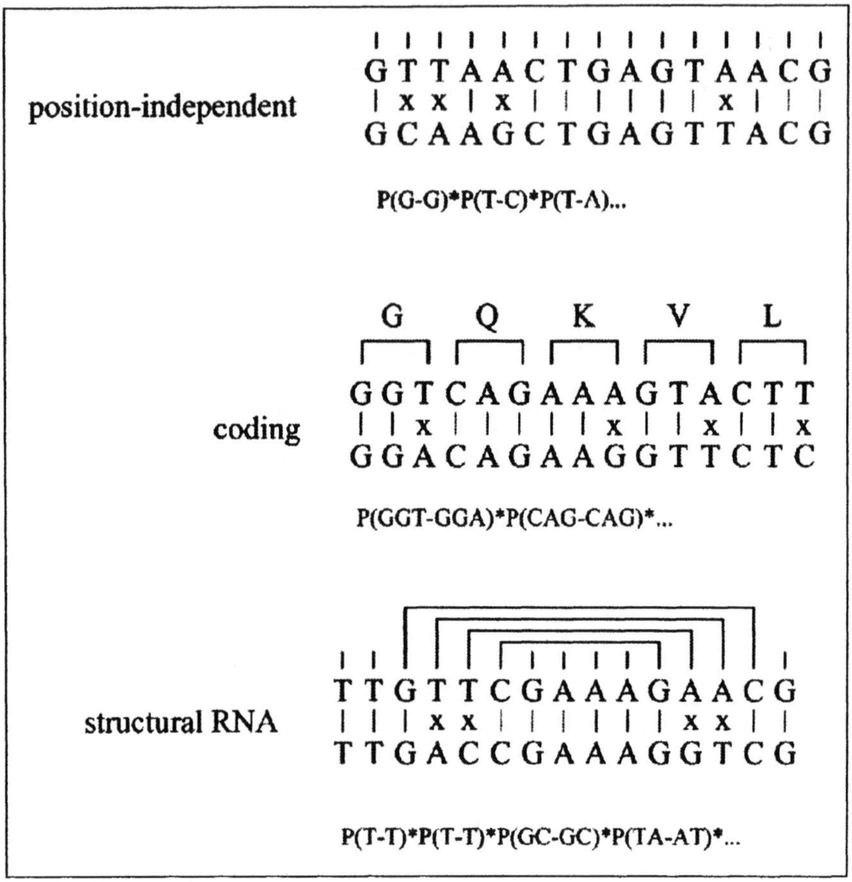

Figure 3. QRNA sequence alignment for protein-coding, structural-RNA-coding and non-coding sequences. Three pairwise alignments of identical composition with identical base substitutions can be classified by distinctive patterns of mutation caused by different selective constraints. The figure indicates how each alignment is scored according to the model that best fits the pattern of mutations: one position at a time for the position-independent model, one codon at a time for the protein-coding model (integrated overall six possible reading frames) and as a combination of base paired positions and single positions for RNA (integrated over all possible secondary structures). For more details see ref. 62. (modified from ref. 62 by permission, copyright 2001 Sean R. Eddy.)

Current Status and Future Prospects for Computational ncRNA Gene-Finding

So just how good are the current computational gene-finders? And what are the prospects for improvement in the near future? As we have seen, in a few cases—such as tRNAs or methylation-guide snoRNAs—current programs work very well. However, in most cases the accomplishments have been more modest. For example, from our knowledge of the number of pseudouridylation sites in eukaryotic rRNA alone, we can be almost certain that there are dozens to hundreds, as yet unidentified, snoRNAs in essentially all eukaryotic genomes. For other classes of RNAs where we have no reliable information as to the total number of RNAs, we simply don't know whether the current programs are performing well or not. For example, were the 31 novel ncRNAs recently found in *E. coli*[63] almost all of the ncRNAs that had previously escaped detection? Or are they only the "tip of the iceberg"? Estimates of the true number of *E. coli* ncRNAs vary, but in reality the performance of current ncRNA gene-finders is still unknown.

In traditional machine-learning analysis, algorithm performance is generally evaluated by dividing the known examples into "training" and "testing" data sets. The program under evaluation is trained solely using the training data set and assessed with the testing data set. However, when few examples are known (the usual situation for ncRNA gene-finders), a modified procedure, "jack-knife testing" is typically used instead. In jack-knife testing, a single known example is sequentially removed from the training set and the program is trained with the remaining data. The program is then tested on its ability to find the omitted example. While eliminating the pitfalls of using the same data for training and testing, jack-knife testing still assumes that the known examples adequately represent the range of targets remaining to be found; otherwise, conclusions from jack-knife testing may be misleading. For example, Carter et al[61] demonstrated using jack-knife testing that RNAGenie had between 90.9% to 93.8% sensitivity in *E. coli* depending on the threshold parameter they chose (see Table 2 of ref. 61). On the other hand, of the 31 novel ncRNAs found by refs. 1, 2, and 64, only 13 were predicted by RNAGenie[61] suggesting a lower sensitivity at finding new ncRNAs.

These observations remind us that only experimental testing can confirm or refute the predictions of a computational gene-finder. Yet even when one uses experimental verification (e.g., Northern analysis or micro-array data) to assess computational gene-finders, one must proceed with caution. A negative result may simply indicate that the ncRNA isn't expressed under the specific cellular environment or the specific tissue type being assayed. On the other hand, even a positive identification on a Northern or micro-array may merely represent the presence of some other stable RNA such as an mRNA leader sequence. Of course, these experimental issues are not insurmountable and can be addressed by careful testing over multiple tissue types and cell environments.[3,4] However, they do remind us that even experimental results—when based on limited data—may not be sufficient to assess the performance of a computational gene-finder.

Despite these caveats, we may speculate a little on the future of ncRNA gene-finding. My belief is that all three classes of gene-finders—the customized gene-finders, the user-configurable motif-finders and the de novo programs—will continue to be important and will be used in a synergistic manner in the next few years. As additional genomes are sequenced, the comparative genomics gene-finders will be able to generate additional candidate ncRNAs. These candidates, along with those identified by the experimental screens,[3,4] will produce additional examples from the known classes of ncRNAs. These new examples will provide additional training data for the custom gene-finders, thereby improving their performance. Meanwhile better RNA-sequence-alignment and structure prediction programs should generate improved models of previously unknown classes of RNAs which can, in turn, be input into the reconfigurable RNA-motif-finders.

So, in the end, how many ncRNAs can we expect to find? One tantalizing hint may have come from the recent publication [65] of human-mouse sequence comparisons. This work showed that 66% of the strongly conserved, syntenic regions between mouse chromosome 16 and the corresponding human chromosomes do not overlap protein-coding exons. Intriguingly, the average length of these conserved syntenic regions is 189 base pairs[65]. How many of them encode ncRNA genes? No one knows. However, with the computational and experimental screening methods already available, the answer should become apparent soon. And if even a fraction do prove to be ncRNAs—as suggested by the recent screens in *E. coli*[1,2,64] and other organisms[3,4]—then locating all these new ncRNAs should provide a window into an exciting modern RNA world far richer than many had previously believed it to be.

Acknowledgements

I am grateful to Dr. Jan Barciszewski for encouraging me to write this article and Dr. Todd Lowe for a critical reading of the manuscript. I would also like to thank Drs. Sean Eddy and Daniel Gautheret for helpful comments on the manuscript.

References

1. Wassarman KM, Repoila F, Rosenow C et al. Identification of novel small RNAs using comparative genomics and microarrays. Genes Dev 2001; 15: 1637-51.
2. Rivas E, Klein RJ, Jones TA, Eddy SR. Computational identification of noncoding RNAs in *E. coli* by comparative genomics. Curr Biol 2001; 11: 1369-73.
3. Huttenhofer A, Kiefmann M, Meier-Ewert S et al. RNomics: an experimental approach that identifies 201 candidates for novel, small, non-messenger RNAs in mouse. EMBO J 2001; 20:2943-53.
4. Tang TH, Bachellerie JP, Rozhdestvensky T et al. Identification of 86 candidates for small non-messenger RNAs from the archaeon Archaeoglobus fulgidus. Proc Natl Acad Sci USA 2002; 99:7536-41.
5. Knudsen B, Hein J. RNA secondary structure prediction using stochastic Context-Free Grammars and evolutionary history. Bioinformatics 1999; 15:446-54.
6. Holmes I, Rubin GM.Pairwise RNA structure comparison with stochastic Context-Free Grammars. Pac Symp Biocomput 2002; 163-74
7. Mathews DH, Turner DH. Dynalign: an algorithm for finding the secondary structure common to two RNA sequences. J Mol Biol 2002; 317:191-203
8. Gorodkin J, Heyer LJ, Stormo GD.Finding the most significant common sequence and structure motifs in a set of RNA sequences. Nucleic Acids Res 1997; 25:3724-32.
9. Mignone F, Gissi C, Liuni S et al. Untranslated regions of mRNAs. Genome Biol 2002; 3(3):
10. Stormo GD. Gene-finding approaches for eukaryotes. Genome Res 2000; 10:394-7.
11. Altschul SF, Gish W, Miller W et al. Basic local alignment search tool. J Mol Biol 1990; 215: 403-10.
12. Burge C, Karlin S. Prediction of complete gene structures in human genomic DNA. J Mol Biol 1997; 268:78-94.
13. Salzberg SL, Delcher AL, Kasif S et al. Microbial gene identification using interpolated Markov models. Nucleic Acids Res 1998; 26:544-8.
14. Kulp D, Haussler D, Reese MG et al. A generalized hidden Markov model for the recognition of human genes in DNA.Proc Int Conf Intell Syst Mol Biol 1996; 4:134-42.
15. Durbin R, Eddy SR, Krogh A et al. Biological Sequence Analysis. Cambridge: Cambridge Unversity Press, 1998.
16. Gribskov M, McLachlan A, Eisenberg D. Profile analysis: detection of distantly related proteins. Proc Natl Acad Sci USA 1987; 84:4355-8.
17. Staden R. A computer program to search for tRNA genes. Nucleic Acids Res 1980; 8:817-25.
18. Margalit H, Shapiro BA, Oppenheim AB et al. Detection of common motifs in RNA secondary structures. Nucleic Acids Res 1989; 17:4829-45.
19. Marvel CC. A program for the identification of tRNA-like structures in DNA sequence data. Nucleic Acids Res 1986; 14:431-5.
20. Fichant GA, Burks C. Identifying potential tRNA genes in genomic DNA sequences. J Mol Biol 1991; 220:659-71.
21. Pavesi A, Conterio F, Bolchi A et al. Identification of new eukaryotic tRNA genes in genomic DNA databases by a multistep weight matrix analysis of transcriptional control regions. Nucleic Acids Res 1994; 22:1247-56.
22. Data taken from Table 1 of ref. 25.
23. Eddy SR, Durbin R. RNA sequence analysis using covariance models. Nucleic Acids Res 1994; 22:2079-88.
24. el-Mabrouk N, Lisacek F. Very fast identification of RNA motifs in genomic DNA. Application to tRNA search in the yeast genome. J Mol Biol 1996; 264:46-55.
25. Lowe TM, Eddy SR. tRNAscan-SE: A program for improved detection of transfer RNA genes in genomic sequence. Nucleic Acids Res 1997; 25:955-64.

26. Sprinzl M, Horn C, Brown M et al. Compilation of tRNA sequences and sequences of tRNA genes. Nucleic Acids Res 1998; 26:148-53
27. Nicoloso M, Qu LH, Michot B et al. Intron-encoded, antisense small nucleolar RNAs: the characterization of nine novel species points to their direct role as guides for the 2'-O-ribose methylation of rRNAs. J Mol Biol 1996; 260:178-95.
28. Lowe TM, Eddy SR. A computational screen for methylation guide snoRNAs in yeast. Science 1999; 283: 1168-71.
29. Omer AD, Lowe TM, Russell AG et al. Homologs of small nucleolar RNAs in Archaea. Science. 2000; 288:517-22.
30. Dandekar T, Sibbald PR. Trans-splicing of pre-mRNA is predicted to occur in a wide range of organisms including vertebrates. Nucleic Acids Res 1990; 1816:4719-25.
31. Lisacek F, Diaz Y, Michel F. Automatic identification of group I intron cores in genomic DNA sequences. J Mol Biol 1994; 2354:1206-17.
31a. Laslett D, Canback B, Anderson S. BRUCE: a program for the detection of transfer-messenger RNA genes in nucleotide sequences. Nucleic Acids Res 2002; 30:3449-53.
32. Staden R. Methods to define and locate patterns of motifs in sequences. Comput Appl Biosci 1988; 4:53-60.
33. Saurin W, Marliere P. Matching relational patterns in nucleic acid sequences. Comput Appl Biosci 1987 Jun3(2):115-20.
34. Gautheret D, Major F, Cedergren R. Pattern searching/alignment with RNA primary and secondary structures: an effective descriptor for tRNA. Comput Appl Biosci 1990; 64:325-31.
35. Laferriere A, Gautheret D, Cedergren R. An RNA pattern matching program with enhanced performance and portability. Comput Appl Biosci 1994; 102:211-2.
36. Eddy SR. RNABOB, http://www.genetics.wustl.edu/eddy/software/#rnabob.
37. Winker S, Overbeek R, Woese CR et al. Structure detection through automated covariance search. Comput Appl Biosci 1990; 64:365-71.
38. Pesole G, Liuni S, D'Souza M. PatSearch: a pattern matcher software that finds functional elements in nucleotide and protein sequences and assesses their statistical significance. Bioinformatics 2000; 16:439-50.
39. Billoud B, Kontic M, Viari A. Palingol: a declarative programming language to describe nucleic acids' secondary structures and to scan sequence database. Nucleic Acids Res 1996; 24:1395-403.
40. Macke TJ, Ecker DJ, Gutell RR et al. RNAMotif, an RNA secondary structure definition and search algorithm. Nucleic Acids Res 2001; 29:4724-35.
41. Brendel V, Busse HG. Genome structure described by formal languages. Nucleic Acids Res 1984; 12:2561-8.
42. Dong S, Searls DB. Gene structure prediction by linguistic methods. Genomics 1994; 23:540-51.
43. Sakakibara Y, Brown M, Hughey R et al. Stochastic Context-Free Grammars for tRNA modeling. Nucleic Acids Res 1994; 2223:5112-20.
44. Ref. 15, chapter 10.
45. Eddy SR. A memory efficient dynamic programming algorithm for optimal alignment of a sequence to an RNA secondary structure. BMC Bioinformatics, 2002; in press.
46. Gautheret D, Lambert A. Direct RNA motif definition and identification from multiple sequence alignments using secondary structure profiles. J Mol Biol 2001; 313:1003-11.
47. Gorodkin J, Stricklin SL, Stormo GD. Discovering common stem-loop motifs in unaligned RNA sequences. Nucleic Acids Res 2001; 29:2135-44.
47a. Gaspin C, Cavaille J, Erauso G et al. Archaeal homologs of eukaryotic methylation guide small nucleolar RNAs: Lessons from the *Pyrococcus* genomes. J Mol Biol 2000; 297:895-906.
48. Ferbeyre G, Smith JM, Cedergren R. Schistosome satellite DNA encodes active hammerhead ribozymes. Mol Cell Biol 1998; Jul18(7):3880-8.
49. Bourdeau V, Ferbeyre G, Pageau M et al. The distribution of RNA motifs in natural sequences. Nucleic Acids Res 1999; 27:4457-67.
50. Ferbeyre G, Bourdeau V, Pageau M et al. Distribution of hammerhead and hammerhead-like RNA motifs through the GenBank. Genome Res 2000; 10:1011-9.
51. Lescure A, Gautheret D, Carbon P et al. Novel selenoproteins identified in silico and in vivo by using a conserved RNA structural motif. J Biol Chem 1999; 274:38147-54.

52. Dandekar T, Hentze MW. Finding the hairpin in the haystack: searching for RNA motifs. Trends Genet 1995; 112:45-50.
53. Le SY, Chen JH, Braun MJ et al. Stability of RNA stem-loop structure and distribution of non-random structure in the human immunodeficiency virus (HIV-I). Nucleic Acids Res 1988; 10:5153-68.
54. Le SY, Chen JH, Currey KM et al. A program for predicting significant RNA secondary structures. Comput Appl Biosci 1988; 4:153-9.
55. Chen JH, Le SY, Maizel JV. A procedure for RNA pseudoknot prediction. Comput Appl Biosci 1992; 8:243-8.
56. Zuker M, Stiegler P. Optimal computer folding of large RNA sequences using thermodynamics and auxiliary information. Nucleic Acids Res 1981; 10(9):133-48.
57. Schuster P, Fontana W, Stadler PF et al. From sequences to shapes and back: a case study in RNA secondary structures. Proc R Soc Lond B Biol Sci 1994; 255:279-84.
58. Rivas E, Eddy SR. Secondary structure alone is generally not statistically significant for the detection of noncoding RNAs. Bioinformatics 2000; 16:583-605.
59. Klein RJ, Misulovin Z, Eddy SR. Noncoding RNA genes identified in AT-rich hyperthermophiles. Proc Natl Acad Sci USA 2002; 99:7542-7.
60. Schattner P. Searching for RNA genes using base-composition statistics. Nucleic Acids Res 2002; 30:2076-82.
61. Carter RJ, Dubchak I, Holbrook SR. A computational approach to identify genes for functional RNAs in genomic sequences. Nucleic Acids Res 2001; Oct 1(2919):3928-38.
62. Rivas E, Eddy SR. Noncoding RNA gene detection using comparative sequence analysis. BMC Bioinformatics 2001; 21:8.
63. Eddy SR. Computational genomics of noncoding RNA genes. Cell 2002; 109:137-40, Table 1.
64. Argaman L, Hershberg R, Vogel J et al. Novel small RNA-encoding genes in the intergenic regions of Escherichia coli. Curr Biol 2001; 11:941-50.
65. Mural R, Adams M, Myers E et al A comparison of whole-genome shotgun-derived mouse chromosome 16 and the human genome. Science 2002; 296:1661-71.

CHAPTER 4

Xist RNA Associates with Chromatin and Causes Gene Silencing

Anton Wutz

Abstract

The mammalian *Xist* gene produces a long, spliced and poly-adenylated noncoding RNA that is uniquely distributed in the nucleus. *Xist* RNA spreads in *cis* from its site of transcription over the entire X-chromosome and mediates X-inactivation, the transitional silencing of one of the two X-chromosomes of female cells. X-inactivation is required for compensation of the dosage difference of X-linked genes between XY males and XX females. A counting and choosing mechanism ensures that *Xist* is expressed from all but one X-chromosome in a random manner. Repression of *Xist* RNA accumulation on the single active X-chromosome is dependent on another large noncoding RNA, the *Tsix* RNA, transcribed in antisense orientation to *Xist*. Initiation of *Xist* expression occurs early in development when cellular differentiation has not yet progressed and involves stabilization of the RNA. Transcriptional repression follows the accumulation of *Xist* RNA with little delay. At a later stage of cell differentiation *Xist* expression does no longer trigger silencing suggesting that *Xist* mediated silencing is restricted to certain cell types which have not yet undergone differentiation. The function of *Xist* RNA has been studied in mouse and humans. Although, it is unclear at present how *Xist* RNA interacts with and spreads over chromatin and effects gene silencing recent studies have begun to shed light on the underlying mechanism. A number of chromosomal proteins and histone modifications have been implicated, in *Xist* mediated gene silencing including variant forms of histone H2A, methylation of H3 and hypoacetylation of H4. The Nova1/2 family of RNA binding proteins has recently been implicated in the function of *Xist* RNA. Chromosomal association has also been observed with the *roX1* and *roX2* RNAs in *Drosophilaee.* In contrast to mammalian *Xist* RNA these RNAs localize to the fly X-chromosome in a sequence directed manner and mediate enhanced transcription of the single male X. Despite mechanistic differences the mammalian *Xist,* and the fly *roX1* and *roX2* RNAs are at present the only RNAs that have been shown to associate with chromatin over the lengths of entire chromosomes and therefore might be viewed as a unique class of chromatin associated RNAs.

Introduction

The problem of sex determination has found many solutions during evolution in diverse organisms. Providing the advantage to enable reshuffling of the genetic information by meiotic recombination during gametogenesis the implementation of particular sex determining systems is driven by a net evolutionary benefit for the advancement of species but draws signifi-

Noncoding RNAs: Molecular Biology and Molecular Medicine, edited by Jan Barciszewski and Volker A. Erdmann.

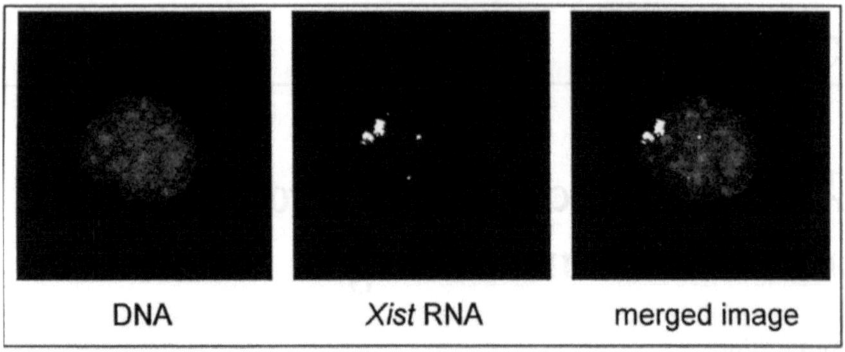

Figure 1. Nuclear localization of mouse *Xist* RNA in the interphase nucleus is visualized by RNA FISH. The RNA forms a cluster over the inactive X in the periphery of the nucleus. The DNA is counter stained to show the extent of the nucleus.

cant cost for the individual organism. Mammalian sex is specified by the unequal genetic constitution of males and females. Males carry the dominant sex determining *SRY* gene on the Y chromosome, which is required for the specification of Sertoli cell fate in the proto-gametogenic organ and forms the basis of the mammalian mode of genetic sex determination (for review see ref. 1). The evolution of the heteromorphic sex chromosomes (X and Y) in mammals results in a difference in the number of genes between XY males and XX females. Thereby, the gene dosage difference of X-linked genes has significantly greater weight as the Y chromosome holds only a small number of genes.[2] To enable successful development a dosage compensation system is required to balance the gene copy of sex linked genes. This is achieved by transcriptional silencing of one of the two X chromosomes in female cells leaving both sexes with one active X.

X-inactivation was encountered for the first time over five decades ago, when a study led by Murray L. Barr aiming at the analyzis of microanatomical changes in feline neurones after stress exposure serendipitously observed a dense staining nuclear structure in female cells that was not present in males. This work was reported in 1949.[3] Although, the meaning of the so-called sex chromatin or Barr body was unclear at the time it spurred follow-up studies. Ten years later it became clear that the Barr body contained a heterochromatic X chromosome and X-inactivation was recognized.[4] In mammals all but one X-chromosome are inactivated per diploid genome and only one X-chromosome is active in XY males as well as XX females (Fig. 2). Thereby, the choice which chromosome is inactivated is random such that each X-chromosome has the chance to become the active one (Xa). Although, the molecular basis for this phenomenon remains mysterious up to present days it is clear that a mechanism must operate early in mammalian embryogenesis that counts the number of X chromosomes, chooses the one-to-be-active, and inactivates all others. Genetic studies based on chromosomal translocations involving the X led to the identification of a single region on human Xq13 that was sufficient and necessary to trigger X inactivation (scored then by the presence of a Barr body). This locus was termed *X inactivation center* (*XIC*). Positional cloning efforts led to the discovery of a gene that mapped to the *XIC* and was expressed exclusively from the inactive X chromosome (Xi). In 1991 the human *XIST* gene mapping to the cytological band Xq13 [5] and mouse *Xist* located on the X chromosome at 42.0 cM [6,7] were reported. *Xist* is an unusual gene. Its product is an untranslated RNA that adheres to chromatin in *cis* of its transcription site and spreads over the entire chromosome causing its heterochromatinisation (Fig. 1). *Xist* expression precedes chromosomal inactivation [8] and gene targeting studies in mouse have demonstrated that *Xist* is required for silencing of the X-chromosome [9,10] clearly establishing *Xist* as a critical

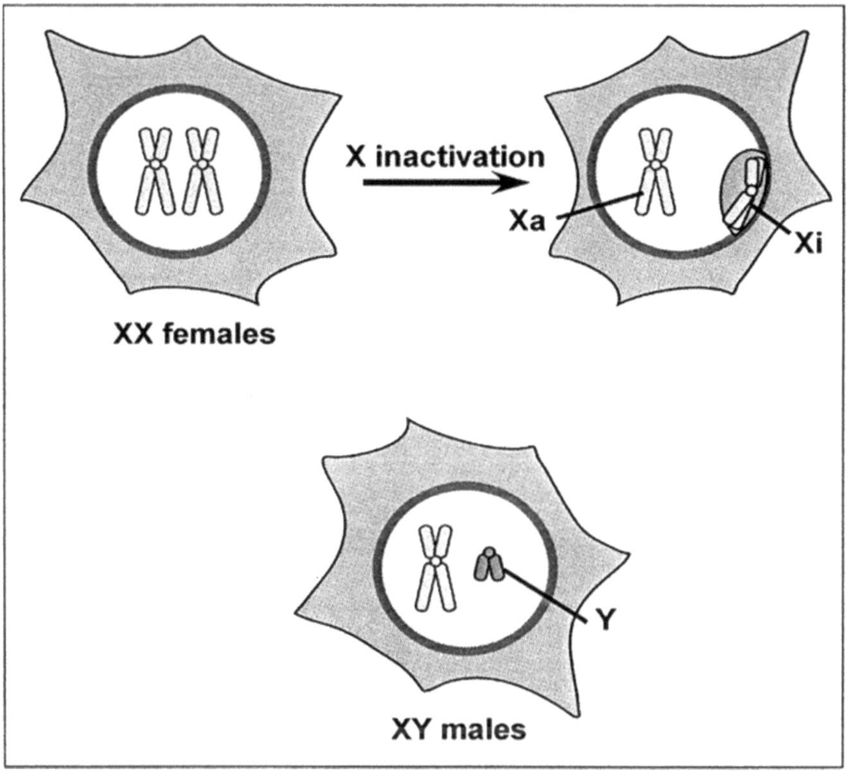

Figure 2. The dosage difference inherent to the genetic constitution of XY males and XX females is balanced by X inactivation in mammalian species. Thereby, one of the two female X chromosomes is inactivated, such that in both sexes one X is active.

regulator. In the same way the discovery of *Xist* put an end to the search for the cause of X-inactivation it raised a set of new questions. Today, *Xist* is studied as a model case for RNA interaction with chromatin and for developmentally regulated changes in chromatin structure. This chapter summarizes what has been learned about *Xist* RNA, what concepts for the understanding of the molecular mechanism have emerged, and what future research will likely focus on.

The *Xist* Gene

The sequences of mouse, human, bovine and vole *Xist* have been determined and the respective gene structures have been established.[6,11-14] *Xist* is 60 percent conserved among the four mammalian species, which is comparable to the average conservation of 5- and 3-prime untranslated regions of mammalian genes but significantly lower than for protein coding sequences. The transcription unit of *Xist* spans approximately 20-30 kilobases of genomic sequence and is split into 8 exons (Fig. 3). The positions of introns are conserved with a few exceptions between different species.[13] The RNA is spliced and poly-adenylated, whereby the human *XIST* gene displays various isoforms due to alternative splicing.[12] Alternative splicing and alternative poly-adenylation site usage generate a number of different transcripts from which the longest are 17 kilobases in mouse and 19 kilobases in humans.[11-14] The large size of

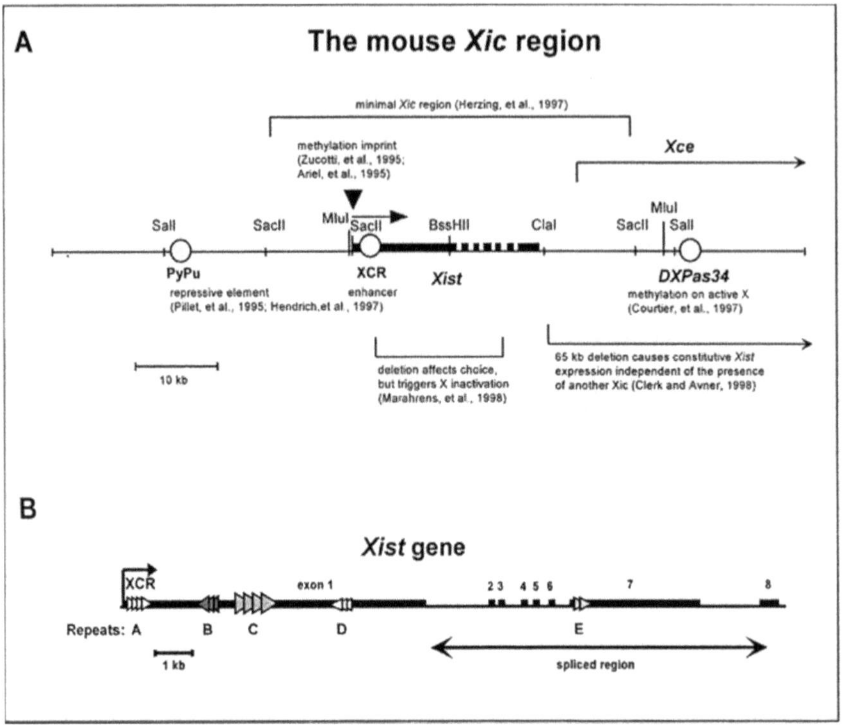

Figure 3. The mouse *Xic* and *Xist* locus are depicted. A) The schematic representation of the mouse *Xic* region shows the *Xist* gene and elements that have been implicated in its regulation. In the *Xist* promoter region a methylation imprint has been identified as a negative regulator and the XCR repeat has been shown to act as a positive regulator in a posttranscriptional manner.[69,70] In the *Xist* upstream region a poly-pyrimidine and polyuridine rich region (PyPu) has been implicated in a repressive function. The *Xce* region extends 3-prime of *Xist* and is a major negative regulator of *Xist* expression. DNA methylation on sequences around the genetic marker DXPas34 has been implicated in this regulation.[71] The location of functionally tested regions are also indicated (see text). B) The diagram shows the mouse *Xist* gene structure. The sequence of *Xist* contains five different repeat elements (repeat A to E) that are conserved between species.

the RNA together with the presence of rare splice variants have led to some confusion about the structure of the gene, which could only be recently clarified taking advantage of the rapidly growing sequence databases.[13] *Xist* RNA lacks significant open reading frames and is not associated with polysomes suggesting that it is unlikely encoding protein.[11,12] A number of repeats (A to E) have been identified in the sequence of *Xist* whose monomer sequences, copynumber and order within *Xist* are conserved among different species (Fig. 3).[11,14] Repeat E is an exception as it is present in high copy number in mouse but only rudiments can be found in the human and vole sequence.

Xist displays a unique pattern of localization inside the nucleus. The RNA associates with chromatin and spreads over the entire chromosome, from which it is transcribed. Fluorescence in situ hybridization (FISH) experiments have made it possible to impressively demonstrate this localization pattern at high resolution (see Fig. 1).[12,15] In mouse *Xist* RNA can be observed attached to metaphase chromosomes in 10 to 20 percent of chromosomal spreads (Note: it is critical that spreads are prepared from cells which have not been arrested with spindle toxins such as Colcemid or Nocodazol). Metaphase localization has not been reported for *XIST* in human cells and it is not clear if there is a genuine difference between the RNAs or it is a consequence of the different cellular context. On mouse metaphase chromosomes *Xist* displays a banding pattern suggesting that its binding is influenced by the regional gene content along the chromosome.[16] This is not unexpected since quantification of *Xist* RNA abundance indicated that there are only 2000 molecules in the cell nucleus, which certainly precludes the possibility that the RNA covers all of the chromosomal DNA.[17] It has been suggested that repetitive elements widely dispersed in the genome might serve as docking stations for *Xist* RNA.[18] However, this possibility has not yet been experimentally tested. At present, it is unclear if *Xist* has specificity or preference for certain chromosomal features. Experiments involving *Xist* transgenes integrated on autosomes have shown clearly that *Xist* RNA can associate in cis with autosomes suggesting that there is no genuine specificity of *Xist* to bind to X chromosomal chromatin. However, it is clear that there are certain chromosomal domains, which are refractory to *Xist* localization. One such example is the pseudoautosomal region of the mouse and human X chromosomes, which is not subject to X-inactivation and from which *Xist* RNA is excluded.[19-21] There are also a number of genes which escape inactivation on Xi. Escape from X-inactivation has been extensively studied in human cells and it appears that such genes are located together in clusters.[21] Other regions have been found with different transgene insertions into autosomes in mouse. These include the very distal part of mouse chromosome 11.[22,23] Further examples include human translocation chromosomes between the X and chromosomes 6 and 16, where *Xist* does not spread into the autosomal region of the chromosome suggesting that it might be refractory to *Xist* localization.[24] These observations clearly document that *Xist* accumulation along the chromosome is influenced by regional differences in chromatin, but the determining factors are unclear at present.

Initiation of *Xist*-Mediated Silencing in the Embryo

Xist is expressed in a developmentally controlled manner. In particular studies in laboratory mice have gained insight into the regulation of *Xist* in exquisite detail. *Xist* transcription can be observed at an early stage in female embryonic development and it appears that X inactivation is initiated in close connection with cellular differentiation in the embryo.[8] In mouse imprinting determines that the male X chromosome is inactivated in the extraembryonic lineages.[25] At the blastocyst stage X inactivation has occurred in the trophectoderm, the first cell lineage entering differentiation in the preimplantation embryo. In the inner cell mass (ICM) of the blastocyst *Xist* expression is low and the RNA accumulates only at the site of transcription, but does not spread over the chromosome.[26] Even in male ICM cells *Xist* expression can be detected as a pinpoint signal using FISH techniques. During epiblast differentiation at the beginning of gastrulation (around 7 to 8 days after fertilization) *Xist* transcription is upregulated on all but one X-chromosomes.[26,27] Thereby, each of the X-chromosomes has a statistical chance of staying active (not expressing *Xist*) which gives rise to a random pattern of X-inactivation in the differentiating cell progeny. It should be noted that once a cell has chosen one X to inactivate all progeny of this cell will maintain the same Xi, such that the female embryo consists of patches of cells with either the paternal or the maternal X inactivated. In human X inactivation is random in the embryonic and extraembryonic lineages. Imprinted X inactivation is only observed in the extraembryonic lineages of rodents and might be an evolutionary relict as also in marsupials the paternal X chromosome is preferentially inactivated [28] (although no orthologs of *Xist* have been identified yet in marsupials).

The process of random X-inactivation can be experimentally recapitulated in mouse embryonic stem (ES) cells.[10,23,26,29] ES cells can be obtained from outgrowths of the ICM. They proliferate indefinitely in culture and when culture conditions are modified they can be induced to differentiate into a variety of embryonic cell types. Female mouse ES cells have two active X-chromosomes similar to the cells of the inner cell mass of blastocysts or the pregastrulation epiblast.[26] Upon differentiation one of the two X-chromosomes becomes inactivated [30] and a number of studies have taken advantage of this system to analyze the chromosomal transition of an active to an inactive X.

Gene targeting experiments in mice and ES cells have shown that *Xist* is required for the initiation of chromosome wide silencing.[9,10] However, once the inactive state has been established *Xist* is no longer required for maintenance in somatic cells. This has been demonstrated by the observation that fragments of the inactive X in somatic cell hybrids which do no longer contain an *Xic* (and thus do not express *Xist*) remain inactive.[31] Deletion of *Xist* by conditional mutagenesis in mice has confirmed this finding.[32] However, *Xist* seems to contribute to the stability of the inactive state, as in somatic cells reactivation frequency of individual genes on the inactive X is elevated, when *Xist* is deleted on the chromosomes.[33] This leads to the interpretation that *Xist* initiates silencing on the X chromosome, but later multiple other chromosomal modifications take over the function and act synergistically. Studies in mouse ES cells have addressed the transition of an active to an inactive X chromosome. It appears that X-inactivation is a stepwise process that involves a number of sequential chromatin adaptations (Fig. 4). *Xist* expression is the initial trigger for inactivation and it has been shown that transcriptional repression follows the localization of *Xist* RNA with little or no delay.[23] In S-phase, the inactive chromosome replicates late compared to other chromosomes, a feature that often correlates with inactive chromatin. At the early stages of inactivation methylation of Lysine 9 has been observed together with acetylation and demethylation of Lysine 4 on histone H3.[34,35] This seems to be the first molecular marker of repressive chromatin. Later yet hypoacetylation of histone H4 can be observed.[30] Then variants of histone H2A are also reshuffled on the inactive X: macroH2A1 a H2A variant with a large C-terminal domain is recruited, whilst H2A.Bbd and H2A.z are both underrepresented on Xi.[36-38] Finally, these modifications culminate in the formation of the Barr body as a heterochromatic structure which localizes to the periphery of the nucleus.

Xic function has been recapitulated outside the X-chromosomal context on autosomes with transgenes containing either genomic sequences [29,39,40] or cDNA sequences of *Xist*.[23] From these studies it has become clear that *Xist* can efficiently associate in *cis* with chromatin, even when expressed form autosomal sites. By using an inducible expression system to regulate *Xist* expression the function of the RNA during ES cell differentiation was analyzed.[23] *Xist* has the capacity to initiate silencing in undifferentiated and early differentiating ES cells. If *Xist* expression is induced later in ES cell differentiation the RNA still localizes to chromatin, but does no longer cause transcriptional repression. This suggests that *Xist* mediated silencing is restricted to cells that have not progressed far in differentiation. This interpretation is further supported by studies carried out in cell lines where *Xist* expression can be induced experimentally by demethylating agents.[41] Therefore, *Xist* is not only required but also sufficient for chromosome wide silencing assuming expression is initiated at the right time in a permissive cellular environment.

When experimentally induced, *Xist* expression causes long range silencing in undifferentiated ES cells. In contrast to female somatic cells, the silent state in ES cells is reversible such that genes are reactivated once *Xist* expression is turned off. At this state inactivation occurs independent of chromosomal modifications like late replication or hypoacetylation of histone H4.[23] This suggests that ES cells represent a model system to study the initiation of X-inactivation and the specific role of *Xist* for causing transcriptional repression. Later in ES cell differentiation the silent state becomes irreversible and independent of *Xist* expression. This transition from a reversible *Xist*-responsive to an irreversible state is accompanied by *Xist* loosing its ability to trigger silencing. The complex kinetics of this transition make it likely that a change in chromatin structure occurs at this point in cellular differentiation which underlies the change

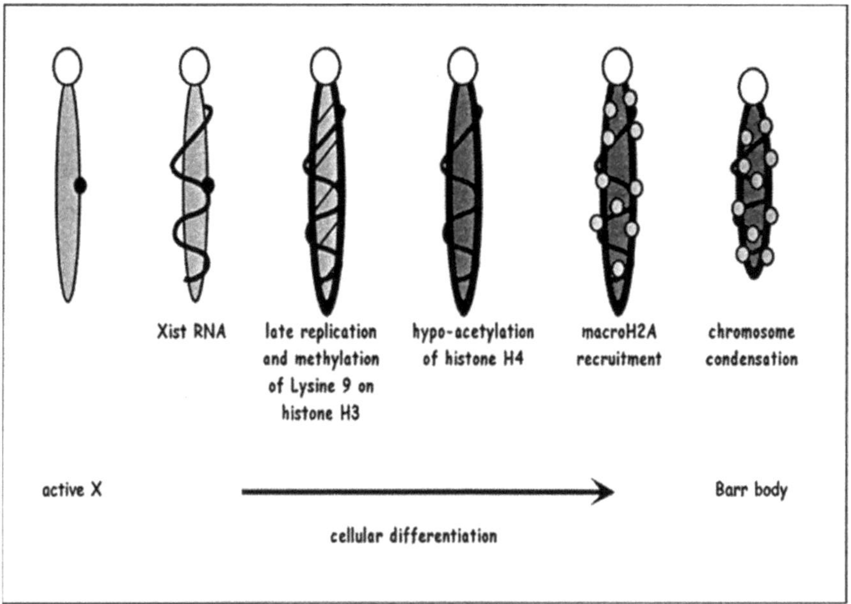

Figure 4. The chromosomal transition of an active to an inactive X is a multistep process and the order of events been studied in differentiating mouse ES cells. Once the cell has chosen, which X chromosome to inactivate, *Xist* expression is upregulated and causes transcriptional repression of X-linked genes. As a consequence transcriptional repression, methylation and deacetylation of Lysine 9, and hypomethylation of Lysine 4 on histone H3 can be observed. Upon further differentiation the chromosome starts to replicate late in S-phase and becomes hypoacetylated on histone H4. Later, H2A variants are reshuffled and macroH2A is recruited to the Xi.

in chromosomal behavior. Notably, X-inactivation has to be initiated at a time when silencing is reversible suggesting that the reversible inactivation is a part of the normal X-inactivation process in female embryos. The phase of reversible silencing might also play an important role for the mechanism of X inactivation (see below).

Regulation of *Xist* Expression—Counting and Choosing the Xs

Xist is regulated by a mechanism ensuring that precisely one X remains active in diploid cells. The number of X-chromosomes is determined before the initiation of X-inactivation by an unknown mechanism (referred to as counting) and a competence factor has been postulated that mediates *Xist* upregulation in cells with more than one X (Fig. 5). The determining factor by which the cell recognizes an X chromosome is the *Xic* region. Thus, the same genomic region that triggers the chromosomal silencing process via *Xist* RNA transcription signals that an X chromosome is present in the nucleus (Fig. 4A). When the number of X-chromosomes exceeds one, *Xist* expression will be triggered from all but one *Xic*. One X is protected and remains the active X chromosome (Xa). For the protection of the future Xa a theoretical activity has been postulated which is referred to as blocking factor. It is clear form studies based on genetically defined deletions in the *Xic* region that *Xist* transcription is influenced by positive and negative regulation and RNA stability has also been shown to be involved in the accumulation of *Xist* RNA.[26,27] The mouse *Xic* region has been confined to a roughly one megabase large region by translocations.[13] In humans, it has been genetically defined to several hundred

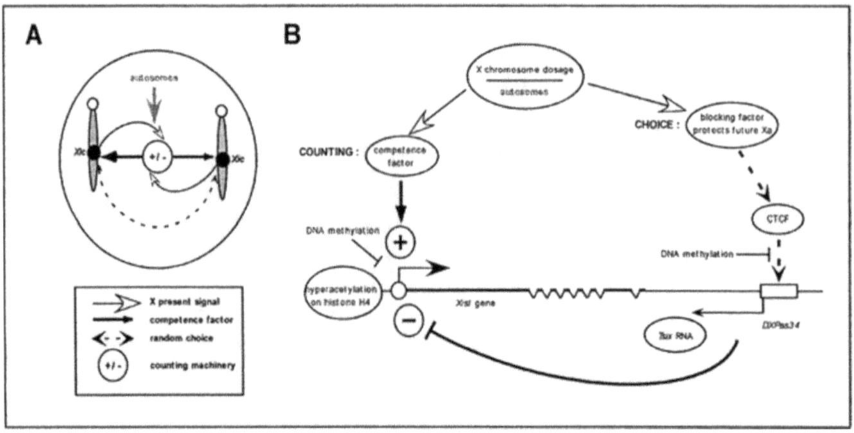

Figure 5. A model of counting and choosing the active X chromosome. A) The diagram shows the signals that are sent from and received by the *Xic*. A counting mechanism measures the number of X chromosomes and determines if inactivation should take place. A hypothetical competence factor has been suggested to mediate this signal. Further regulation is required to ensure that one X chromosome is protected and stays active. B) Schematic representation of elements involved in the regulation of *Xist* (see also (Fig. 3A) and text).

kilobases in size.[42] By studying a panel of autosomal insertions of transgenes containing the genomic region of the mouse *Xic* in mouse ES cells it was concluded that the relevant elements for signalling the presence of an additional *Xic* are confined within 80 kilobases.[22] Although, these transgenes work in high copynumber, a single copy integration of a 450 kb *Xic* YAC transgene will not reliably trigger X-inactivation suggesting that there are other X chromosomal elements that maybe enable or enhance *Xic* function.[40]

Gene targeting experiments in mouse ES cells have further probed the function of different elements in the *Xic* region. Deletions of part of the *Xist* transcription unit do not abolish counting of the mutant chromosome.[9,10] This shows that *Xist* sequences that are clearly crucial for chromosomal silencing are not required for counting. However, deletions in *Xist* can alter the choice of the chromosome for inactivation.[43] A deletion of *Xist* including the promoter was reported to neither affect counting nor choice suggesting that *Xist* transcription per se is not required for counting or choosing.[10] Still a nonrandom pattern of X-inactivation is observed in mutant ES cells due to cell selection and loss that occurs secondarily in the differentiating cell population. This indicates that the presence of two active X chromosomes compromises proliferation or survival of cells during differentiation.

Choice - Xce - Tsix

Once the number of X chromosomes has been determined a mechanism (referred to as choosing) has to decide which of these to inactivate. In mouse, the X-chromosome to stay active is chosen randomly but the likelihood with which a particular X is inactivated is influenced by a genetically defined region including sequences 3-prime to *Xist*, which has been termed X controlling element (*Xce*).[44] Although, no human counterpart has been identified to date studies of the *Xce* region have provided important insight into the workings of the choosing mechanism. A large deletion encompassing sequences of 65 kilobases 3-prime of *Xist* leads to constitutive inactivation of the mutant X chromosome even if no other *Xic* is present in the cell.[45] This indicates that sequences that normally ensure repression of *Xist* are located in this 3-prime region.

One component of the *Xce* is the noncoding *Tsix* RNA, which has been identified as a regulator of *Xist* expression in cis.[46,47] *Tsix* is transcribed in antisense orientation to *Xist* from a promoter 20 kb downstream of the 3-prime end of *Xist* and spans the entire *Xist* transcription unit including the promoter. The idea has been put forward that *Tsix* prevents the accumulation of *Xist* on the future active X.[48] Although, this interpretation is supported by the expression profile of mouse *Tsix* there are several points to be clarified. *Tsix* has only a marked effect in cells that have more than one X chromosome. It appears that interference with *Tsix* transcription renders the mutant chromosome more likely to become inactivated implicating *Tsix* as a negative regulator of *Xist* expression. However, deletion of *Tsix* in male cells does not lead to increased *Xist* expression.[47,49,50] This shows that *Tsix* is not absolutely required to repress *Xist*. To get around this observation a competence mechanism has been envisioned according to which *Xist* expression is actively licensed once more than one X are present. Consequently, *Tsix* is only required in female cells to repress *Xist*.[47] The general relevance of *Tsix* has been questioned as it appears to be poorly conserved between other mammals. A human *TSIX* gene has been described, but it has important differences to the mouse gene that make it unlikely to be a relevant player in regulating *XIST* expression.[51] This could indicate that *Tsix* is a mediator of the *Xce* effect in mice but future work is clearly needed to reconcile the different observations with a central role of *Tsix* in choosing and counting.

Autosome Number

It is assumed that X inactivation regulates the dosage of X linked genes relative to the autosomal gene content or nuclear volume, but at present the nature of the reference for counting is unknown. The molecular basis of counting is complex indeed. This is exemplified by the observation of X-inactivation patterns in tetraploid cells. As expected, studies in tetraploid conceptuses in mouse and human show that two X-chromosomes per tetraploid autosome set are kept active (in analogy to one per diploid). But this observation is deceiving, as it is strongly biased by cells selection. Dosage of two active X chromosomes in tetraploid nuclei is probably the genetically least compromised setting that gives such cells a much greater capability to contribute to the conceptus compared to cells with aberrant numbers of active X chromosomes. The actual effect of tetraploidy on counting might be quite different. In tetraploid mouse embryos, cells with abnormal numbers of active and inactive X-chromosomes have been observed.[52] The situation is even more dramatic in differentiating tetraploid embryonal carcinoma cell lines [53] suggesting that the molecular mechanism for counting and choice is severely compromised in cells with tetraploid genome content. The autosomal influence on counting remains therefore unclear, although, for theoretical reasons some connection is likely to exist. Notably, some insight into how autosomal genes influence X chromosome choice has been gained. A dominant mutation implicated in *Xce* function has been recently mapped genetically to a region on the middle of chromosome 15 [54] and the DNA binding factor CTCF has been implicated in the regulation in *Tsix* expression.[55] However, functional studies of these genes will be needed to clarify their roles.

The Mechanism of *Xist* Function

Factors Interacting with Xist

Despite over a decade of intense study the molecular players that interact with *Xist* to achieve chromosomal silencing are still poorly characterized. There are a number of predictions that can be made. First RNA binding proteins must recognize *Xist* RNA. Second, from *Xist* localization it can be inferred that the RNA has to associate with chromosomal proteins and this interaction also needs to effect silencing in the end. And third, an interface layer of proteins would hardly be surprising considering how complex biological systems are in general. Today, a few proteins have been implicated in the process. Biochemical studies have implicated general RNA binding proteins as interactors of *Xist*. Heterogeneous RNA binding protein (hnRNP) C1 and C2 have been shown to interact with *Xist* RNA in vitro.[56] However, the significance of this observation remains unclear, as these proteins will bind virtually any RNA. Other factors include the Nova family of RNA binding proteins, which have been implicated indirectly, by interference studies employing antisense peptide nucleic acids.[57] For both RNA binding proteins it is presently not clear if they are recruited in *Xist* dependent manner to the Xi, and their role for the mechanism remains unclear.

Factors that are recruited to the Xi late in cellular differentiation have been identified. This is exemplified by the localization of the histone H2A variant macroH2A1.[36] MacoH2A localizes to the Xi in strictly *Xist* dependent manner. When *Xist* is deleted from the Xi by conditional mutagenesis in mouse somatic cells, enrichment for macroH2A is lost despite the chromosome staying transcriptionally repressed, late replicating and hypoacetylated on histone H4.[32] Similarly, induction of *Xist* expression in somatic cells leads to accumulation of macroH2A, despite transcriptional silencing is not effected.[58,59] This demonstrates that macroH2A is recruited dependent on *Xist* RNA, but independent of silencing. However, the function of macroH2A in the X inactivation process remains unclear, because it is localized to the Xi at a very late stage in cellular differentiation when normally X-inactivation is largely completed. Undoubtedly, macroH2A could be one of the redundant factors that are involved in the maintenance of the silent state.

A number of chromosomal modifications occur on the inactive X, and it stands to reason that some of them might be affected by a general program of the cell that leads to progressive heterochromatinisation once the process has been initiated by *Xist* expression. The existence of such an autonomous program can be envisioned by looking on the centromeric heterochromatin or the Y chromosome, which is by and large heterochromatic. In ES cells both regions are methylated on Lysine 9 on histone H3 and replicate late, but acetylation of H4 is comparable to autosomal active chromatin. Upon differentiation hypoacetylation of H4 occurs, probably as a consequence of the preexisting chromatin modifications or the repressed state. To view certain modifications on the chromatin of the Barr body as effects of its silent state is a useful concept to separate the cause and consequence of transcriptional repression. But, it should be noted that the idea of an autonomous heterochromatinisation program is rather theoretical and it is likely flawed at higher resolution as heterochromatin certainly reveals some specificity making the generality of such a program questionable.

The fact that factors such as macroH2A are clearly recruited independent of the autonomous silencing program and relatively late in the process of X inactivation suggests that *Xist* RNA acts at multiple stages to effect a variety of molecular transformation. Future, exploration might turn up a multitude of new markers for the Barr body and in fact a number of markers have recently been encountered. It is presently unclear if the absence of H2A.z and H2A.Bbd are caused by silencing or by *Xist* localization and macroH2A1 recruitment.[38] Also, an autoimmune serum (referred to as serum 154) has been identified that contains antibodies that specifically stain the Barr body at later stages of differentiation.[60] Although, the protein recognized by the serum has not yet been identified it heralds new insight in the molecular signature of the Xi.

The critical proteins with which *Xist* interacts in the initiation of X inactivation remain to be determined. However, it is clear that there are at least two classes of factors that decorate the Barr body. First, factors are localized by interaction with *Xist* RNA in a direct or indirect manner, these also include factors needed to initiate silencing. A second type of modifications and structural components of heterochromatin are consequences of the repressed chromatin state and are effected by default.

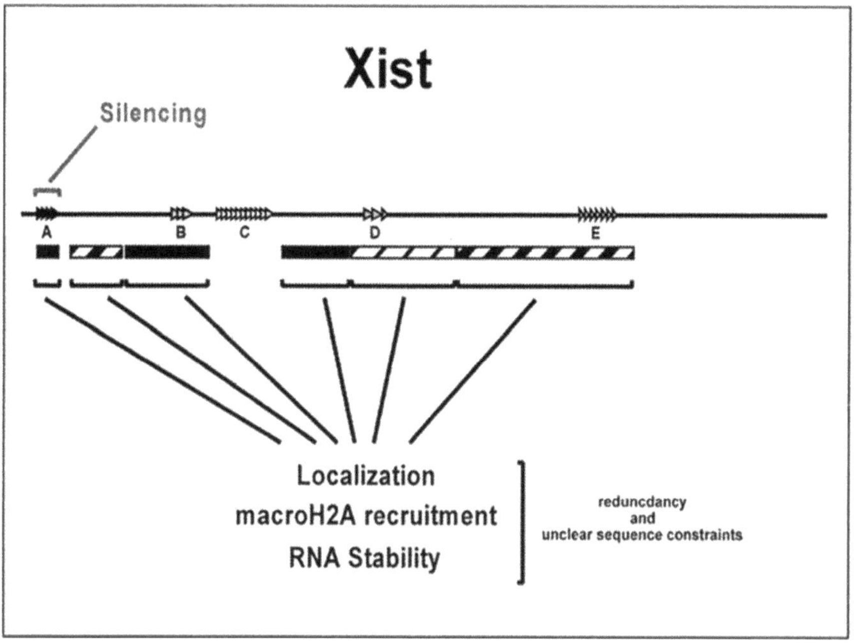

Figure 6. Sequences of *Xist* required for chromosomal localisation and silencing have been identified. Repeat A is crucial for silencing, whilst other sequences of *Xist* (indicated by shaded boxes beneath, the darkness of shading corresponds to relative importance) are required for localization of the RNA to chromatin and spreading in cis.

The Structure of Xist RNA

By looking at the sequence of the *Xist* RNA insight has been gain into the mechanism of *Xist* function. *Xist* is a large RNA, which makes computational structure predictions difficult. However, it is clear from the sequence that *Xist* is not likely to form a compact folded RNA. *Xist* might have some local folded domains that are linked together by long regions with undefined structure. The sequence requirements for silencing and localization have been explored by a deletion analysis of *Xist* RNA in mouse ES cells.[59] Specific mutants of *Xist* RNA have been generated that are able to localize to chromatin in cis and spread over the chromosome, but are not causing gene silencing. This shows that localization and transcriptional repression can be mechanistically separated and has important implications for models of *Xist* RNA localization, as silencing and hence chromatin condensation is unlikely to be involved. The first repeat on the 5-prime end of *Xist* (repeat A also named XCR for *Xist*-conserved-repeat, see Fig. 6) was identified as a critical sequence to effect silencing. This repeat consists of seven to eight copies of a sequence motif that can fold into an RNA structure comprising two stem loops (Fig. 7). These might be structural features that are recognized by factors that mediate transcriptional repression.

Repeat A only functions together with other sequences of *Xist* that mediate RNA localization, demonstrating that RNA localization is a prerequisite for silencing. Multiple spatially separated sequences within *Xist* that are redundant and act cooperatively mediate localization to chromatin and spreading of the RNA over the chromosome (Fig. 6). No motifs have been identified between these redundant sequences yet, suggesting that either heteromeric

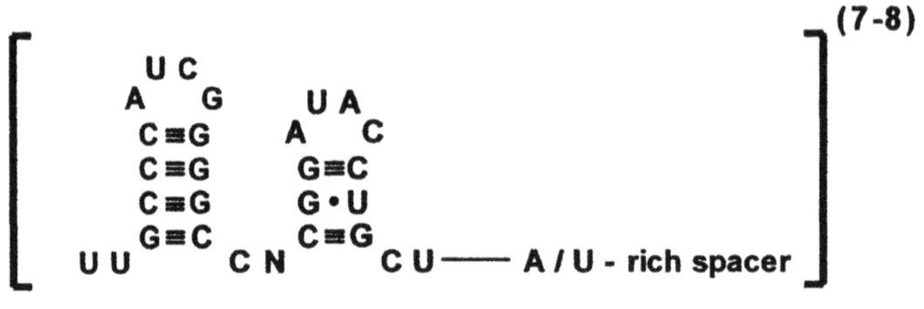

Figure 7. A critical sequence motif of repeat A is predicted to fold into two stem loops, which could be binding sites for factors that repress transcription. If these loops are mutated the sequence does no longer effect silencing.

complexes are formed that function in a redundant manner or the sequences of binding sites on *Xist* RNA are not heavily constrained. The latter could point towards the presence of low affinity binding sites. Such sites could be efficiently occupied by factors only, if a mechanism for cooperative binding exists such as to endow the final assembled complex with a significant stability.

A Model for *Xist* Function

Lessons Learned and Open Questions about the Silencing Mechanism

The mechanism by which *Xist* RNA spreads from its site of synthesis over the entire chromosome is still unclear. Among the possible scenarios are the two orthogonal alternatives, namely specific transport or restricted diffusion. Specific transport of the RNA requires the existence of a machinery that is capable of transporting the RNA over the chromosome, which has not been describes thus far. In addition such a mechanism predicts a defined sequence responsible for localization that acts as a docking site for the transport machinery and a bimodal mutant phenotype, namely either transport or no localization. In contrast, a panel of different *Xist* mutants displays a graded potential to localize to chromatin and to spread over the chromosome. Generally, an inverse correlation between the size of the deletion and the ability to localize is followed. Additionally, the sequence requirements for localization are not clearly defined suggesting that multiple proteins bind to *Xist* and heteromorphic complexes cooperate to achieve localization. These findings suggest that localization of *Xist* RNA to the chromosome might be achieved by a mechanism based on restricted diffusion of *Xist* RNA, cooperative binding of *Xist* containing complexes to each other and to chromatin, and stability of the RNA (Fig. 8). This model is supported by a number of observations but present data cannot exclude the possibility that specific transport is involved. Notably, *Xist* RNA mutants that fail to localize to chromatin do not give rise to stable RNA. This could be a way of avoiding inappropriate *Xist* localization to chromatin at remote sites in the nucleus. A number of different functionally redundant sequences or modules is required for efficient localization. This suggests that more than one interaction site per RNA molecule are needed to interact in all three dimensions for binding to chromatin and to other *Xist* RNA complexes. After its transcription, *Xist* will bind to factors that mediate its chromosomal attachment ensuring that the RNA is not freely diffusing away from its site of production but allowing it to travel a decent distance. Binding to other *Xist* containing complexes will enable more efficient binding to chromatin, establishing a mechanism of cooperative binding. Thereby, the stability of the

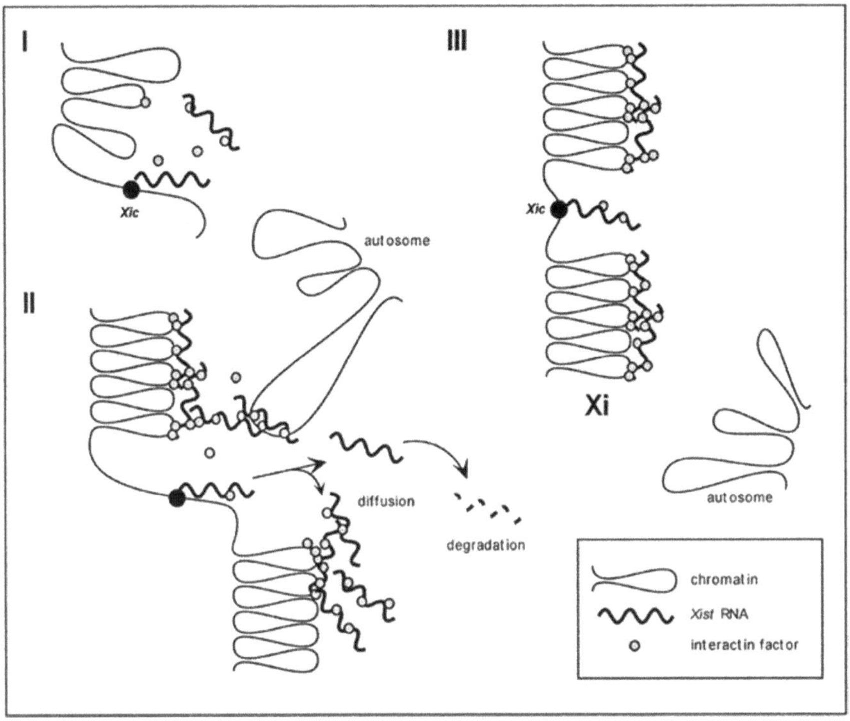

Figure 8. A model for *Xist* RNA localization to chromatin in cis is depicted. Different stages (I to III) are shown to clarify important features. (I) *Xist* RNA is transcribed off the *Xic* and associates with binding proteins. The complexes formed are unstable due to the presence of only low affinity binding sites. (II) When the concentration of *Xist* is sufficiently high a cooperative binding mechanism ensures complex stability. One of the factors in the cooperative binding process are docking sites on chromatin and *Xist* will preferentially localize to chromatin in proximity to its site of synthesis (which potentially could also include regions of other chromosomes which are by chance in close proximity to the *Xic*). Complexes that fail to associate with chromatin diffuse away and are degraded. (III) *Xist*-mediated silencing is initially reversible and upon cell division inadvertently silenced parts of autosomes will be reactivated, because they are unlinked to the *Xic* and are therefore unlikely to remain in close distance. This adjusts inactivation pattern before it becomes irreversible and explains the precise pattern normally found in cells.

complex is dependent on the local abundance of factors. Conversely, once they fail to attach to chromatin, freely diffusible *Xist* complexes are inherently unstable avoiding silencing of unrelated chromosomes. Proximity to both chromatin and other complexes is a prerequisite for efficient binding, thereby, mediating RNA stability. Indeed, stabilization of *Xist* RNA has been documented at the onset of X-inactivation.[26,27]

The proposed mechanism of *Xist* localization ensures preferential inactivation of chromatin in proximity to the transcription site of *Xist* and, hence, of the X compared to other chromosomes, but it can hardly explain the high fidelity of the naturally observed phenomenon. An additional feature of X-inactivation might come into play here. *Xist*-mediated silencing is reversible early in ES cell differentiation. Reversibility of silencing provides for the possibility to correct for inadvertently silenced regions of other chromosomes. Notably, a window corresponding to roughly the time of one cell division exists in ES cell differentiation within which silencing becomes irreversible and *Xist* loses its ability to initiate silencing.[23] This transition might allow the cell to reactivate inappropriately silenced genes in the course of a cell division cycle and effectively carry out the "proofing" of the silencing pattern before making it irreversible and committing to it for good.

Evolutionary Considerations

Mammals have evolved a unique dosage compensation system for which no parallel exists in other phyla and orthologs of *Xist* have not been identified in other vertebrates. Yet the components of the X-inactivation machinery are unlikely to have arisen de novo in evolution. Conceivably, proteins involved in other cellular pathways have been adapted for the purpose of dosage compensation. RNA interaction with chromatin might in fact be an evolutionary old scheme. Two noncoding RNAs have been found involved in dosage compensation in the fruit fly *Drosophilae melanogaster.*[61,62] These RNAs, *roX1* and *roX2*, localize to the male X chromosome in flies and are required for spreading of the dosage compensation complex.[63,64] In contrast to *Xist* these RNAs can act in trans, such that their recruitment to the X is determined by features on the X-chromosome rather than their transcription site. Also, fly dosage compensation leads to enhanced transcription of the single male X to adjust for the presence of two X chromosomes in female flies.[65] Despite this mechanistically quite different effect, it is stunning to see that chromatin associated RNAs are essential components of dosage compensation systems in diverse phyla as mammals and flies.[66] Recently, RNA has been implicated in the structure of the pericentric heterochromatin in mice,[67] giving yet another example of a function of RNA in chromatin. Noncoding RNAs also function in the regulation of expression domains and gene clusters. The *Air* RNA has been shown to be required for repression of genes in the imprinted gene cluster on mouse chromosome 17 including the *Igf2r* gene.[68] In this case the number of genes and the extent of the chromosomal domain regulated is small, but *Air* leads to repression in *cis* exclusively on the paternally inherited chromosome from which it is transcribed. This may suggest a similarity to *Xist* function in X-inactivation. The involvement of RNA in chromatin structure and gene regulation might point to a common feature of RNA that makes it so useful for the regulation of chromatin domains. Also, the variety and abundance of RNA binding proteins in the cell might serve as a basis for the evolutionary adaptation for specific processes such as dosage compensation.

Concluding Remarks

Functional studies of the X inactivation mechanism will allow insight into the epigenetic regulation of gene expression in mammals. It appears that *Xist* mediated silencing of the X chromosome is a paradigm for a powerful epigenetic system that is capable of heterochromatizing an entire chromosome and determine its specific nuclear localization. It is expected that similar interactions underlie regulation of other genes - however, with less dramatic consequences. The involvement of an RNA in chromatin formation is an intriguing finding, and might also pinpoint a general process by which transcription feeds back onto chromatin structure. The outline of the mechanism of *Xist* function has been determined in the past, now it needs to be filled with molecular details. Therefore, the isolation of interacting factors is a major focus of ongoing research. Equally important is the understanding of the molecular basis of counting and the stunning changes in chromatin behavior during cellular differentiation. Much about the workings of the nucleus will certainly be learned in the near future and noncoding RNAs hold great promise to yield information of how the cell achieves the regulation of chromatin domains.

References

1. Goodfellow PN, Lovell-Badge R. SRY and sex determination in mammals. Annu Rev Genet 1993; 27:71-92.
2. Lahn BT, Page DC. Four evolutionary strata on the human X chromosome. Science 1999; 286(5441):964-967.
3. Barr ML, Bertram EG. A morphological distinction between neurons of the male and female, and the behavior of the nucleolar satellite during accelerated nucleoprotein synthesis. Nature 1949; 163:676-677.
4. Lyon MF. Gene action in the X-chromosome of the mouse (Mus musculus L.). Nature 1961; 190:372-373.
5. Brown CJ, Ballabio A, Rupert JL et al. A gene from the region of the human X inactivation centre is expressed exclusively from the inactive X chromosome. Nature 1991; 349(6304):38-44.
6. Borsani G, Tonlorenzi R, Simmler MC et al. Characterization of a murine gene expressed from the inactive X chromosome. Nature 1991; 351(6324):325-329.
7. Brockdorff N, Ashworth A, Kay GF et al. Conservation of position and exclusive expression of mouse *Xist* from the inactive X chromosome. Nature 1991; 351(6324):329-331.
8. Kay GF, Penny GD, Patel D et al. Expression of *Xist* during mouse development suggests a role in the initiation of X chromosome inactivation. Cell 1993; 72(2):171-182.
9. Marahrens Y, Panning B, Dausman J et al. *Xist*-deficient mice are defective in dosage compensation but not spermatogenesis. Genes Dev 1997; 11(2):156-166.
10. Penny GD, Kay GF, Sheardown SA et al. Requirement for *Xist* in X chromosome inactivation. Nature 1996; 379(6561):131-137.
11. Brockdorff N, Ashworth A, Kay GF et al. The product of the mouse *Xist* gene is a 15 kb inactive X-specific transcript containing no conserved ORF and located in the nucleus. Cell 1992; 71(3):515-526.
12. Brown CJ, Hendrich BD, Rupert JL et al. The human *XIST* gene: analyzis of a 17 kb inactive X-specific RNA that contains conserved repeats and is highly localized within the nucleus. Cell 1992; 71(3):527-542.
13. Chureau C, Prissette M, Bourdet A et al. Comparative sequence analyzis of the x-inactivation center region in mouse, human, and bovine. Genome Res 2002; 12(6):894-908.
14. Nesterova TB, Slobodyanyuk SY, Elisaphenko EA et al. Characterization of the genomic *Xist* locus in rodents reveals conservation of overall gene structure and tandem repeats but rapid evolution of unique sequence. Genome Res 2001; 11(5):833-849.
15. Clemson CM, McNeil JA, Willard HF et al. *XIST* RNA paints the inactive X chromosome at interphase: evidence for a novel RNA involved in nuclear/chromosome structure. J Cell Biol 1996; 132(3):259-275.
16. Duthie SM, Nesterova TB, Formstone EJ et al. *Xist* RNA exhibits a banded localization on the inactive X chromosome and is excluded from autosomal material in cis. Hum Mol Genet 1999; 8(2):195-204.
17. Buzin CH, Mann JR, Singer-Sam J. Quantitative RT-PCR assays show *Xist* RNA levels are low in mouse female adult tissue, embryos and embryoid bodies. Development 1994; 120(12):3529-3536.
18. Lyon MF. X-chromosome inactivation: a repeat hypothesis. Cytogenet Cell Genet 1998; 80(1-4):133-137.
19. Perry J, Palmer S, Gabriel A et al. A short pseudoautosomal region in laboratory mice. Genome Res 2001; 11(11):1826-1832.
20. Tsuchiya KD, Willard HF. Chromosomal domains and escape from X inactivation: comparative X inactivation analyzis in mouse and human. Mamm Genome 2000; 11(10):849-854.
21. Carrel L, Cottle AA, Goglin KC et al. A first-generation X-inactivation profile of the human X chromosome. Proc Natl Acad Sci USA 1999; 96(25):14440-14444.
22. Lee JT, Lu N, Han Y. Genetic analyzis of the mouse *X inactivation center* defines an 80-kb multifunction domain. Proc Natl Acad Sci USA 1999; 96(7):3836-3841.
23. Wutz A, Jaenisch R. A shift from reversible to irreversible X inactivation is triggered during ES cell differentiation. Mol Cell 2000; 5(4):695-705.

24. Keohane AM, Barlow AL, Waters J et al. H4 acetylation, *XIST* RNA and replication timing are coincident and define x;autosome boundaries in two abnormal X chromosomes. Hum Mol Genet 1999; 8(2):377-383.
25. Kay GF, Barton SC, Surani MA et al. Imprinting and X chromosome counting mechanisms determine *Xist* expression in early mouse development. Cell 1994; 77(5):639-650.
26. Panning B, Dausman J, Jaenisch R. X chromosome inactivation is mediated by *Xist* RNA stabilization. Cell 1997; 90(5):907-916.
27. Sheardown SA, Duthie SM, Johnston CM et al. Stabilization of *Xist* RNA mediates initiation of X chromosome inactivation. Cell 1997; 91(1):99-107.
28. Wakefield MJ, Keohane AM, Turner BM et al. Histone underacetylation is an ancient component of mammalian X chromosome inactivation. Proc Natl Acad Sci USA 1997; 94(18):9665-9668.
29. Lee JT, Strauss WM, Dausman JA et al. A 450 kb transgene displays properties of the mammalian X-inactivation center. Cell 1996; 86(1):83-94.
30. Keohane AM, O'Neill L P, Belyaev ND et al. X-Inactivation and histone H4 acetylation in embryonic stem cells. Dev Biol 1996; 180(2):618-630.
31. Brown CJ, Willard HF. The human X-inactivation centre is not required for maintenance of X-chromosome inactivation. Nature 1994; 368(6467):154-6.
32. Csankovszki G, Panning B, Bates B et al. Conditional deletion of *Xist* disrupts histone macroH2A localization but not maintenance of X inactivation. Nat Genet 1999; 22(4):323-324.
33. Csankovszki G, Nagy A, Jaenisch R. Synergism of *Xist* RNA, DNA methylation, and histone hypoacetylation in maintaining X chromosome inactivation. J Cell Biol 2001; 153(4):773-784.
34. Peters AH, Mermoud JE, O'Carroll D et al. Histone H3 lysine 9 methylation is an epigenetic imprint of facultative heterochromatin. Nat Genet 2002; 30(1):77-80.
35. Heard E, Rougeulle C, Arnaud D et al. Methylation of histone H3 at Lys-9 is an early mark on the X chromosome during X inactivation. Cell 2001; 107(6):727-738.
36. Costanzi C, Pehrson JR. Histone macroH2A1 is concentrated in the inactive X chromosome of female mammals. Nature 1998; 393(6685):599-601.
37. Mermoud JE, Costanzi C, Pehrson JR et al. Histone macroH2A1.2 relocates to the inactive X chromosome after initiation and propagation of X-inactivation. J Cell Biol 1999; 147(7):1399-1408.
38. Chadwick BP, Willard HF. Histone H2A variants and the inactive X chromosome: identification of a second macroH2A variant. Hum Mol Genet 2001; 10(10):1101-1113.
39. Herzing LB, Romer JT, Horn JM et al. *Xist* has properties of the X-chromosome inactivation centre. Nature 1997; 386(6622):272-275.
40. Heard E, Mongelard F, Arnaud D et al. *Xist* yeast artificial chromosome transgenes function as X-inactivation centers only in multicopy arrays and not as single copies. Mol Cell Biol 1999; 19(4):3156-3166.
41. Clemson CM, Chow JC, Brown CJ et al. Stabilization and localization of *Xist* RNA are controlled by separate mechanisms and are not sufficient for X inactivation. J Cell Biol 1998; 142(1):13-23.
42. Brown CJ, Lafreniere RG, Powers VE et al. Localization of the X inactivation centre on the human X chromosome in Xq13. Nature 1991; 349(6304):82-84.
43. Marahrens Y, Loring J, Jaenisch R. Role of the *Xist* gene in X chromosome choosing. Cell 1998; 92(5):657-664.
44. Simmler MC, Cattanach BM, Rasberry C et al. Mapping the murine *Xce* locus with (CA)n repeats. Mamm Genome 1993; 4(9):523-530.
45. Clerc P, Avner P. Role of the region 3' to *Xist* exon 6 in the counting process of X- chromosome inactivation. Nat Genet 1998; 19(3):249-253.
46. Lee JT, Davidow LS, Warshawsky D. *Tsix*, a gene antisense to *Xist* at the X-inactivation centre. Nat Genet 1999; 21(4):400-404.
47. Lee JT, Lu N. Targeted mutagenesis of *Tsix* leads to nonrandom X inactivation. Cell 1999; 99(1):47-57.
48. Stavropoulos N, Lu N, Lee JT. A functional role for *Tsix* transcription in blocking *Xist* RNA accumulation but not in X-chromosome choice. Proc Natl Acad Sci USA 2001; 98(18):10232-10237.

49. Luikenhuis S, Wutz A, Jaenisch R. Antisense transcription through the *Xist* locus mediates *Tsix* function in embryonic stem cells. Mol Cell Biol 2001; 21(24):8512-8520.
50. Sado T, Wang Z, Sasaki H et al. Regulation of imprinted X-chromosome inactivation in mice by *Tsix*. Development 2001; 128(8):1275-1286.
51. Migeon BR, Lee CH, Chowdhury AK et al. Species differences in *TSIX/Tsix* genes reveal the roles of these genes in X-Chromosome inactivation. Am J Hum Genet 2002; 71(2):286-293.
52. Webb S, de Vries TJ, Kaufman MH. The differential staining pattern of the X chromosome in the embryonic and extraembryonic tissues of postimplantation homozygous tetraploid mouse embryos. Genet Res 1992; 59(3):205-214.
53. Takagi N. Variable X chromosome inactivation patterns in near-tetraploid murine EC x somatic cell hybrid cells differentiated in vitro. Genetica 1993; 88(2-3):107-117.
54. Percec I, Plenge RM, Nadeau JH et al. Autosomal dominant mutations affecting X inactivation choice in the mouse. Science 2002; 296(5570):1136-1139.
55. Chao W, Huynh KD, Spencer RJ et al. CTCF, a candidate trans-acting factor for X-inactivation choice. Science 2002; 295(5553):345-347.
56. Brown CJ, Baldry SE. Evidence that heteronuclear proteins interact with *XIST* RNA in vitro. Somat Cell Mol Genet 1996; 22(5):403-417.
57. Beletskii A, Hong YK, Pehrson J et al. PNA interference mapping demonstrates functional domains in the noncoding RNA *Xist*. Proc Natl Acad Sci USA 2001; 98(16):9215-9220.
58. Rasmussen TP, Wutz AP, Pehrson JR et al. Expression of *Xist* RNA is sufficient to initiate macrochromatin body formation. Chromosoma 2001; 110(6):411-420.
59. Wutz A, Rasmussen TP, Jaenisch R. Chromosomal silencing and localization are mediated by different domains of *Xist* RNA. Nat Genet 2002; 30(2):167-174.
60. Hong B, Reeves P, Panning B et al. Identification of an autoimmune serum containing antibodies against the barr body. Proc Natl Acad Sci USA 2001; 98(15):8703-8708.
61. Amrein H, Axel R. Genes expressed in neurons of adult male *Drosophilae*. Cell 1997; 88(4):459-469.
62. Meller VH, Wu KH, Roman G et al. *roX1* RNA paints the X chromosome of male *Drosophilae* and is regulated by the dosage compensation system. Cell 1997; 88(4):445-457.
63. Kageyama Y, Mengus G, Gilfillan G et al. Association and spreading of the *Drosophilae* dosage compensation complex from a discrete *roX1* chromatin entry site. Embo J 2001; 20(9):2236-2245.
64. Meller VH, Rattner BP. The roX genes encode redundant male-specific lethal transcripts required for targeting of the MSL complex. Embo J 2002; 21(5):1084-1091.
65. Baker BS, Gorman M, Marin I. Dosage compensation in *Drosophilae*. Annu Rev Genet 1994; 28:491-521.
66. Willard HF, Salz HK. Remodelling chromatin with RNA. Nature 1997; 386(6622):228-229.
67. Maison C, Bailly D, Peters AH et al. Higher-order structure in pericentric heterochromatin involves a distinct pattern of histone modification and an RNA component. Nat Genet 2002; 30(3):329-334.
68. Sleutels F, Zwart R, Barlow DP. The noncoding *Air* RNA is required for silencing autosomal imprinted genes. Nature 2002; 415(6873):810-813.
69. Pillet N, Bonny C, Schorderet DF. Characterization of the promoter region of the mouse *Xist* gene. Proc Natl Acad Sci USA 1995; 92(26):12515-12519.
70. Hendrich BD, Plenge RM, Willard HF. Identification and characterization of the human *XIST* gene promoter: implications for models of X chromosome inactivation. Nucleic Acids Res 1997; 25(13):2661-2671.
71. Courtier B, Heard E, Avner P. *Xce* haplotypes show modified methylation in a region of the active X chromosome lying 3' to *Xist*. Proc Natl Acad Sci USA 1995; 92(8):3531-3535.

Chapter 5

Dosage Compensation in *Drosophila*:

A Ribonucleoprotein Complex Mediates Transcriptional Up-Regulation

Dianne Kindel and Hubert Amrein

Summary

Sex-specific chromosomes (commonly referred to as X and Y) provide the basis for sex determination in many animal species. However, this difference in karyotype has drastic consequences for the quantitative set-up of the genome of the two sexes, as sex chromosomes harbor thousands of genes with no role in sex determination but important functions in basic cellular and organismal processes. Therefore, heterogametic animal species have acquired distinct mechanisms, commonly referred to as dosage compensation, to equalize the transcriptional output of different copy numbers of X–linked genes. Because sex chromosomes evolved numerous times in life history, many different mechanisms of dosage compensation have been established in various animal phyla. Common to all these mechanisms are two features: First, dosage compensation operates in one sex only, requiring a reliable regulatory mechanism that distinguishes the two karyotypes. Second, dosage compensation acts on one (or a pair) of sex chromosomes, but does not affect any of the far more numerous autosomes in the nucleus, demanding a need for chromosome recognition by the compensation machinery.

In the fruit fly *Drosophila melanogaster*, dosage compensation is achieved in the heterogametic (XY) male by two-fold hyper-transcription of virtually all genes on the single X chromosome. The dosage compensation complex (DCC) associates with the male X chromosome, leading to chromosome wide acetylation of histone H4, and perhaps other, yet unknown chromatin modification events. The five proteins that form the DCC, generally referred to as the male-specific lethal (MSL) proteins, have distinct essential roles, which is revealed by the lethality of males mutant for any of the *msl* genes. Assembly of the MSLs occurs in a stepwise fashion at about 35 so-called chromatin entry sites that are evenly distributed on the X chromosome and also involves two regulatory RNAs, *roX1* and *roX2* (*RNA on the X*). Interestingly, the two *roX* genes themselves are two of these 35 sites. The chromatin entry sites also appear to serve as bases for the distribution of the DCCs, which spread into neighboring chromatin regions where they catalyze chromatin modification events that lead to the two-fold up-regulation of gene expression for most X-linked genes.

From the day of its discovery, first in *Drosophila* more than half a century ago[1,2] and soon after in mammals,[3] the phenomenon of dosage compensation has intrigued researchers in an increasing number of fields including genetics, cytology, gene regulation, chromatin structure, sex determination, epigenetics and RNA biology. Thus, one could view dosage compensation

Noncoding RNAs: Molecular Biology and Molecular Medicine, edited by Jan Barciszewski and Volker A. Erdmann.

as a process at the intersection of numerous biological disciplines. One of its hallmarks is the variety of mechanisms that have evolved in different animal systems. The last few years of research in dosage compensation have provided some new insights into this intriguing and complex problem. The purpose of this review is to present the novel findings of these studies in the fruit fly, *Drosophila melanogaster.*

Different Mechanisms of Dosage Compensation

The need for dosage compensation is a direct consequence of the evolution of sex chromosomes, whose purpose is the implementation of genetic control of sexual development and differentiation.[4-7] This control is mediated by various different mechanisms, but usually involves specific genes located on one of the sex chromosomes themselves. The different copy number of gene-rich chromosomes such as the X in mammals, *Drosophila*, and *C. elegans* causes a large-scale genetic imbalance. For example, about 3300 of the roughly 14,000 genes in *Drosophila* map to the X chromosome,[8] which means that almost one quarter of all gene products would be twice as abundant in XX females compared to XY males in the absence of a compensation mechanism. Such a bias, if unattended, would have severe consequences during development and ultimately lead to lethality. The *Drosophila* Y chromosome contains only a few genes, most of which are essential for male fertility.

Four gene regulatory solutions can be envisioned to deal with large-scale genetic imbalances and at least three of these solutions are used in various animal systems.[7,9,10] The first is implemented in mammals, where an entire X chromosome in all somatic cells of the homogametic (XX) female is silenced (X-inactivation).[11] A second solution, the down-regulation of gene expression by 50% in the homogametic (XX) hermaphrodite is realized in the nematode *C. elegans.*[12] A third mechanism, two-fold up-regulation of all genes on the X chromosome, is implemented in the heterogametic (XY) *Drosophila* male.[13-15] Common to all three mechanisms is that virtually all genes on the affected chromosome are transcriptionally regulated in a chromosome wide manner. Finally, one can imagine a fourth process in which the expression of genes with crucial functions (during development or in regulatory networks) would be selectively adjusted. Recent studies in birds, long thought not to dosage compensate their sex-chromosome-linked genes, provide evidence that some but not all genes on the gene-rich Z sex-chromosome are compensated.[16,17] Whether the noncompensated Z-linked genes are just rare exceptions of an established mechanism, as in mammalian X inactivation, where some genes found in the pseudo-autosomal region escape inactivation, or whether they reflect a truly distinct mechanism of dosage compensation remains to be seen.

Basic Principles of Dosage Compensation

Regardless of the specific mechanism, dosage compensation must adhere to several important principles: First, it must be sex-specific (i.e., dosage compensation is implemented only in one sex). Second, at least one component (core component) of the compensation machinery must specifically recognize the sex chromosome. Third, the various dosage compensation components must assemble into functional complexes (the dosage compensation complexes or DCCs). And finally, DCCs need to spread to chromatin containing regulatory (DNA) elements of (almost) all genes on that chromosome.

Male-Specific Hyper-Transcription Through Chromatin Modification

In *Drosophila melanogaster*, early cytological investigations of polytene chromosomes from salivary gland nuclei of male and female larvae revealed that the single male X chromosome appears broader than the two paired female Xs,[18,19] indicating that the chromatin on the male

X chromosome is less densely packed than that of the two female X chromosomes. The first direct measurements of RNA synthesis of X-linked genes confirmed that the transcriptional output of the two gene copies on the female X chromosomes is identical to that of the single copy on the male X.[20,21] Taken together, these studies pointed to a male compensation mechanism via hyper-transcription rather than female hypo-transcription. It was only after genetic studies identified several male-specific lethal mutations that dosage compensation was more directly linked to the male sex.[22-25] Genetic and molecular analysis of these genes, which became known as the *male-specific lethal* (*msl*) genes - *msl1*, *msl2*, *msl3*, *maleless* (*mle*) and *males-absent on the first* (*mof*) - showed that mutations in any of the them produced virtually identical phenotypes: male-specific lethality during late third instar larval and early pupal stages, when a major reorganization of the animal's body plan from larva to fly begins. The similarity of the mutant phenotypes indicated that the MSL proteins are likely to function in the same process. Protein localization studies using antibodies against each of the MSLs showed that all five proteins colocalize to hundreds of sites along the male X chromosome, providing the first evidence for a dosage compensation complex (DCC).[26-31] A molecular link between dosage compensation and changes in chromatin composition/structure of the male X chromosome was provided by Turner and coworkers about ten years ago, when they noted that a hyper acetylated histone H4 isoform, H4Ac16, is highly enriched on the male X chromosome, but not on those of females or on autosomes.[32] Indeed, H4Ac16 colocalizes with the MSL proteins to the hundreds of discrete bands on the salivary gland chromosomes.[26]

Male-Specific Implementation of Dosage Compensation Is Dependent on the Absence of SXL Repressor

In *Drosophila*, the regulatory control mechanism that implements dosage compensation in males is well understood.[33] Dosage compensation is established in males because they lack the Sex-lethal (SXL) repressor, an RNA binding protein that inhibits translation of the *male-specific lethal 2* (*msl2*) mRNA, which encodes a key component of the DCC.[34-37]

Female-specific activation of *Sxl* is directly linked to the highly sensitive nature of its promoters to the concentration of two helix-loop-helix transcriptional regulators, SIS-A and SIS-B, in the early embryo.[38-41] The *sis-A* and *sis-B* genes are located on the X-chromosome; consequently, their protein products are twice as abundant in female as opposed to male embryos, a difference that results in the activation of the *Sxl* promoter in females and lack of activation in males. Hence, only the young female embryo produces "early" SXL protein. A few hours later in development, a promoter switch occurs. SIS-A and SIS-B are no longer expressed and the *sis-A*/*sis-B* dependent *Sxl* promoter is shut down; instead, a constitutive promoter is activated,[42] which produces *Sxl* transcripts in both sexes. However, proper splicing of this new *Sxl* transcript requires "early" female-specific SXL protein.[43,44] This ensures that additional SXL protein is produced in females throughout development and in adult life. In males, SXL protein is never produced because functional *Sxl* mRNA splicing fails to occur due to the absence of "early" SXL protein. Lack of SXL allows the translation of *msl2* RNA and production of MSL2 protein, a crucial component of the dosage compensation machinery. Thus, lack of dosage compensation during early embryonic development is used as a tool for its implementation in males only.

It should be noted that some X-linked genes essential for early development are transiently dosage-compensated in females by SXL directly at the level of RNA translation/stability (for a review see ref. 15). However, we shall focus on the male-specific dosage compensation process that is controlled at the transcriptional level and is an indirect consequence of the absence of SXL in males. This mechanism is implemented after embryogenesis and is maintained throughout the life of the animal.

The Components of the DCC

The lethal phenotype of males mutant for any one of the *msl* genes clearly established that a full complement of MSL proteins is essential for dosage compensation and indicated that each MSL protein serves a unique function in this process.[22-25] Not surprisingly, the five MSL proteins share no similarity with each other and belong to distinct gene families or are novel proteins (Table 1).

MSL1 is the only protein that cannot be associated with a gene family and it contains no discernable structural motifs other than its amino terminal acidic stretches.[31]

MSL2 and MSL3 harbor sequence motifs that are known to mediate DNA binding or chromatin association in other proteins. MSL2 contains a RING finger, a motif that has been implicated in protein-protein interactions and/or nucleic acid or chromatin binding. [28,34,36,45] The RING finger of MSL2 is essential for dosage compensation.[28,46]

MSL3 and its mammalian and yeast orthologs, MRG15 and Eaf3p, respectively, feature two chromodomains,[47-49] regions that have been implicated in chromatin remodeling and transcriptional control. Chromodomains are a common motif found in a variety of other chromatin-associated proteins, including HP1.[50]

MLE and MOF are two other proteins that have highly similar counterparts in mammals. MLE is the ortholog of human RNA helicase A, with which it shares about 70% sequence similarity.[29,51] In addition to a large (~ 600 aa) central DExH helicase homology domain,[52] both MLE and RNA helicase A contain two double-stranded RNA binding motifs of about 60 amino acids (aa) at the amino terminus and a glycine/tyrosine/arginine rich carboxy terminus.[29,53,54] The RNA substrate for the double stranded RNA binding domain of MLE is not known. MLE likely binds RNA through another motif, the carboxy terminal glycine-rich heptad repeat.[55] These findings suggest that MLE protein may directly interact with nascent transcripts of X-linked genes and/or some integral component of the DCC such as the *roX* RNAs. The *roX* RNAs may form stable secondary structures that could interact with the double stranded RNA binding domain of MLE, whereas the helicase domain might be involved in relieving RNA secondary structures in nascent transcripts.

MOF is a member of the MYST family of histone acetyltransferases, which includes its closest relative, the human Tip60 protein.[25] MOF has a significantly larger amino-terminus than Tip60, but both proteins contain a single chromodomain in this region that is followed by a long (300 aa) conserved carboxy terminus (the HAT domain), which includes a single zinc finger and an acetyl-CoA binding site. The HAT domain is well conserved between the two proteins (70% sequence similarity). The identification of the MOF histone acetyltransferase suggested a direct link between the DCC and the reported chromatin modification of the X chromosome, hyperacetylated histone H4 at lysine residue 16 (H4Ac16).[26,32] Numerous studies in mammals and yeast have shown that histone acetyltransferases are associated with transcriptional activation.[56]

JIL-1 kinase, which phosphorylates histone H3 at Serine 10 (H3Ser10) in vitro[57] and is essential for general viability and establishment and/or maintenance of higher order chromatin structure,[58] might also be involved in chromatin modification events associated with male dosage compensation. JIL-1 is enriched approximately two-fold on the male X chromosome [57] and colocalizes with the MSLs.[59] In addition, there is both in vivo and in vitro evidence that JIL-1 associates with the DCC.[59]

The identification of *roX1* and *roX2* (*RNA on the X*) was an important breakthrough because it provided a new viewpoint for how the X chromosome might be recognized by the DCC.[60,61] As their names imply, the *roX* genes are located on the X chromosome and encode regulatory nuclear RNAs that associate specifically with the X chromosome. They were isolated somewhat serendipitously in a molecular screen for genes sex-specifically expressed in the brain

Table 1. Important features of DCC components

Features/ DCC Component	Overall Structure	Structural Motifs/ Domains	Implicated Function	Demonstrated Function or Activity	Putative Function in Dosage Compensation	Orthologs	Original Reference
MSL1	Novel acidic protein	Two acidic stretches in N-terminus	Chromatin remodeling and/or transcription	N-terminus is essential for localization to X	Localization of DCC core to chromatin entry sites; stabilization of DCC	None	31
MSL2	Hydrophilic, acidic protein	RING finger near N-terminus	Protein-protein interactions and/or nucleic acid binding	RING finger is essential for dosage compensation N-terminus is essential for localization to X	Localization of DCC core to chromatin entry sites; stabilization of DCC	None	28, 34 and 36
		Metallothionein-like domain	Regulating zinc availability to RING finger	None			
		Coiled-coil domain	Protein-protein interactions	None			

Continued on next page

Table 1. Important features of DCC components (continued)

Features/ DCC Component	Overall Structure	Structural Motifs/ Domains	Implicated Function	Demonstrated Function or Activity	Putative Function in Dosage Compensation	Orthologs	Original Reference
MLE	Member of DExH subfamily of NTPase and/or helicase proteins	NTP-binding/ Helicase domain	Unwinding nucleic acids using NTP hydrolysis	RNA/DNA binding and helicase; ATPase/helicase function is essential for dosage compensation	Destabilize chromatin Stimulate transcription by modifying structure of nascent RNA Interact with *roX* RNAs	RNA helicase A (human) Nuclear DNA helicase II (bovine)	29
		Double stranded RNA binding domains	Binding double stranded RNA	None			
		Glycine-rich repeats	Nucleic acid binding; Protein-protein interactions	None			
MSL3	Member of novel family of chromodomain proteins	Chromodomain at each terminus	Chromatin remodeling; Transcriptional regulation	Interacts with RNA nonspecifically in vitro	Interact with *roX* RNAs	MRG15 (human) Eaf3 (yeast)	27

Continued on next page

Table 1. Important features of DCC components (continued)

Features/ DCC Component	Overall Structure	Structural Motifs/ Domains	Implicated Function	Demonstrated Function or Activity	Putative Function in Dosage Compensation	Orthologs	Original Reference
MOF	Member of MYST family of acetylases	Histone acetyltransferase (HAT) domain near N terminus	Chromatin remodeling by histone acetylation	Specifically monoacetylates Lysine 16 on histone H4	Hyperacetylates histone H4 on the X chromosome Associates with *roX2* RNA in vivo	Tip60 (human) MOZ (human)	25
		Chromodomain near N terminus	Chromatin remodeling; Transcriptional regulation	Interacts with RNA nonspecifically in vitro			
		C_2HC-type zinc finger	RNA binding	Affects HAT activity and chromatin binding			
roX1	Novel male-specific,	*roX* genes share virtually	None	At least one *roX* RNA is necessary	Unknown; DCC	None	60 and 61

Continued on next page

Table 1. Important features of DCC components (continued)

Features/ DCC Component	Overall Structure	Structural Motifs/ Domains	Implicated Function	Demonstrated Function or Activity	Putative Function in Dosage Compensation	Orthologs	Original Reference
roX2	noncoding RNAs 3.7 kb (*roX1*) and 1.2 kb (*roX2*) in length	no sequence homology with each other or other genes secondary structure not known		for DCC spreading from chromatin entry sites and/or maintenance of the DCC on the X chromosome	spreading from chromatin entry sites and/or maintenance of the DCC on the X chromosome?		
JIL-1	Novel tandem kinase	Two Ser/Thr kinase domains	Chromatin remodeling by histone phosphorylation	Phosphorylates Ser10 on histone H3 viability and	Unknown JIL-1 is essential for chromatin structure in both males and females	None	57

All components that are known to be involved in dosage compensation are listed. With the exception of *Jil-1*, the genes have no other known, established functions in *Drosophila*. Only the original reference describing the identification of each gene is cited. For more details see text.

and in an enhancer trap analysis. Both *roX* RNAs are stably expressed only in males. The first evidence that they function in dosage compensation was revealed by their mode of regulation. Male-specific expression of the *roX* genes is dependent on the entire set of MSL proteins.[60] Moreover, the *roX* RNAs colocalize with the MSL proteins on the male X chromosome.[62] Despite the absence of any significant sequence similarity between *roX1* (3.7 kb) and *roX2* (1.2 kb),[60] the two RNAs are functionally redundant,[63,64] which prevented their discovery in the genetic screens that identified the *msl* genes.

The *roX* RNAs Are Chromatin Entry Sites

The identification of the *roX* genes prompted new speculation about how the MSL proteins identify the appropriate (X) chromosome in the nucleoplasm and how they reach the hundreds of sites along the X chromatin. For many years, the search for a specific 'mark' on the X chromosome that would allow the MSL proteins to distinguish it from the autosomes focused on specific DNA sequence motifs. However, numerous experimental approaches as well as the completion of the *Drosophila* genome sequence did not reveal any sequences specific for the X. The X-linked *roX* genes/RNAs provided an alternative to a DNA-mediated recognition motif, namely X-chromatin-associated RNAs. It suggested that male-specific expression and chromatin association of the *roX* RNAs could allow the recruitment of MSL complexes to the two *roX* regions on the X chromosome. From these and other chromatin entry sites (possibly expressing additional *roX* RNAs), the *roX*/MSL complex might spread along the entire length of the X chromosome in cis, similar to the *Xist* RNA in mammals, which spreads from the X inactivation center to the rest of the X chromosome.[10,65]

Evidence that the *roX* genes/RNAs recruit the DCC was obtained by immunolocalization studies of the MSL proteins in males that contained autosomal *roX* transgenes. For example, MSL proteins and endogenous *roX* RNAs are recruited to *roX* transgenes inserted on autosomes whether or not the *roX* transgenes are transcribed.[62,66,67] Moreover, the recruited DCCs can spread in cis along flanking autosomal chromatin for distances up to a few hundred kilobases and acetylate histone H4, suggesting that these DCCs are fully functional. In addition, *roX1* RNA transcribed from an autosomal transgene or translocated *roX*1 gene assembles into DCCs, which can relocate onto the X chromosome.[61,66] Taken together, these findings indicate that the *roX* genes function as chromatin entry sites for complete DCC assembly and that these DCCs can spread into neighboring chromatin in cis as well as exchange freely between different chromosomes.

Ordered Assembly of the DCC

The study of the assembly process of the DCC components has been greatly facilitated by the availability of polytene chromosomes in *Drosophila* larval salivary gland cells and antibodies against each of the MSL proteins. In these studies, the distribution of the MSL proteins was determined in *msl* mutant larvae of males or females ectopically expressing MSL2 (Fig. 1).

The most intriguing observation is that loss of either MSL1 or MSL2 prevents association of any of the remaining MSL proteins with the male X chromosome, indicating that these two proteins form the core during the assembly process.[27,46,68] In males that lack MLE, only MSL1 and MSL2 are found on the X, but not MSL3 or MOF.[27,30,46,68] Interestingly, the number of MSL1 and MSL2 containing sites is significantly reduced from several hundred to approximately 35 X-linked sites, the so-called chromatin entry sites. Similarly, MSL1/2 localization is restricted to the entry sites in flies lacking MSL3 or MOF.[25,27,30,46,68] These studies indicate that MSL1 and MSL2 form a core that assembles in a first step on the X, and that the other components enter the complex later. MLE appears to enter the complex in a second phase, as absence of MSL3 or MOF allows weak association of MLE to the entry sites.[25,27,30] Incorporation

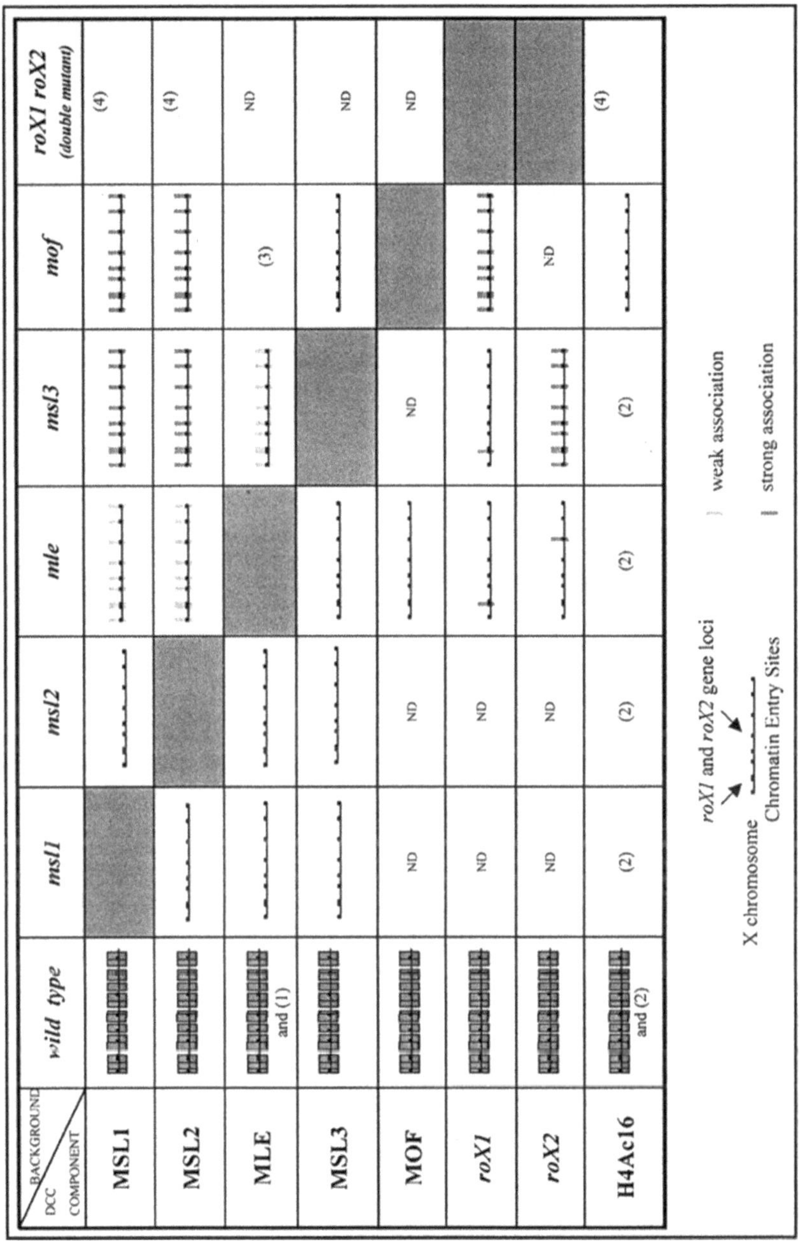

Figure 1. Localization of DCC components in male X chromosome in wild type and various mutant backgrounds. Mutant backgrounds that do not affect localization (*roX1*⁻ and *roX2*⁻) are not shown. ND = not determined. (1) some autosomal sites are also observed. (2) weaker association with all autosomes is observed as well. (3) reduced association on X and mislocalization to some sites on autosomes. (4) severely reduced association on X and mislocalization to several autosomal sites and the chromocenter.

of MSL-3 and MOF appears to be dependent on MLE, as absence of MLE completely abolishes MSL3 and MOF association with the entry sites.[27,30]

A subtle, but possibly significant difference in localization to the chromatin entry sites of the core proteins (MSL1 and MSL2) is observed in flies that lack MLE or MOF/MSL3. Whereas all 35 chromatin entry sites including *roX1* and *roX2* were occupied when MOF or MSL3 is absent, all but the two *roX* entry sites are occupied in flies lacking MLE.[67] This suggests a qualitative difference in core protein recruitment between transcribed and nontranscribed chromatin entry sites.

The distribution of *roX* RNAs has also been investigated in various *msl* mutant animals. In the absence of MSL1, MLS2 or MLE, both *roX* RNAs are found only at their sites of synthesis (i.e., the location of the *roX1* and *roX2* genes).[62] Interestingly, in the absence of MSL3, *roX2* is observed at the 35 chromatin entry sites, colocalizing with MSL1 and MSL2, whereas *roX1* is still only found at its site of synthesis. This difference in localization pattern was interpreted as evidence that *roX2* RNA enters the complex before *roX1* does, and it also suggested that MLE and MSL3 are needed for the incorporation of *roX2* and *roX1* RNAs, respectively. However, the order by which the *roX* RNAs are incorporated into the DCC cannot be essential, since functional DCCs are formed in the presence of only one *roX* RNA.[64] Because partially assembled DCCs (i.e., lacking any of the MSL proteins or both *roX* RNAs) are never observed at the hundreds of sites characteristic of wild type males, it is thought that spreading from the chromatin entry sites into neighboring chromatin requires a complete DCC.

JIL-1 localizes to all chromosomes in both sexes, but it is enriched approximately two-fold on the male X chromosome,[57] where it colocalizes with the MSLs.[59] However, the MSL proteins and H4Ac16 properly localize to the polytene X chromosome in *Jil-1* mutant larvae, indicating that JIL-1 is not necessary for assembly of the DCC.[58]

Specific Functions of and Interactions Among MSL Proteins

In addition to genetic investigations, the specific functions of the MSL proteins have also been characterized by coimmunoprecipitation studies, in vitro protein-RNA interaction experiments and analyses using the yeast two-hybrid system. For example, it was shown that the two proteins forming the core during the assembly, MSL1 and MSL2, can be coimmunoprecipitated from larval and S2 (*Drosophila* cell line) protein extracts.[36] Moreover, it was shown that this interaction is mediated through the RING finger of MSL2.[69] In addition, MSL1 also interacts with MSL3 as assayed by coimmunoprecipitation experiments. Deletion analysis of MSL1 in the yeast two hybrid system showed that the amino terminus of this protein mediates interaction with MSL2, whereas the carboxy terminus is involved in binding to MSL3 and MOF.[70] Finally, anti-MOF antibody coimmunoprecipitates all other MSL proteins.[71,72]

MLE, like its human counterpart, the DNA/RNA helicase A, shows strong unwinding activity on RNA/RNA and RNA/DNA duplexes.[51] Unwinding is directional (3'-> 5') and NTP dependent, as mutations in the ATP binding site reduce or abolish unwinding. The NTPase/helicase activities of MLE are required for proper MLE function in dosage compensation in vivo, but these activities are not required for MSL1, MSL3, MLE, or MOF localization to the chromatin entry sites.[30,51]

There is clear evidence that MOF is responsible for the enrichment of the histone isoform H4Ac16 on the X chromosome, as a single amino acid substitution in the acetyl-CoA binding site of MOF abolishes histone acetyltransferase activity in vivo and in vitro. In such *mof* mutant males, H4Ac16 is not found on the X chromosome[25] and DCC spreading along the X chromosome does not occur.[30,73] MOF specifically acetylates histone H4 at the lysine 16 residue in vitro.[71,74] Both MOF and MSL3 have nonspecific RNA binding activity in vitro,[72] which is supported by data showing that *roX2* can be amplified by RT-PCR from immunoprecipitates obtained from S2 cell extracts using MSL antibodies.[62,71,72] MOF binds *roX2* RNA in vivo through its chromodomain; this interaction may be necessary for stable integration of MOF into the DCC.[72] Since MLE is necessary for *roX2* entry into the DCC [62] and MOF can bind to *roX2*, it is possible that these two proteins do not contact each other directly, but are linked to each other via an RNA bridge; indeed, MLE-MOF interaction is lost under conditions of high ionic strength, suggesting a loose or indirect interaction between the acetyltransferase and the helicase.[72] The chromodomains of HP1 and other proteins have been shown to interact with histones, particularly H3 and H4.[50] There is good evidence that MOF interacts with histone H4 directly. Whether and with what kind of histone the chromodomain of MSL3 interacts is not known.

As a word of caution, we should note that no in vitro protein-protein binding studies have been performed with purified MSL proteins. Thus, it is still not clear for many of these interactions whether they are direct or whether they are mediated through other protein or RNA components associated with the DCC.

The *roX* RNAs Are Essential for Dosage Compensation

Although in vivo and in vitro data have clearly established that the *roX* genes function as chromatin entry sites[62,66,67] and that the *roX* RNAs are components of the DCC,[62,71,72] their specific roles in dosage compensation are still poorly understood. However, the essential requirement for the *roX* genes in dosage compensation was shown recently in genetic studies.[64] Whereas males lacking either the *roX1* or *roX2* gene are completely viable and show normal DCC distribution on the X chromosome, males lacking both *roX* genes are almost completely lethal and the dying male larvae show aberrant MSL1 and MSL2 localization: weak association with the chromatin entry sites on the X and concomitant ectopic association with the chromocenter and some autosomal sites. Thus, it appears that *roX1* and *roX2* have redundant functions, which might be surprising considering the lack of significant sequence similarity between the two genes. However, the two *roX* RNAs are likely to fold into secondary and/or tertiary structures, which might be similar and far more relevant to their function than primary sequence similarity. Unfortunately, the size of the RNAs is so large that too many possible/probable secondary structures are predicted for experimental testing.

In contrast to the lethal phenotype of males lacking any one of the *msl* genes[22-25] a small percentage of *roX1*$^-$*roX2*$^-$ double mutant males survive to adulthood, albeit with developmental delay.[64] Surprisingly, these *roX1*$^-$*roX2*$^-$ males are quite healthy and fertile, which can be interpreted as evidence that the *roX* genes, although essential for efficient dosage compensation, are not absolutely necessary for the recruitment and distribution of the DCC throughout the X chromosome. In addition, when males lacking functional *roX* genes on the X are provided with an autosomal transgene that produces *roX1* or *roX2* RNA, viability is completely restored. Thus, the recruitment of the DCC to the X chromosome can occur without the local accumulation of either *roX* RNA from its endogenous locus, questioning the relevance of the X chromosomal location of the *roX* genes altogether. However, a third (or more?) *roX*-like gene might exist on the X chromosome, and its location on the X as well as the RNA it produces might explain both incomplete lethality of *roX1*$^-$*roX2*$^-$ mutant males as well as proper localization of the DCC on the X in males whose only other source of *roX* RNA is an autosomal *roX1* or *roX2* transgene. Regardless, the genetic analyses revealed that the presence of at least one of the two known *roX* RNAs is necessary for normal establishment of dosage compensation.

Models and Outlook

What are the roles of the *roX* RNAs/genes in dosage compensation? The four crucial events are: recognition of the chromatin entry sites located on the X (1), stepwise assembly of the DCC at these sites (2), spreading of the DCC into the adjacent chromatin (3) and modification of chromatin structure (4). In principle, the *roX* genes/RNAs could be involved in one, several or all of these processes (Fig. 2). A role for the *roX* genes in recruiting the core proteins onto the X is indicated by the observations that the *roX* genes are located at two of the 35 chromatin entry sites and that autosomal *roX* transgenes can recruit and support the assembly of entire DCCs.[62,66,67] However, such a role is not essential, because in males lacking both *roX1* and *roX2* chromatin entry sites, DCCs assemble normally on the X in the presence of an autosomal source of either *roX* RNA.[64]

Involvement in DCC assembly is probably the best-established role of the *roX* RNAs.[62,71,72] Whereas the *roX* RNAs appear not to be necessary for the association of the core of the DCC (consisting of MSL1 and MSL2) with the chromatin entry sites, at least one of them is required for the transition into a fully functional DCC.[64] The role of the *roX* RNAs in assembly may also be linked to a function in stabilization of the DCC. For example, it is possible that the *roX* RNAs might support weak interactions between MSLs.[75] Several of the MSLs (MOF, MSL3 and MLE) have putative or known RNA binding properties (Table 1), and thus the *roX* RNAs might strengthen weak protein/protein interactions in the form of an RNA string that 'runs through' the different MSL proteins.

Does spreading from the chromatin entry sites to the hundreds of sites along the X depend on the *roX* RNAs? There is no direct experimental data that would suggest so. At the very least, the *roX* RNAs might play a permissive role in this process by virtue of their role in stabilizing the DCC. Only complete DCC but not partial complexes (e.g., lacking both *roX* RNAs or any of the MSL proteins) appear to be able to spread and thus, at least an indirect role for the *roX* RNAs in this process is likely.

It is well established that MOF is the major component involved in chromatin modification of the male X chromosome. Since the *roX* RNAs are probably associated with MOF, it seems possible that they play a direct role in chromatin modification as well. A precedent for a noncoding RNA that is involved in chromatin remodeling is the human SRA (Steroid receptor RNA Activator) transcript, which selectively enhances steroid receptor-mediated transactivation and acts as part of a ribonucleoprotein complex containing SRC-1.[76]

What is the structure of the *roX* RNAs? Their nucleotide sequence has not given us any clue about important regions of the RNA and it is likely that secondary and tertiary structures are more important in defining their function. Evolutionary sequence comparisons of *roX* genes from diverse *Drosophila* species might be valuable in this regard as comparative phylogenetic studies have contributed significantly to our understanding of many structural RNAs, particularly the ribosomal RNAs. Another intriguing question is the different phenotypes of males with mutations in individual *msl* genes and males lacking both *roX* genes. How do surviving *roX1⁻roX2⁻* males distribute their DCCs on the X in the absence of any *roX* RNA? Or is there perhaps still another *roX* gene (*roX3*) to be discovered, which provides a small, but sufficient "*roX-like*" activity to allow occasional survival of these males? Finally, if localization of the *roX* genes on the X chromosome is not necessary for DCC spreading, what is? Is it just the concentration of 35 sites on one chromosome, spread more or less evenly along the 20 MB of DNA, or are there qualitatively distinct sites whose presence on the X is necessary for DCC recruitment? And what is the nature of these different types of chromatin entry sites?

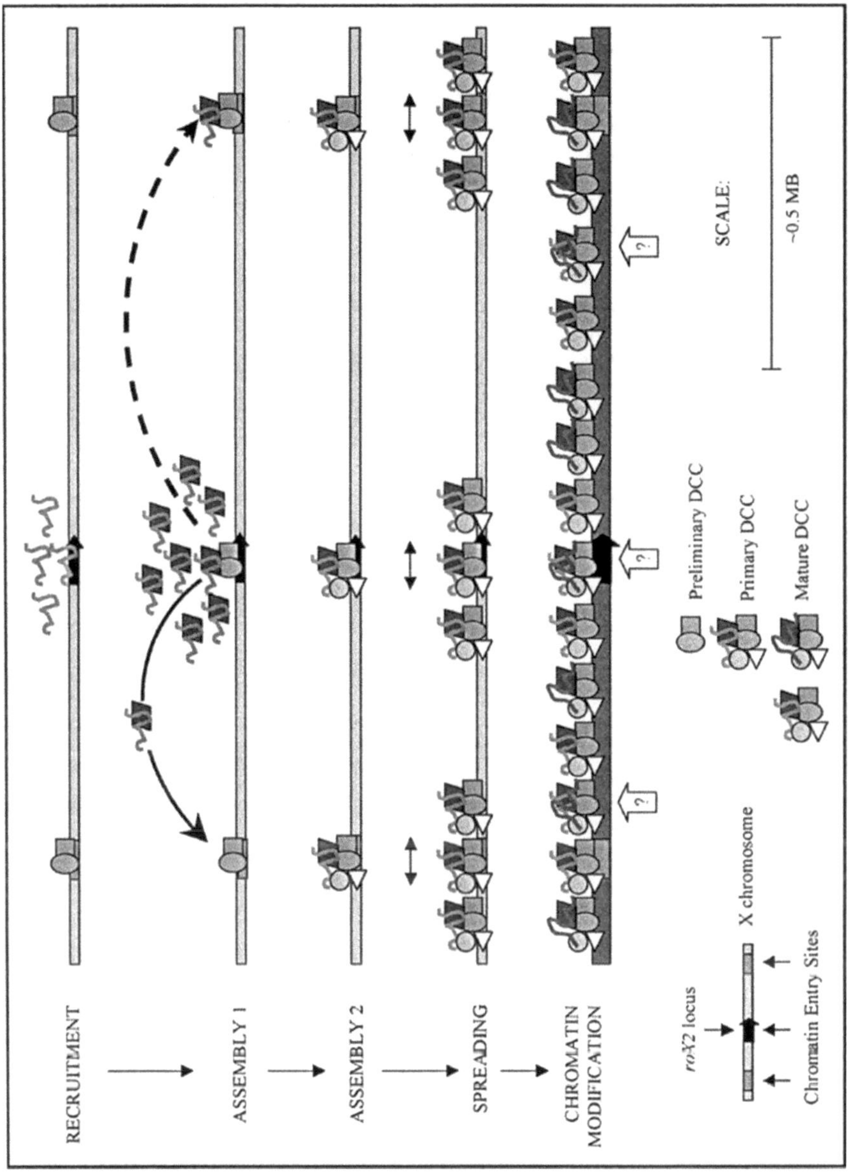

Dosage compensation will remain a fascinating problem in the field of chromatin structure and transcriptional regulation. It is safe to say that most crucial components have been identified, their specific roles in this process have been established and some of their enzymatic activities and binding properties have been resolved. The future will have to look at questions old and new including the still unclear recruitment of the DCC (or its core components) to the X chromosome, the fascinating process of spreading of the DCC from sites of assembly to sites of activity and additional chromatin and DNA modification events that are the cause for up-regulation.

Figure 2, at left. Assembly, spreading and action of the DCC (Model) Assembly, spreading and action of DCCs in a region of the X chromosome with three chromatin entry sites, including *roX2*, is shown (assembly in regions closer to the *roX1* gene might be similar, but involve *roX1* RNA instead). The recruitment of the core proteins to form the preliminary (pre)DCC is probably *roX* RNA independent and might occur preferentially at nontranscribed entry sites. The first assembly step involves the addition of a *roX2* RNA and MLE. MLE/*roX2* RNA might first form at the *roX2* entry site and then rapidly be translocated to nontranscribed entry sites (either via chromatin/DNA looping or through a free intermediate). MSL1/MSL2/MLE and *roX2* form a primary (prim)DCC, that is more stable than the core; this complex, however, is still unable to spread. Addition of MSL3 and MOF (and possibly JIL-1) ultimately leads to the formation of the mature complex, which has the capability to spread along the X and reach DNA between entry sites. There, MOF and JIL-1 (and possibly other enzymes) alter chromatin structure via histone acetylation/phosphorylation (darker coloration of the broader X chromosome). It is not clear whether individual DCCs in wild type males contain one or both *roX* RNAs, but it is known that DCCs function properly with only one of them. DCCs containing *roX1*, *roX2* or both RNAs are shown.

References

1. Muller H. Further studies on the nature and causes of gene mutations. Proceedings of the 6th International Congress on Genetics 1932; 1:213-255.
2. Muller H. Evidence of the precision of genetic adaptation. Harvey Lecture Series 1950; XLIII:165-229.
3. Lyon MF. Gene action in the X chromosome of the mouse (*Mus musculus L.*). Nature 1961; 190:372-372.
4. Lucchesi JC. On the origin of sex chromosomes. Bioessays 1999; 21(3):188-90.
5. Marin I, Siegal ML, Baker BS. The evolution of dosage-compensation mechanisms. Bioessays 2000; 22(12):1106-14.
6. Kelley RL, Kuroda MI. The role of chromosomal RNAs in marking the X for dosage compensation. Curr Opin Genet Dev 2000; 10(5):555-61.
7. Meller VH. Dosage compensation: making 1X equal 2X. Trends Cell Biol 2000; 10(2):54-9.
8. Adams MD, Celniker SE, Holt RA et al. The genome sequence of *Drosophila melanogaster*. Science 2000; 287(5461):2185-95.
9. Lucchesi JC. Dosage compensation in flies and worms: the ups and downs of X-chromosome regulation. Curr Opin Genet Dev 1998; 8(2):179-84.
10. Park Y, Kuroda MI. Epigenetic aspects of X-chromosome dosage compensation. Science 2001; 293(5532):1083-5.
11. Brockdorff N. X-chromosome inactivation: closing in on proteins that bind *Xist* RNA. Trends Genet 2002; 18(7):352-8.
12. Meyer BJ. Sex in the worm: counting and compensating X-chromosome dose. Trends Genet 2000; 16(6):247-53.
13. Stuckenholz C, Kageyama Y, Kuroda MI. Guilt by association: noncoding RNAs, chromosome-specific proteins and dosage compensation in *Drosophila*. Trends Genet 1999; 15(11):454-8.
14. Meller VH, Kuroda MI. Sex and the single chromosome. Adv Genet 2002; 46:1-24.
15. Kelley RL, Kuroda MI. Equality for X chromosomes. Science 1995; 270(5242):1607-10.
16. McQueen HA, McBride D, Miele G et al. Dosage compensation in birds. Curr Biol 2001; 11(4):253-7.
17. Ellegren H. Dosage compensation: do birds do it as well? Trends Genet 2002; 18(1):25-8.
18. Aronson J, Rudkin G, Schultz J. A comparison of giant X-chromosomes in male and female *Drosophila melanogaster* by cytophotometry in the ultraviolet. Cytochemistry 1954; 2:458-459.
19. Dobzhansky T. The X-chromosome in the larval salivary glands of hybrids *Drosophila insularis x Drosophila tropicalis*. Chromosoma 1957; 8:691-698.
20. Mukherjee AS, Beermann W. Synthesis of ribonucleic acid by the X-chromosomes of *Drosophila melanogaster* and the problem of dosage compensation. Nature 1965; 207(998):785-6.
21. Mukherjee AS. Dosage compensation in *Drosophila*; an autoradiographic study. The Nucleus 1966; 9:83-96.
22. Belote JM, Lucchesi JC. Male-specific lethal mutations of *Drosophila melanogaster*. Genetics 1980; 96(1):165-86.

23. Uchida S, Uenoyama T, Oishi K. Studies on the sex-specific lethals of *Drosophila melanogaster*. III. A third chromosome male-specific lethal mutant. The Japan Journal of Genetics 1981; 56:523-527.
24. Fukunaga A, Tanaka A, Oishi K. *Maleless*, a recessive autosomal mutant of *Drosophila melanogaster* that specifically kills male zygotes. Genetics 1975; 81:135-141.
25. Hilfiker A, Hilfiker-Kleiner D, Pannuti A et al. *mof*, a putative acetyl transferase gene related to the Tip60 and MOZ human genes and to the SAS genes of yeast, is required for dosage compensation in *Drosophila*. EMBO J 1997; 16(8):2054-60.
26. Bone JR, Lavender J, Richman R et al. Acetylated histone H4 on the male X chromosome is associated with dosage compensation in *Drosophila*. Genes Dev 1994; 8(1):96-104.
27. Gorman M, Franke A, Baker BS. Molecular characterization of the *male-specific lethal-3* gene and investigations of the regulation of dosage compensation in *Drosophila*. Development 1995; 121(2):463-75.
28. Zhou S, Yang Y, Scott MJ et al. *Male-specific lethal 2*, a dosage compensation gene of *Drosophila*, undergoes sex-specific regulation and encodes a protein with a RING finger and a mettallothionein-like cysteine cluster. EMBO J 1995; 14(12):2884-95.
29. Kuroda MI, Kernan MJ, Kreber R et al. The *maleless* protein associates with the X chromosome to regulate dosage compensation in *Drosophila*. Cell 1991; 66(5):935-47.
30. Gu W, Szauter P, Lucchesi JC. Targeting of MOF, a putative histone acetyl transferase, to the X chromosome of *Drosophila melanogaster*. Dev Genet 1998; 22(1):56-64.
31. Palmer MJ, Mergner VA, Richman R et al. The *male-specific lethal-one* (*msl-1*) gene of *Drosophila melanogaster* encodes a novel protein that associates with the X chromosome in males. Genetics 1993; 134(2):545-57.
32. Turner BM, Birley AJ, Lavender J. Histone H4 isoforms acetylated at specific lysine residues define individual chromosomes and chromatin domains in *Drosophila* polytene nuclei. Cell 1992; 69(2):375-84.
33. Cline TW, Meyer BJ. Vive la difference: males vs females in flies vs worms. Annu Rev Genet 1996; 30:637-702.
34. Bashaw GJ, Baker BS. The *msl-2* dosage compensation gene of *Drosophila* encodes a putative DNA-binding protein whose expression is sex specifically regulated by *Sex-lethal*. Development 1995; 121(10):3245-58.
35. Bashaw GJ, Baker BS. The regulation of the *Drosophila msl-2* gene reveals a function for *Sex-lethal* in translational control. Cell 1997; 89(5):789-98.
36. Kelley RL, Solovyeva I, Lyman LM et al. Expression of *msl-2* causes assembly of dosage compensation regulators on the X chromosomes and female lethality in *Drosophila*. Cell 1995; 81(6):867-77.
37. Kelley RL, Wang J, Bell L et al. *Sex lethal* controls dosage compensation in *Drosophila* by a nonsplicing mechanism. Nature 1997; 387(6629):195-9.
38. Cline TW. Evidence that *sisterless-a* and *sisterless-b* are two of several discrete "numerator elements" of the X/A sex determination signal in *Drosophila* that switch Sxl between two alternative stable expression states. Genetics 1988; 119(4):829-62.
39. Bopp D, Bell LR, Cline TW et al. Developmental distribution of female-specific *Sex-lethal* proteins in *Drosophila melanogaster*. Genes Dev 1991; 5(3):403 15.
40. Erickson JW, Cline TW. A bZIP protein, *sisterless-a*, collaborates with bHLH transcription factors early in *Drosophila* development to determine sex. Genes Dev 1993; 7(9):1688-702.
41. Parkhurst SM, Bopp D, Ish-Horowicz D. X:A ratio, the primary sex-determining signal in *Drosophila*, is transduced by helix-loop-helix proteins. Cell 1990; 63(6):1179-91.
42. Keyes LN, Cline TW, Schedl P. The primary sex determination signal of *Drosophila* acts at the level of transcription. Cell 1992; 68(5):933-43.
43. Bell LR, Maine EM, Schedl P et al. *Sex-lethal*, a *Drosophila* sex determination switch gene, exhibits sex-specific RNA splicing and sequence similarity to RNA binding proteins. Cell 1988; 55(6):1037-46.
44. Bell LR, Horabin JI, Schedl P et al. Positive autoregulation of *Sex-lethal* by alternative splicing maintains the female determined state in *Drosophila*. Cell 1991; 65(2):229-39.
45. Saurin AJ, Borden KL, Boddy MN et al. Does this have a familiar RING? Trends Biochem Sci 1996; 21(6):208-14.

46. Lyman LM, Copps K, Rastelli L et al. *Drosophila* Male-specific lethal-2 protein: structure/function analysis and dependence on MSL-1 for chromosome association. Genetics 1997; 147(4):1743-53.
47. Koonin EV, Zhou S, Lucchesi JC. The chromo superfamily: new members, duplication of the chromo domain and possible role in delivering transcription regulators to chromatin. Nucleic Acids Res 1995; 23(21):4229-33.
48. Eisen A, Utley RT, Nourani A et al. The yeast NuA4 and *Drosophila* MSL complexes contain homologous subunits important for transcription regulation. J Biol Chem 2001; 276(5):3484-91.
49. Bertram MJ, Berube NG, Hang-Swanson X et al. Identification of a gene that reverses the immortal phenotype of a subset of cells and is a member of a novel family of transcription factor-like genes. Mol Cell Biol 1999; 19(2):1479-85.
50. Eissenberg JC. Molecular biology of the chromo domain: an ancient chromatin module comes of age. Gene 2001; 275(1):19-29.
51. Lee CG, Chang KA, Kuroda MI et al. The NTPase/helicase activities of *Drosophila maleless*, an essential factor in dosage compensation. EMBO J 1997; 16(10):2671-81.
52. Tanner NK, Linder P. DExD/H box RNA helicases: from generic motors to specific dissociation functions. Mol Cell 2001; 8(2):251-62.
53. Lee CG, Hurwitz J. Human RNA helicase A is homologous to the *maleless* protein of *Drosophila*. J Biol Chem 1993; 268(22):16822-30.
54. Gibson TJ, Thompson JD. Detection of dsRNA-binding domains in RNA helicase A and *Drosophila maleless:* implications for monomeric RNA helicases. Nucleic Acids Res 1994; 22(13):2552-6.
55. Richter L, Bone JR, Kuroda MI. RNA-dependent association of the *Drosophila maleless* protein with the male X chromosome. Genes Cells 1996; 1(3):325-36.
56. Brown CE, Lechner T, Howe L et al. The many HATs of transcription coactivators. Trends Biochem Sci 2000; 25(1):15-9.
57. Jin Y, Wang Y, Walker DL et al. JIL-1: a novel chromosomal tandem kinase implicated in transcriptional regulation in *Drosophila*. Mol Cell 1999; 4(1):129-35.
58. Wang Y, Zhang W, Jin Y et al. The JIL-1 tandem kinase mediates histone H3 phosphorylation and is required for maintenance of chromatin structure in *Drosophila*. Cell 2001; 105(4):433-43.
59. Jin Y, Wang Y, Johansen J et al. JIL-1, a chromosomal kinase implicated in regulation of chromatin structure, associates with the male specific lethal (MSL) dosage compensation complex. J Cell Biol 2000; 149(5):1005-10.
60. Amrein H, Axel R. Genes expressed in neurons of adult male *Drosophila*. Cell 1997; 88(4):459-69.
61. Meller VH, Wu KH, Roman G et al. *roX1* RNA paints the X chromosome of male *Drosophila* and is regulated by the dosage compensation system. Cell 1997; 88(4):445-57.
62. Meller VH, Gordadze PR, Park Y et al. Ordered assembly of *roX* RNAs into MSL complexes on the dosage-compensated X chromosome in *Drosophila*. Curr Biol 2000; 10(3):136-43.
63. Franke A, Baker BS. The *rox1* and *rox2* RNAs are essential components of the compensasome, which mediates dosage compensation in *Drosophila*. Mol Cell 1999; 4(1):117-22.
64. Meller VH, Rattner BP. The *roX* genes encode redundant male-specific lethal transcripts required for targeting of the MSL complex. EMBO J 2002; 21(5):1084-91.
65. Kelley RL, Kuroda MI. Noncoding RNA genes in dosage compensation and imprinting. Cell 2000; 103(1):9-12.
66. Kelley RL, Meller VH, Gordadze PR et al. Epigenetic spreading of the *Drosophila* dosage compensation complex from *roX* RNA genes into flanking chromatin. Cell 1999; 98(4):513-22.
67. Kageyama Y, Mengus G, Gilfillan G et al. Association and spreading of the *Drosophila* dosage compensation complex from a discrete *roX1* chromatin entry site. EMBO J 2001; 20(9):2236-45.
68. Palmer MJ, Richman R, Richter L et al. Sex-specific regulation of the *male-specific lethal-1* dosage compensation gene in *Drosophila*. Genes Dev 1994; 8(6):698-706.
69. Copps K, Richman R, Lyman LM et al. Complex formation by the *Drosophila* MSL proteins: role of the MSL2 RING finger in protein complex assembly. EMBO J 1998; 17(18):5409-17.
70. Scott MJ, Pan LL, Cleland SB et al. MSL1 plays a central role in assembly of the MSL complex, essential for dosage compensation in *Drosophila*. EMBO J 2000; 19(1):144-55.
71. Smith ER, Pannuti A, Gu W et al. The *Drosophila* MSL complex acetylates histone H4 at lysine 16, a chromatin modification linked to dosage compensation. Mol Cell Biol 2000; 20(1):312-8.

72. Akhtar A, Zink D, Becker PB. Chromodomains are protein-RNA interaction modules. Nature 2000; 407(6802):405-9.
73. Gu W, Wei X, Pannuti A et al. Targeting the chromatin-remodeling MSL complex of *Drosophila* to its sites of action on the X chromosome requires both acetyl transferase and ATPase activities. EMBO J 2000; 19(19):5202-11.
74. Akhtar A, Becker PB. Activation of transcription through histone H4 acetylation by MOF, an acetyltransferase essential for dosage compensation in *Drosophila*. Mol Cell 2000; 5(2):367-75.
75. Amrein H. Multiple RNA-protein interactions in *Drosophila* dosage compensation. Genome Biol 2000; 1(6): 1030.1-1030.5.
76. Lanz RB, McKenna NJ, Onate SA et al. A steroid receptor coactivator, SRA, functions as an RNA and is present in an SRC-1 complex. Cell 1999; 97(1):17-27.

Chapter 6

The Structure, Regulation and Function of the Imprinted *H19* RNA

Raluca I. Verona and Marisa S. Bartolomei

Abstract

H19 is a member of a small subset of genes that are subject to the parent-of-origin dependent expression known as genomic imprinting. The *H19* gene is transcribed exclusively from the maternal allele, and this paternal imprinting is evolutionarily conserved in mammals. In addition to its imprinting, *H19* displays a complex expression profile, with high levels present during embryonic development in a spatially- and temporally-restricted manner. The product of the *H19* gene is a fully processed RNA that lacks any significant open reading frames. Hence, it is thought that *H19* exerts its functions at the RNA level, as a regulatory RNA or riboregulator. Over the years, *H19* has been proposed to function in many different processes, ranging from transcriptional regulation to tumor suppression and oncogenesis. However, the exact biological functions, if any, of *H19* remain to be defined. This chapter reviews important aspects of *H19* properties, regulation, and potential functions. First, we describe the highly complex regulation of *H19* imprinting and tissue-specific expression, including cis-acting elements that are important for these processes. We also describe the sequence conservation of the *H19* locus, including the striking secondary structure conservation of the *H19* RNA across different species. Finally, we discuss possible biological roles for the *H19* gene product, using data that have emerged from human cancer patients, mouse models, as well as cell culture studies.

Introduction

Genomic imprinting is a mechanism of transcriptional regulation through which expression of a gene occurs solely from its maternal or paternal allele.[1] A small subset of genes in mammals is subject to genomic imprinting, and, interestingly, many of these genes regulate cell proliferation and differentiation. The importance of proper imprinted expression for normal growth and development is underscored by two observations. First, embryos containing only maternal or paternal genomes fail to develop normally.[2,3] Second, loss of imprinting plays a causative role in a significant number of inherited human diseases and cancers.[4,5]

The monoallelic expression of imprinted genes is believed to result from epigenetic modifications that differentially mark the parental alleles during gametogenesis.[1] The two copies of an imprinted gene often exhibit differential chromatin structure (DNA methylation, histone modifications and hypersensitive sites), as well as asynchronous replication.[6] While the exact nature of the epigenetic mark remains to be established fully, and may in fact be gene-specific, DNA methylation of cytosine residues in CpG dinucleotides is a strong candidate for such a

Noncoding RNAs: Molecular Biology and Molecular Medicine, edited by Jan Barciszewski and Volker A. Erdmann.

mark. DNA methylation has been shown to silence gene expression either by directly inhibiting transcription factor binding or through recruitment of repressive chromatin remodeling complexes.[7] DNA methylation fulfills many of the requirements needed for an epigenetic mark—it is stably maintained during replication, and hence heritable through multiple cell divisions, and it can also be erased, as would be needed when imprints are reset in the germ line.[1] Most of the imprinted genes identified to date contain regions that exhibit differential methylation of the parental alleles. In addition, embryos deficient in DNA methyltransferase family members exhibit defects in imprinted gene expression.[8-12] Hence, DNA methylation has been implicated strongly in the regulation of imprinting and is believed to play a crucial role in the establishment and/or maintenance of monoallelic gene expression.[13]

The *H19* gene is one of the most extensively studied imprinted genes. *H19* was identified more than fifteen years ago in independent differential cDNA screens for genes regulated by the *alpha-fetoprotein* regulator *raf* or genes that are activated upon cellular differentiation.[14-17] *H19* is transcribed by RNA polymerase II, and the RNA is spliced and polyadenylated.[18] However, despite these features, the *H19* transcript is not translated.[18] Sequence analysis reveals the presence of several short open reading frames (ORFs), none of which exhibit any conservation across species. Furthermore, antibodies raised against the longest of these ORFs have failed to detect the protein product in vivo.[18] Early cell fractionation studies showed that the *H19* RNA is cytoplasmic, where it is associated with other proteins but not associated with ribosomes.[18] Together, these features suggest that *H19* exerts its function at the RNA level, as a regulatory RNA or riboregulator.

Riboregulatory RNAs

H19 is part of a growing class of RNA polymerase II-transcribed noncoding RNAs that includes Air, Kcnq1ot1, and Xist. Many of these noncoding RNAs occur at imprinted loci, are themselves imprinted, and function as riboregulators. Most commonly, these RNAs function to negatively regulate gene expression in cis (Table 1).

The examples listed in Table 1 illustrate the fact that noncoding RNAs are prevalent at imprinted loci. In many cases, noncoding RNAs are expressed in an antisense orientation to a sense, protein-coding gene. Interestingly, the sense and antisense RNAs are often oppositely imprinted and expressed from different parental alleles (Table 1). Furthermore, methylation appears to be preferentially associated with one of the chromosomes, providing a possible mechanism for the regulation of allele-specific transcription. For example, at the mouse *Insulin-like growth factor type 2 receptor* (*Igf2r*) imprinted cluster on chromosome 17, the paternally-expressed noncoding Air transcript is expressed from a promoter located in intron 2 of the maternally-expressed *Igf2r* and is transcribed in antisense orientation to *Igf2r*.[19,20] Maternal-specific methylation of the Air promoter is believed to be important for silencing of the Air transcript on this allele.[20-22]

Expression of the antisense RNA often correlates with repression of the sense gene at many different loci (Table 1). Hence, it is generally thought that the noncoding RNA plays a role in silencing of the sense gene. The mechanisms through which this regulation occurs remain poorly understood. One possibility is that the antisense RNA successfully competes for components of the transcriptional machinery or enhancer elements, thus preventing expression of the sense gene. Alternatively, the antisense RNA could directly block transcription of the sense gene, perhaps by interfering with RNA polymerase II, or could alternatively act at a post-transcriptional level, by base-pairing with the sense RNA and either sequestering it or causing its degradation through an RNAi-like mechanism. At the *Igf2r* locus, expression of Air silences *Igf2r* expression in cis.[22,23] A recent study showing that a truncated Air transcript loses its ability to silence transcription suggests that the full length RNA, or simply transcription through the region, is required for its function.[23]

Table 1. Properties of imprinted noncoding RNAs

Noncoding RNA	Expression	Overlapping Sense Gene	Expression of Sense Gene	Locus Methylation	Function of Nocoding RNA	Evolutionary Conservation#	References
Air	Paternal	*Igf2r*	Maternal	Maternal (*Air* promoter)	Repression of *Igf2r, Slc22a2, Slc22a3*	Mouse only	20-22, 25, 26
COPG2IT1 (CIT1) / Copg2as	Paternal	*COP2G*	Maternal (mouse only)	N/D	Unknown	Mouse and human	82, 83
Gtl2	Maternal	None	None	Paternal	Unknown	Mouse and human	84-88
H19	Maternal	None	None	Paternal (5′ region)	Unknown	Mouse and human	see text
Igf2as	Paternal	*Igf2*	Paternal	Paternal	Unknown	Mouse (human antisense RNA has ORF)	89, 90
Kcnq1ot1	Paternal	*Kcnq1*	Maternal	Maternal (*Kcnq1ot1* promoter)	Repression of *Kcnq1, Cdkn1c, Slc22a1l, Tssc3*	Mouse and human	91-93
Nespas/GNAS1AS	Paternal	*Nesp*	Maternal	Maternal (5′ of transcript)	Regulation of *Nesp*?	Mouse and human	94-96
PEG1-AS/MESTIT1	Paternal	*PEG1/MEST*	Paternal (isoform-specific)	Maternal	Regulation of *PEG1/MEST* expression during development?	Human, mouse N/D	97, 98
Tsix	Maternal*	*Xist*	Paternal*	Paternal (*Tsix* promoter)	Silencing of *Xist*	Mouse only	99-101
Ube3a antisense	Paternal	*Ube3a*	Maternal	Maternal	Silencing of *Ube3a* and potentially of *ATP10C*	Mouse and human	102, 103
Xist	Paternal*	*Tsix*	Maternal*	Paternal (*Tsix* promoter)	Silencing of X-linked genes	Mouse only	99-101
Znf127AS	Paternal	*Znf127*	Paternal	Maternal	Unknown	Mouse and human	104, 105

#Conservation of noncoding RNA between mice and humans; *only in extra-embryonic tissues; N/D=not determined.

Another interesting property of these riboregulatory RNAs is that they can regulate the transcription not only of overlapping sense genes, but also of genes that are more distantly located. Once silencing of the overlapping gene is established, and the surrounding DNA has become heterochromatic, it could spread in both directions to genes located at a distance 5' and 3', resulting in transcriptional silencing. Deletion of the Air promoter or truncation of its RNA leads to derepression not only of *Igf2r*, but also of the distant *Slc22a2* and *Slc22a3* genes that are not overlapped by Air.[23,24] Similarly, at both the mouse and human *Cdkn1c* ($p57^{Kip2}$)/ *Kcnq1* imprinted gene clusters, expression of noncoding Kcnq1ot1 RNA on the paternal chromosome is proposed to negatively regulate the expression of the overlapping *Kcnq1*, as well as of the more distant *Cdkn1c, Slc22a1l* (*Impt1*), and *Tssc3* (*Ipl*) genes.[25,26]

Many of the noncoding RNAs have been implicated in transcriptional regulation., With a few exceptions, however, the in vivo functions of these noncoding RNAs remain to be established and await the generation of deletions at the endogenous loci in mice. Although it is likely that in many cases noncoding RNAs function to silence gene expression, some of the noncoding RNAs that have been identified may not be functional and could be transcribed simply as the result of a favorable open chromatin configuration of the surrounding DNA. Finally, whereas many noncoding RNAs have been found at imprinted loci, noncoding RNAs may be just as prevalent at nonimprinted loci.

The *H19* RNA is yet another example of an imprinted noncoding RNA that has been implicated in transcriptional control, as well as in tumorigenesis. As discussed below, *H19* expression is highly complex and tightly regulated. Although the control of *H19* imprinted expression is becoming very well understood, the function of this transcript remains enigmatic.

Imprinting of *H19*

H19 is paternally imprinted and its exclusive maternal expression is conserved among mammals.[27-29] *H19* is located within a cluster of imprinted genes on human chromosome 11p15.5 and the corresponding syntenic region at the distal end of mouse chromosome 7. In both human and mouse, *H19* is located 3' to the maternally imprinted gene Insulin-like growth factor 2 (*Igf2*).[30-32] *H19* and *Igf2* are not simply neighbors, but share significant aspects of their regulation through common DNA elements.

Clues about the imprinted regulation of the mouse *H19* gene have emerged from studies carried out using three different approaches: (1) analysis of methylation of the parental alleles, (2) determination of minimal sequence requirements for imprinting of *H19* transgenes, and (3) examination of the effect of *H19* deletions at the endogenous locus. Together, these studies have shown clearly that a 2 kilobase (kb) region located 2 kb upstream of the *H19* promoter regulates imprinting of both *H19* and *Igf2*. This 2 kb region is known as the differentially methylated domain (DMD) or as the imprinting control region (ICR) (Fig. 1).

Methylation

In contrast to most imprinted genes that are characterized by regions with predominantly maternal methylation, *H19* exhibits paternal-specific methylation. In fact, the DMD has been proposed to harbor the imprinting mark that distinguishes the parental alleles of *H19*. Consistent with this hypothesis, the two *H19* alleles exhibit striking differences in their methylation profiles, with the paternal DMD being hypermethylated relative to its maternal counterpart.[33-35] The differential methylation of the DMD is present in the early embryo, is resistant to the genome-wide demethylation that occurs during preimplantation, and persists throughout subsequent development.[33,34] The paternal allele is also hypermethylated at the *H19* promoter, consistent with its silent transcriptional status. However, methylation in this region occurs after implantation and is believed to occur as a result of spreading of methylation from the DMD sequence into the promoter region. Loss of *H19* methylation caused by absence of the

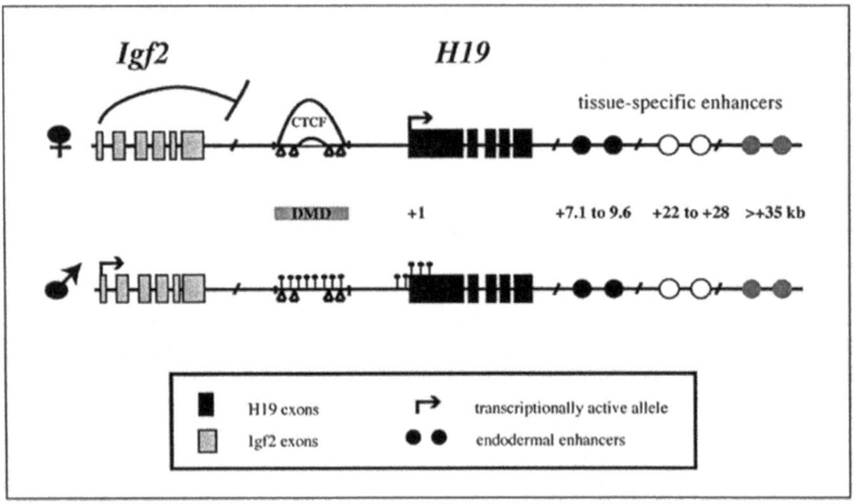

Figure 1. The *H19/Igf2* locus at the distal end of mouse chromosome 7. *H19* and *Igf2* exons are marked with black and grey rectangles, respectively, and the black arrows denote the starts of transcription. The four CTCF binding sites within the differentially methylated region (DMD) are shown as triangles. Regions of *H19* paternal specific-methylation in the DMD and promoter region are marked by filled lollipops. The location of the endodermal enhancers and the proposed location of the mesodermal enhancers are also indicated.

DNA methyltransferase1 (Dnmt1) leads to biallelic expression, supporting a role for methylation in regulating mono-allelic expression at this locus.[11]

The DMD is unmethylated in oocytes.[33,36,37] In contrast, this region is methylated in sperm.[33,34] The paternal-specific methylation of the DMD is established during male gametogenesis, after erasure of methylation at this locus in primordial germ cells.[38,39] Interestingly, the two parental alleles are methylated at different times during male germ cell development, suggesting that an epigenetic signal other than methylation plays a role in differentiating the parental chromosomes at this time.[38,39] The characterization of *H19* methylation in germ cells has raised many important questions about the exact nature of the early imprinting mark in the DMD region. Understanding how the two parental copies are distinguished in the absence of methylation is critical for elucidating how the *H19* imprint is established in the germ line.

Transgenes

Studies using mouse *H19* transgenes have been instrumental in establishing the role of the DMD in *H19* imprinting and in defining the minimal sequence requirements for proper imprinting. An *H19* transgenic construct containing 4 kb of 5' *H19* sequence, an internally-deleted *H19* structural gene, and 8 kb of 3' sequence (RRSRdBam) mimicks the imprinting of the endogenous *H19* gene.[35,40] RRSRdBam is expressed and hypomethylated when maternally inherited; it is repressed and hypermethylated in the 5' flanking region and gene body upon paternal transmission. Deletion analysis of sequences contained within the RRSRdBam construct demonstrated that the DMD is required for transgene imprinting (BRSRdBam).[40] However, in the context of these transgenes that lack the 5'-most DMD sequence, it appears that *H19* 3' sequence is also important for proper imprinting.[41] Together, these data show that multiple sequences both 5' and 3' to *H19* can cooperate to regulate imprinting of this gene.

Deletions of H19 Sequence at the Endogenous Locus

Mutation of the DMD at the endogenous locus has a profound effect on imprinting. Paternal inheritance of a 2 kb DMD deletion leads to loss of paternal-specific *H19* methylation and to activation of this normally silent allele.[42,43] Conditional deletion of the DMD specifically in terminally differentiated muscle cells shows that, once imprinting is established, the DMD is no longer required for *H19* imprinting,[44] presumably because once methylation spreads from the DMD to the *H19* promoter, a silent transcriptional state is locked in at this allele. Together, these studies show that the DMD region is required to establish the paternal imprinting of *H19*.

Conversely, maternal inheritance of a DMD deletion activates the silent *Igf2* allele,[42,43] demonstrating that the DMD is required for the parental-specific expression of both genes. However, the DMD regulates imprinting of *H19* and *Igf2* through distinct mechanisms (see below). Consistent with this idea, the DMD region is continuously required to maintain the proper imprinting of *Igf2*.[44]

Deletion of only 1.2 kb of the DMD causes loss of *H19* imprinting on the paternal allele without affecting *Igf2* on the maternal chromosome.[45] This result suggests that distinct regions within the DMD are required for the silencing of *H19* and *Igf2*. More precise deletions and mutations in this region are needed to determine the specific regions within the DMD that function to regulate the reciprocal imprinting of *H19* and *Igf2*.

Model for H19 and Igf2 Imprinting

Recent studies have suggested that the DMD can act as an insulator element in regulating the imprinting of *Igf2*. Insulators are sequences that prevent promoter-enhancer interactions in a directional manner.[46] The DMD was shown to contain four putative binding sites for the vertebrate insulator factor CTCF, and binding of CTCF to these elements in vitro is inhibited by methylation of the target sequence.[47-51] The DMD can also act as an insulator in transfection assays and in transgenic mice.[48,50-52]

Together, these data have led to the following model. On the maternal allele, the DMD is unmethylated and *H19* is transcribed. CTCF binding to the unmethylated DMD is hypothesized to insulate the *Igf2* promoter from enhancers located 3' to *H19*, allowing *H19* exclusive access to these elements. On the paternal allele, the DMD methylation spreads to the *H19* promoter region, leading to the silencing of *H19*. The hypermethylated DMD cannot be bound by CTCF, thereby preventing the formation of an insulator and allowing *Igf2* exclusive access to the enhancers.

The recent identification of the CTCF binding sites within the *H19* DMD and the emergence of the CTCF model for *Igf2* regulation have been very exciting developments in the *H19*/*Igf2* field. The exact role, if any, that CTCF plays in the establishment of parental imprints remains to be elucidated.[53] One possibility is that CTCF binds to the *H19* ICR in the female germ line and protects this region from methylation, thus establishing a maternal imprint. Alternatively, binding of CTCF could occur post-fertilization, when the imprints have already been established. The methylation of the paternal allele in the male germ line would prevent CTCF binding on this allele; however, CTCF could bind to the unmethylated maternal allele at this time. Elucidating the contribution of CTCF in the establishment and/or maintenance of imprinting at this locus will be critical for the full understanding of allele-specific *H19* and *Igf2* expression.

Tissue-Specific Expression of *H19*

H19 displays a complex expression profile that is highly regulated both spatially and temporally. The *H19* transcript is present at high levels, representing one of the most abundant RNAs in the developing mouse embryo. *H19* is transcribed in tissues of endodermal and mesodermal origin. Expression of *H19* begins at the late blastocyst stage and persists throughout embryogenesis. After birth, *H19* transcription is turned off gradually within a few weeks in most tissues, with the exception of muscle where low levels of expression are found through adulthood.[54]

Two elements located between +7.0 and +10 kb relative to the start of *H19* transcription mediate endodermal expression of *H19* [55,56] (Fig. 1). Deletion of sequence spanning these elements results in loss of *H19* expression in liver, gut, kidney, and lung.[55] Interestingly, these enhancers were also shown to be important for endodermal expression of *Igf2*,[55] consistent with its expression pattern in the same tissue types as *H19*.[35,57,58]

Studies using yeast and bacterial artificial chromosomes (YACs and BACs) have examined what sequences are required for expression of *H19* in mesodermal tissues. An 130 kb YAC transgene spanning 5' to *Igf2* through 35 kb 3' to *H19* exhibits expression in endodermal tissues, as expected, but in only a subset of mesodermal tissues, namely skeletal but not cardiac muscle.[59,60] In contrast, a BAC transgene spanning *H19* sequence from –7 to +130 kb is expressed in all endodermal and mesodermal tissues.[52] Results from these transgenic studies, together with in vitro mapping experiments using transient transfection assays, suggest that the skeletal muscle enhancers are located between +22 to +28 kb downstream of the *H19* transcription start, whereas the cardiac muscle enhancers reside between +35 and +130 (Fig. 1).

Mutations at the endogenous *H19* locus have supported this preliminary mapping of the mesodermal enhancers. Maternal inheritance of a +10.7 to +34.7 kb deletion results in a five- to six-fold decrease in *H19* expression in skeletal muscle, without a change in cardiac muscle expression.[61] As in the case of the shared endodermal enhancers, it appears that the same sequences mediate mesodermal expression of both *H19* and *Igf2*, as paternal inheritance of this deletion leads to a three- to four-fold reduction of *Igf2* skeletal muscle expression.[61] The location of these elements 3' of the DMD is consistent with the insulator model of *Igf2* imprinting and with the DMD deletions that result in loss of *Igf2* imprinting in mesodermal tissues.[43] More precise mapping of the mesodermal enhancers awaits the generation of smaller deletions at this locus.

Deletion of sequence within the DMD also reduces the level of expression of the normally active *H19* and *Igf2* alleles. Interestingly, different tissues are affected to varying degrees by these deletions.[42,43] In liver, maternal transmission of the deletion causes *H19* expression to be reduced by 50% relative to the wild type maternal allele, whereas in some mesodermal tissues such as kidney and heart *H19* is drastically down regulated to 10-26% of wild type maternal expression levels.[43] When the same deletion is inherited paternally, the expression level of the normally active *Igf2* allele is also decreased, most significantly in liver.[43] These results suggest that the DMD could harbor sequences that are important for full tissue-specific expression of the normally active allele. Although the mechanism through which this occurs remains unknown, it is possible that the DMD interacts differentially with the tissue-specific enhancers.

Discrete sequences located between *H19* and *Igf2* have also been implicated in the regulation of tissue-specific *H19* transcription. Two 400-bp elements upstream of *H19* (HUC1 and HUC2) have been proposed recently to play a role in driving mesodermal expression of *H19*.[62] These sequences do not exhibit allele-specific methylation, are biallelically transcribed, and are believed to be noncoding RNAs. The function of these regions in the control of *H19* expression at the endogenous locus has not yet been determined, although it is unclear how they might influence allele- and tissue- specific transcription of *H19*. However, it is interesting to note that in the context of a larger deletion that includes HUC1, mesodermal expression of both *H19* and *Igf2* can still occur.[52]

It is likely that more elements important for *H19* tissue-specific expression remain to be identified at this locus. The mechanism by which imprinted tissue-specific expression occurs may be dependent on specific trans-acting factors as well as multiple cis-acting DNA elements.

Sequence Conservation of *H19*

H19 is very well conserved between mice and humans, exhibiting 77% overall sequence identity, with some regions as high as 86%.[18] The genomic organization consisting of five exons and four small introns is also conserved through evolution.[18]

Additional comparative analyses between human and mouse *H19* sequences have revealed the presence of discrete regions of homology at this locus outside the *H19* gene body (Fig. 2). These homologous regions include elements located both 5' and 3' to *H19*.[56,62-65] The high degree of sequence conservation of these elements through evolution strongly suggests that they have important regulatory functions.

Interestingly, the sequence conservation in the DMD region 5' to *H19* is limited to 21-base pair (bp) repeats that are present four times in the mouse and rat genes and six times in the human sequence.[63] Since these 21-bp repeats are the CTCF binding sites within the DMD region, it is clear that the mechanism of reciprocal imprinting of *H19* and *Igf2* remains very well conserved between mice and humans, despite little overall sequence similarity in this region.

Several conserved sequence blocks are present 3' of the *H19* transcription unit.[64] Not surprisingly, among these are the two endodermal enhancer elements. Therefore, it is likely that other conserved elements in this region may be functionally important. Consistent with this hypothesis, some of these elements exhibit enhancer function in transgenic mice.[64]

The high degree of *H19* sequence conservation offers a useful tool to identify elements that are important for its regulation. Testing the role of conserved elements at the endogenous locus in mice will likely uncover many new cis-acting DNA elements that function in both tissue-specific transcription and imprinting.

H19 and Growth Control

H19 is located on human chromosome 11p15.5, a region commonly mutated in tumorigenesis. Loss of *H19* expression frequently occurs in many pediatric cancers, including Wilms' tumor, adrenocortical carcinoma, hepatoblastoma, and embryonal rhabdomyosarcoma.[66] Loss of heterozygosity (LOH) of this region almost always involves loss of the maternal chromosome. In tumors that retain both alleles, *H19* is silenced and hypermethylated on both parental chromosomes.[66] These data support a role for *H19* as a tumor suppressor.

Interestingly, *H19* has also been proposed to function to promote tumorigenesis, as an oncofetal gene—a gene usually expressed exclusively during fetal development that is abnormally activated in tumors.[67] Loss of imprinting of *H19* resulting in biallelic expression has been observed in adult cancers including lung, esophageal, cervical, bladder, prostate, testicular germ cell tumors (TGCTs) and choriocarcinoma.[68]

Furthermore, mutation at the *H19* locus that results in loss of *IGF2* imprinting is thought to contribute to the etiology of Beckwith-Wiedemann syndrome, a disorder characterized by somatic overgrowth and increased childhood tumor susceptibility.[66,69,70] As *H19* and *IGF2* are both closely linked and oppositely imprinted, a single genetic or epigenetic mutational event can simultaneously act to inactivate *H19* and overexpress *IGF2*.[66] Indeed, analysis of Wilms' tumors, a common cancer of BWS patients, revealed that loss of *H19* is concurrent with activation of *IGF2* in 75% of cases.[70]

Although deregulation of *H19* expression is a frequent event in cancer, it is still unclear whether these transcriptional changes directly contribute to neoplasia. Before the roles of *H19* and *IGF2* in tumorigenesis can be elucidated, it is essential to understand the regulation of these genes in normal cells.

Is the *H19* RNA Functional?

Whether *H19* encodes a functional gene product remains an intensely controversial subject. At the sequence level, the *H19* gene is very well conserved between mouse and human, as

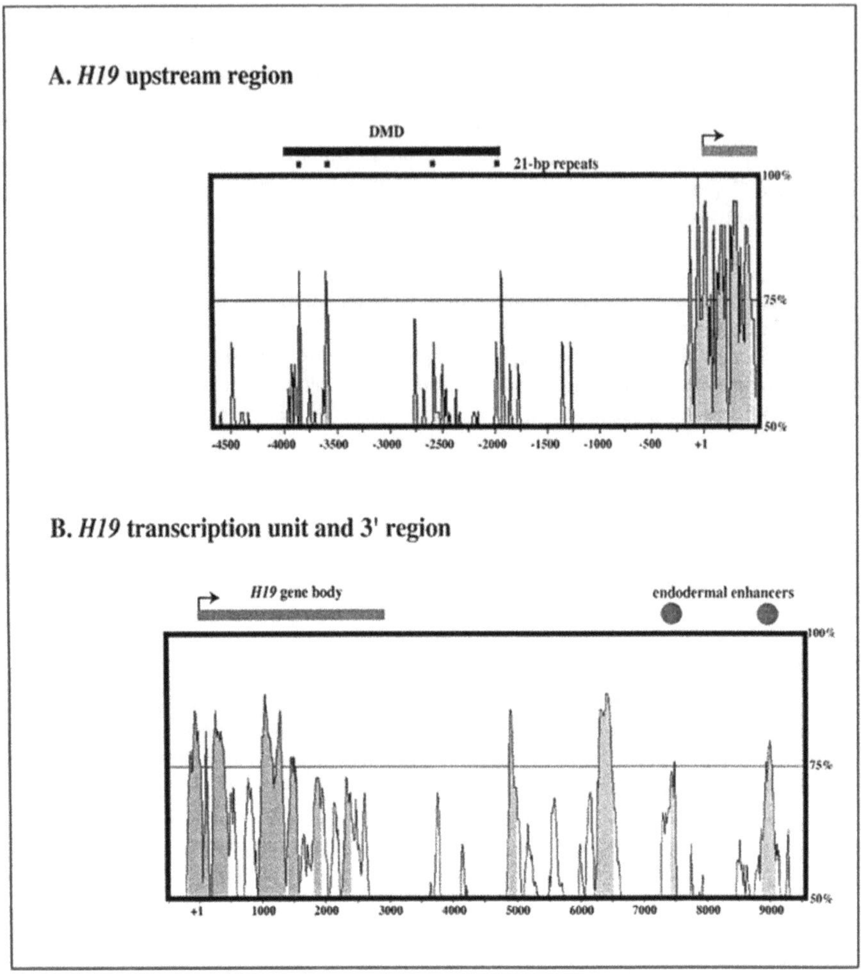

Figure 2. Evolutionary conservation between mouse and human *H19* sequence. Mouse and human *H19* sequences were compared using VISTA (http://www-gsd.lbl.gov/vista).[80,81] Sequence identity plots are shown in (A) for *H19* upstream sequence and (B) for *H19* gene body and 3' region. Mouse sequence is displayed on the X-axis and percentage identity is indicated on the Y-axis. The window length used in the comparison for percent identity was 21 bp for (A) and 100 bp for (B). Shaded areas represent regions with sequence conservation of 75% or above for the specific window length. The location of the DMD region between -2000 and -4000 relative to the start of transcription is indicated in (A). The 21-bp repeats marked below the DMD are located at -2101, -2556, -3587 and -3835.[63] The 3 kb *H19* transcription unit (starting at +1) and the two endodermal enhancer elements are indicated in (B).

well as among other mammals.[18] In addition, the *H19* RNA exhibits striking conservation of secondary structure among different species ranging from mouse and human to rat, orangutan, cat, and rabbit.[71] Furthermore, comparison of the rates of evolution between mouse and rat *H19* genes shows significant stabilizing selection.[72] Together, these properties strongly support a functional role for the *H19* RNA.

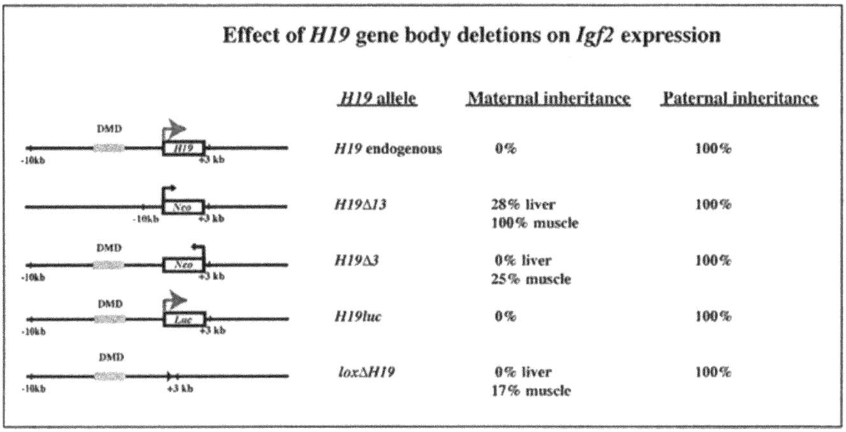

Figure 3. Effect of *H19* deletions on *Igf2* expression. *Igf2* expression upon maternal or paternal inheritance of distinct *H19* alleles is shown for liver and muscle tissues, unless otherwise indicated. The *H19* promoter is marked by a large grey arrow, whereas the *neomycin* gene promoter is represented by a small black arrow. The gray rectangle denotes the DMD. The $H19^{\Delta 13}$ allele contains a deletion of the *H19* transcription unit and 10kb upstream, including the *H19* promoter and the DMD.[76] $H19^{\Delta 3}$ harbors a deletion of the *H19* promoter and gene body.[74] In $H19^{luc}$, the H19 transcription unit has been replaced by the *luciferase* gene (*luc*), leaving the *H19* promoter intact.[75] *LoxΔH19* contains a deletion of both the *H19* promoter and gene body, leaving a loxP site in their place.[73]

Over the past decade, *H19* has been proposed to function in transcriptional regulation, as well as in tumorigenesis, as either a tumor suppressor or as an oncofetal gene. However, experiments in the mouse have failed to define a role for *H19* in any of these processes.

Despite the extremely high levels of its transcript throughout embryogenesis, *H19* appears to be dispensable for normal growth and development in mice.[73-75] Although in some cases *H19* mutant mice exhibit an overgrowth phenotype, this phenotype is attributed to upregulation of *Igf2* caused by the mutation of *H19* upstream sequence.[76] Overexpression of *H19* can be growth-suppressive in cancer cell lines,[77] supporting the hypothesis that *H19* functions as a tumor suppressor. However, none of the mice deficient for *H19* develop tumors.

Given that *H19* and *Igf2* are expressed from opposite parental alleles and share many of their control elements, one of the initial models postulated that the *H19* RNA functions to regulate the imprinted expression of *Igf2*. Specifically, the maternally expressed *H19* RNA was hypothesized to inhibit *Igf2* expression in cis. This proposed function for *H19* was tested in several studies in which the *H19* transcription unit was deleted at the endogenous locus in mice (Fig. 3). Although the $H19^{\Delta 13}$ allele caused a derepression of *Igf2* expression, this result was subsequently explained by the absence of the DMD region.[76] Deletion of the *H19* transcription unit alone, leaving the promoter intact, does not affect *Igf2* imprinting,[75] whereas the absence of both the promoter region and *H19* transcription unit leads to a slight relaxation of *Igf2* imprinting in neonatal muscle but not in any other tissues.[73,74] Thus, data from the *H19* deletion mouse strains suggest that the *H19* RNA does not play a major role in regulating *Igf2* expression.

Nevertheless, several in vitro studies using cell lines have proposed a role for *H19* in negatively regulating *Igf2* levels. One report concluded that the *H19* RNA is associated with polysomes and affects *Igf2* translatability.[78] Another study using *H19* sense and antisense stable transfection assays proposed that *H19* negatively affects *Igf2* transcription in cis.[79] However, the in vivo relevance of these results remains unclear.

Surprisingly, few details about the function of the *H19* RNA have emerged over the past twelve years. Although the degree of sequence conservation at both the DNA and RNA level is very striking, the function of this abundantly expressed transcript remains elusive.

References

1. Bartolomei MS, Tilghman SM. Genomic imprinting in mammals. Annu Rev Genet 1997; 31:493-525.
2. Surani MAH, Barton SC, Norris ML. Development of reconstituted mouse eggs suggest imprinting of the genome during gametogenesis. Nature 1984; 308:548-550.
3. McGrath J, Solter D. Completion of mouse embryogenesis requires both the maternal and paternal genomes. Cell 1984; 37:179-183.
4. Joyce JA, Schofield PN. Genomic imprinting and cancer. J Clin Path: Mol Path 1998; 51:185-190.
5. Lyle R. Gametic imprinting in development and disease. J of Endocrinology 1997; 155:1-12.
6. Brannan CI, Bartolomei MS. Mechanisms of genomic imprinting. Curr Opin Genet Dev 1999; 9(2):164-170.
7. Kass SU, Pruss D, Wolffe AP. How does DNA methylation repress transcription?. Trends Genet 1997; 13(11):444-449.
8. Okano M, Takebayashi S, Okumura K et al. Assignment of cytosine-5 DNA methyltransferases Dnmt3a and Dnmt3b to mouse chromosome bands 12A2-A3 and 2H1 by in situ hybridization. Cytogenet Cell Genet 1999; 86(3-4):333-334.
9. Howell CY, Bestor TH, Ding F et al. Genomic imprinting disrupted by a maternal effect mutation in the Dnmt1 gene. Cell 2001; 104(6):829-838.
10. Hata K, Okano M, Lei H et al. Dnmt3L cooperates with the Dnmt3 family of de novo DNA methyltransferases to establish maternal imprints in mice. Development 2002; 129(8):1983-1993.
11. Li E, Beard C, Jaenisch R. Role for DNA methylation in genomic imprinting. Nature 1993; 366(6453):362-365.
12. Bourc'his D, Xu GL, Lin CS et al. Dnmt3L and the establishment of maternal genomic imprints. Science 2001; 294(5551):2536-2539.
13. Feil R, Khosla S. Genomic imprinting in mammals: an interplay between chromatin and DNA methylation?. Trends in Genetics 1999; 15(11):431-435.
14. Davis RL, Weintraub H, Lassar AB. Expression of a single transfected cDNA converts fibroblasts to myoblasts. Cell 1987; 51:987-100.
15. Pachnis V, Belayew A, Tilghman SM. Locus unlinked to α-fetoprotein under the control of the murine raf and Rif genes. Proc Natl Acad Sci USA 1984; 81:5523-5527.
16. Pachnis V, Brannan CI, Tilghman SM. The structure and expression of a novel gene activated in early mouse embryogenesis. EMBO J 1988; 7:673-681.
17. Poirier F, Chan CT, Timmons PM et al. The murine *H19* gene is activated during embryonic stem cell differentiation in vitro and at the time of implantation in the developing embryo. Development 1991; 113(4):1105-1114.
18. Brannan CI, Dees EC, Ingram RS et al. The product of the *H19* gene may function as an RNA. Mol Cell Biol 1990; 10(1):28-36.
19. Lyle R, Watanabe D, te Vruchte D et al. The imprinted antisense RNA at the *Igf2r* locus overlaps but does not imprint Mas1. Nat Genet 2000; 25(1):19-21.
20. Wutz A, Smrzka OW, Schweifer N et al. Imprinted expression of the *Igf2r* gene depends on an intronic CpG island. Nature 1997; 389(6652):745-749.
21. Stöger R, Kubicka P, Liu C-G et al. Maternal-specific methylation of the imprinted mouse *Igf2r* locus identifies the expressed locus as carrying the imprinting signal Cell 1993; 73:61-71.
22. Wutz A, Theussl HC, Dausman J et al. Nonimprinted *Igf2r* expression decreases growth and rescues the Tme mutation in mice. Development 2001; 128(10):1881-1887.
23. Sleutels F, Zwart R, Barlow DP. The noncoding Air RNA is required for silencing autosomal imprinted genes. Nature 2002; 415(6873):810-813.

24. Zwart R, Sleutels F, Wutz A et al. Bi-directional action of the *Igf2r* imprint control element on upstream and downstream imprinted genes. Genes Dev 2001; 15(18):2361-2366.
25. Cleary MA, van Raamsdonk CD, Levorse J et al. Disruption of an imprinted gene cluster by a targeted chromosomal translocation in mice. Nat Genet 2001; 29(1):78-82.
26. Horike S, Mitsuya K, Meguro M et al. Targeted disruption of the human LIT1 locus defines a putative imprinting control element playing an essential role in Beckwith- Wiedemann syndrome. Hum Mol Genet 2000; 9(14):2075-2083.
27. Bartolomei MS, Zemel S, Tilghman SM. Parental imprinting of the mouse *H19* gene. Nature 1991; 351(6322):153-155.
28. Rachmilewitz J, Goshen R, Ariel I et al. Parental imprinting of the human *H19* gene. FEBS Lett 1992; 309(1):25-28.
29. Overall M, Bakker M, Spencer J et al. Genomic imprinting in the rat: linkage of *Igf2* and *H19* genes and opposite parental allele-specific expression during embryogenesis. Genomics 1997; 45(2):416-420.
30. DeChiara TM, Robertson EJ, Efstratiadis A. Parental imprinting of the mouse insulin-like growth factor II gene. Cell 1991; 64:849-859.
31. Giannoukakis N, Deal C, Paquette J et al. Parental genomic imprinting of the human *IGF2* gene. Nature Genet 1993; 4:98-101.
32. Ohlsson R, Nystrom A, Pfeifer-Ohlsson S et al. *IGF2* is parentally imprinted during human embryogenesis and in the Beckwith-Wiedemann syndrome. Nature Genet 1993; 4:94-97.
33. Tremblay KD, Duran KL, Bartolomei MS. A 5' 2-kilobase-pair region of the imprinted mouse *H19* gene exhibits exclusive paternal methylation throughout development. Mol Cell Biol 1997; 17(8):4322-4329.
34. Tremblay KD, Saam JR, Ingram RS et al. A paternal-specific methylation imprint marks the alleles of the mouse *H19* gene. Nat Genet 1995; 9(4):407-413.
35. Bartolomei MS, Webber AL, Brunkow ME et al. Epigenetic mechanisms underlying the imprinting of the mouse *H19* gene. Genes Dev 1993; 7(9):1663-1673.
36. Lucifero D, Mertineit C, Clarke HJ et al. Methylation dynamics of imprinted genes in mouse germ cells. Genomics 2002; 79(4):530-538.
37. Olek A, Walter J. The preimplantation ontogeny of the *H19* methylation imprint. Nat Genet 1997; 17(3):275-276.
38. Davis TL, Trasler JM, Moss SB et al. Acquisition of the *H19* methylation imprint occurs differentially on the parental alleles during spermatogenesis. Genomics 1999; 58(1):18-28.
39. Davis TL, Yang GJ, McCarrey JR et al. The *H19* methylation imprint is erased and reestablished differentially on the parental alleles during male germ cell development. Hum Mol Genet 2000; 9(19):2885-2894.
40. Elson DA, Bartolomei MS. A 5' differentially methylated sequence and the 3' flanking region are necessary for *H19* transgene imprinting. Mol Cell Biol 1997; 17:309-317.
41. Cranston MJ, Spinka TL, Elson DA et al. Elucidation of the minimal sequence required to imprint *H19* transgenes. Genomics 2001; 73(1):98-107.
42. Thorvaldsen JL, Duran KL, Bartolomei MS. Deletion of the *H19* differentially methylated domain results in loss of imprinted expression of *H19* and *Igf2*. Genes Dev 1998; 12(23):3693-3702.
43. Thorvaldsen JL, Mann MR, Nwoko O et al. Analysis of sequence upstream of the endogenous *H19* gene reveals elements both essential and dispensable for imprinting. Mol Cell Biol 2002; 22(8):2450-2462.
44. Srivastava M, Hsieh S, Grinberg A et al. *H19* and *Igf2* monoallelic expression is regulated in two distinct ways by a shared cis acting regulatory region upstream of *H19*. Genes Dev 2000; 14(10):1186-1195.
45. Drewell RA, Brenton JD, Ainscough JF et al. Deletion of a silencer element disrupts *H19* imprinting independently of a DNA methylation epigenetic switch. Development 2000; 127(16):3419-3428.
46. Bell AC, Felsenfeld G. Stopped at the border: boundaries and insulators. Curr Opin Genet Devel 1999; 9:191-198.
47. Kanduri C, Holmgren C, Pilartz M et al. The 5' flank of mouse *H19* in an unusual chromatin conformation unidirectionally blocks enhancer-promoter communication. Curr Biol 2000; 10(8):449-457.

48. Kanduri C, Pant V, Loukinov D et al. Functional association of CTCF with the insulator upstream of the *H19* gene is parent of origin-specific and methylation-sensitive. Curr Biol 2000; 10(14):853-856.
49. Szabo PE, Tang S-H, Rentsendorj A et al. Maternal-specific footprints at putative CTCF sites in the *H19* imprinting control region give evidence for insulator function. Current Biol 2000; 10:607-610.
50. Hark AT, Schoenherr CJ, Katz DJ et al. CTCF mediates methylation-sensitive enhancer-blocking activity at the *H19/Igf2* locus. Nature 2000; 405(6785):486-489.
51. Bell AC, Felsenfeld G. Methylation of a CTCF-dependent boundary controls imprinted expression of the *Igf2* gene. Nature 2000; 405(6785):482-485.
52. Kaffer CR, Srivastava M, Park KY et al. A transcriptional insulator at the imprinted *H19/Igf2* locus. Genes Dev 2000; 14(15):1908-1919.
53. Thorvaldsen JL, Bartolomei MS. Mothers setting boundaries. Science 2000; 288:2145-2146.
54. Tilghman SM, Brunkow ME, Brannan CI et al. The mouse *H19* gene: its structure and function in mouse development. In: Sharp PA, ed. Nuclear Processes and Oncogenes. New York: Academic Press, 1992:188-200.
55. Leighton PA, Saam JR, Ingram RS et al. An enhancer deletion affects both *H19* and *Igf2* expression. Genes Dev 1995; 9(17):2079-2089.
56. Yoo-Warren H, Pachnis V, Ingram RS et al. Two regulatory domains flank the mouse *H19* gene. Mol Cell Biol 1988; 8(11):4707-4715.
57. Zemel S, Bartolomei MS, Tilghman SM. Physical linkage of two mammalian imprinted genes, *H19* and insulin-like growth factor 2. Nat Genet 1992; 2(1):61-65.
58. Bartolomei MS, Tilghman SM. Parental imprinting of mouse chromosome 7. Sem Dev Biol Sem 1992; 3:107-117.
59. Ainscough JF, Dandolo L, Surani MA. Appropriate expression of the mouse *H19* gene utilises three or more distinct enhancer regions spread over more than 130 kb. Mech Dev 2000; 91(1-2):365-368.
60. Ainscough JF, John RM, Surani MA. Mechanism of imprinting on mouse distal chromosome 7. Genet Res 1998; 72(3):237-245.
61. Kaffer CR, Grinberg A, Pfeifer K. Regulatory mechanisms at the mouse *Igf2/H19* locus. Mol Cell Biol 2001; 21(23):8189-8196.
62. Drewell RA, Arney KL, Arima T et al. Novel conserved elements upstream of the *H19* gene are transcribed and act as mesodermal enhancers. Development 2002; 129(5):1205-1213.
63. Stadnick MP, Pieracci FM, Cranston MJ et al. Role of a 461-bp G-rich repetitive element in *H19* transgene imprinting. Dev Genes Evol 1999; 209(4):239-248.
64. Ishihara K, Hatano N, Furuumi H et al. Comparative genomic sequencing identifies novel tissue-specific enhancers and sequence elements for methylation-sensitive factors implicated in *Igf2/H19* imprinting. Genome Res 2000; 10(5):664-671.
65. Koide T, Ainscough J, Wijgerde M et al. Comparative analysis of Igf-2/*H19* imprinted domain: identification of a highly conserved intergenic DNase I hypersensitive region. Genomics 1994; 24(1):1-8.
66. Tycko B. Genomic imprinting and cancer. Results Probl Cell Differ 1999; 25:133-169.
67. Ariel I, Ayesh S, Perlman EJ et al. The product of the imprinted *H19* gene is an oncofetal RNA. Mol Pathol 1997; 50(1):34-44.
68. Looijenga LH, Verkerk AJ, De Groot N et al. *H19* in normal development and neoplasia. Mol Reprod Dev 1997; 46(3):419-439.
69. Maher ER, Reik W. Beckwith-Wiedemann syndrome: imprinting in clusters revisited. J Clin Invest 2000; 105:247-252.
70. Tycko B. Genomic imprinting and human neoplasia. In: Ehrlich M, ed. DNA alterations in cancer. Natick, MA: Eaton Publishing, 2000:333-349.
71. Juan V, Crain C, Wilson C. Evidence for evolutionarily conserved secondary structure in the *H19* tumor suppressor RNA. Nucleic Acids Res 2000; 28(5):1221-1227.
72. Hurst LD, Smith NG. Molecular evolutionary evidence that *H19* mRNA is functional. Trends Genet 1999; 15(4):134-135.

73. Schmidt JV, Levorse JM, Tilghman SM. Enhancer competition between *H19* and *Igf2* does not mediate their imprinting. Proc Natl Acad Sci USA 1999; 96(17):9733-9738.
74. Ripoche MA, Kress C, Poirier F et al. Deletion of the *H19* transcription unit reveals the existence of a putative imprinting control element. Genes Dev 1997; 11(12):1596-1604.
75. Jones BK, Levorse JM, Tilghman SM. *Igf2* imprinting does not require its own DNA methylation or *H19* RNA. Genes Dev 1998; 12(14):2200-2207.
76. Leighton PA, Ingram RS, Eggenschwiler J et al. Disruption of imprinting caused by deletion of the *H19* gene region in mice. Nature 1995; 375(6526):34-39.
77. Hao Y, Crenshaw T, Moulton T et al. Tumour-suppressor activity of *H19* RNA. Nature 1993; 365(6448):764-767.
78. Li YM, Franklin G, Cui HM et al. The *H19* transcript is associated with polysomes and may regulate *IGF2* expression in trans. J Biol Chem 1998; 273(43):28247-28252.
79. Wilkin F, Paquette J, Ledru E et al. *H19* sense and antisense transgenes modify insulin-like growth factor-II mRNA levels. Eur J Biochem 2000; 267(13):4020-4027.
80. Dubchak I, Brudno M, Loots GG et al. Active conservation of noncoding sequences revealed by three-way species comparisons. Genome Res 2000; 10(9):1304-1306.
81. Mayor C, Brudno M, Schwartz JR et al. VISTA : visualizing global DNA sequence alignments of arbitrary length. Bioinformatics 2000; 16(11):1046-1047.
82. Lee YJ, Park CW, Hahn Y et al. Mit1/Lb9 and Copg2, new members of mouse imprinted genes closely linked to Peg1/Mest(1). FEBS Lett 2000; 472(2-3):230-234.
83. Yamasaki K, Hayashida S, Miura K et al. The novel gene, gamma2-COP (COPG2), in the 7q32 imprinted domain escapes genomic imprinting. Genomics 2000; 68(3):330-335.
84. Schmidt JV, Matteson PG, Jones BK et al. The Dlk1 and Gtl2 genes are linked and reciprocally imprinted. Genes Dev 2000; 14(16):1997-2002.
85. Takada S, Tevendale M, Baker J et al. Delta-like and gtl2 are reciprocally expressed, differentially methylated linked imprinted genes on mouse chromosome 12. Curr Biol 2000; 10(18):1135-1138.
86. Takada S, Paulsen M, Tevendale M et al. Epigenetic analysis of the Dlk1-Gtl2 imprinted domain on mouse chromosome 12: implications for imprinting control from comparison with *Igf2-H19*. Hum Mol Genet 2002; 11(1):77-86.
87. Paulsen M, Takada S, Youngson NA et al. Comparative sequence analysis of the imprinted Dlk1-Gtl2 locus in three mammalian species reveals highly conserved genomic elements and refines comparison with the *Igf2-H19* region. Genome Res 2001; 11(12):2085-2094.
88. Wylie AA, Murphy SK, Orton TC et al. Novel imprinted DLK1/GTL2 domain on human chromosome 14 contains motifs that mimic those implicated in *IGF2/H19* regulation. Genome Res 2000; 10(11):1711-1718.
89. Moore T, Constancia M, Zubair M et al. Multiple imprinted sense and antisense transcripts, differential methylation and tandem repeats in a putative imprinting control region upstream of mouse *Igf2*. Proc Natl Acad Sci USA 1997; 94(23):12509-12514.
90. Okutsu T, Kuroiwa Y, Kagitani F et al. Expression and imprinting status of human PEG8/*IGF2*AS, a paternally expressed antisense transcript from the *IGF2* locus, in Wilms' tumors. J Biochem (Tokyo) 2000, 127(3):475-483.
91. Smilinich NJ, Day CD, Fitzpatrick GV et al. A maternally methylated CpG island in KvLQT1 is associated with an antisense paternal transcript and loss of imprinting in Beckwith- Wiedemann syndrome. Proc Natl Acad Sci USA 1999; 96(14):8064-8069.
92. Mitsuya K, Meguro M, Lee MP et al. LIT1, an imprinted antisense RNA in the human KvLQT1 locus identified by screening for differentially expressed transcripts using monochromosomal hybrids. Hum Mol Genet 1999; 8(7):1209-1217.
93. Engemann S, Strodicke M, Paulsen M et al. Sequence and functional comparison in the Beckwith-Wiedemann region: implications for a novel imprinting centre and extended imprinting. Hum Mol Genet 2000; 9(18):2691-2706.
94. Hayward BE, Bonthron DT. An imprinted antisense transcript at the human GNAS1 locus. Hum Mol Genet 2000; 9(5):835-841.
95. Wroe SF, Kelsey G, Skinner JA et al. An imprinted transcript, antisense to Nesp, adds complexity to the cluster of imprinted genes at the mouse Gnas locus. Proc Natl Acad Sci USA 2000; 97(7):3342-3346.

96. Williamson CM, Skinner JA, Kelsey G et al. Alternative noncoding splice variants of Nespas, an imprinted gene antisense to Nesp in the Gnas imprinting cluster. Mamm Genome 2002; 13(2):74-79.
97. Nakabayashi K, Bentley L, Hitchins MP et al. Identification and characterization of an imprinted antisense RNA (MESTIT1) in the human MEST locus on chromosome 7q32. Hum Mol Genet 2002; 11(15):1743-1756.
98. Li T, Vu TH, Lee KO et al. An imprinted PEG1/MEST antisense expressed predominantly in human testis and in mature spermatozoa. J Biol Chem 2002; 277(16):13518-13527.
99. Mlynarczyk SK, Panning B. X inactivation: Tsix and Xist as yin and yang. Curr Biol 2000; 10(24):R899-903.
100. Percec I, Bartolomei MS. Genetics. Do X chromosomes set boundaries? Science 2002; 295(5553):287-288.
101. Avner P, Heard E. X-chromosome inactivation: counting, choice and initiation. Nat Rev Genet 2001; 2(1):59-67.
102. Runte M, Huttenhofer A, Gross S et al. The IC-SNURF-SNRPN transcript serves as a host for multiple small nucleolar RNA species and as an antisense RNA for UBE3A. Hum Mol Genet 2001; 10(23):2687-2700.
103. Chamberlain SJ, Brannan CI. The Prader-Willi syndrome imprinting center activates the paternally expressed murine Ube3a antisense transcript but represses paternal Ube3a. Genomics 2001; 73(3):316-322.
104. Jong MT, Carey AH, Caldwell KA et al. Imprinting of a RING zinc-finger encoding gene in the mouse chromosome region homologous to the Prader-Willi syndrome genetic region. Hum Mol Genet 1999; 8:795-803.
105. Jong MT, Gray TA, Ji Y et al. A novel imprinted gene, encoding a RING zinc-finger protein, and overlapping antisense transcript in the Prader-Willi syndrome critical region. Hum Mol Genet 1999; 8:783-793.

CHAPTER 7

MicroRNAs

Eric G. Moss

Abstract

MicroRNAs are the smallest functional noncoding RNAs of plants and animals. They are about 22 nucleotides in length with no common structural or sequence features. Some are conserved across great evolutionary distances, indicating that their sequence is not arbitrary. A single organism may have hundreds of distinct microRNAs, some of which are expressed in stage-, tissue- or cell type-specific patterns. MicroRNAs are believed to act through specific complementary sites in target mRNAs to inhibit gene expression post-transcriptionally. Because of their size and potential to exist in almost any sequence, microRNAs are a versatile means for controlling gene expression both in nature and artificially.

Introduction

Naturally-occurring microRNAs (miRNAs) are 19 to 25 nucleotides (nt) long, cleaved from partially duplexed precursors, and are explicitly encoded in the genomes of plants and animals. They have essentially no sequence features in common, although some families of miRNAs are highly related in sequence. Of the hundreds known from a handful of species, a few show broad phylogenetic conservation. Some miRNAs are expressed widely in the organism, others display restricted expression in developmental or tissue-specific patterns. All are assumed to be post-transcriptional regulators of gene expression of specific target mRNAs. Most of our understanding about the mechanism of action of miRNAs comes from work on *lin-4*, a regulator of developmental timing in *C. elegans* which appears to function at the ribosome. Genetic studies of plant and animal proteins involved in miRNA processing and function have suggested that miRNAs have many fundamental roles in development. Because of their extreme size and sequence diversity miRNAs are likely to evolve rapidly. Their existence in both plant and animal kingdoms suggests that have existed since the origin of multicellular life.

miRNAs are similar to small interfering RNAs (siRNAs) that are generated during RNA-induced gene silencing, or RNA interference (RNAi). siRNAs are approximately 21 to 25nt and generated by cleavage of double-stranded RNA (dsRNA) by a multidomain ribonuclease, named Dicer in animals.[1,2] The siRNAs, which are themselves double-stranded, are then incorporated into an RNA-induced silencing complex (RISC) which targets mRNAs for cleavage based on exact complementarity to one of the two siRNA strands.[3,4] This process of gene silencing occurs in animals, plants and some fungi. It may have evolved in response to viruses, transposable elements, or other repetitive sequences, and the double-stranded RNA may be endogenous or exogenous. Processing and function of miRNAs and siRNAs requires some of the same kinds of proteins, but whereas miRNAs are explicitly encoded in the genome and have coevolved with their targets sites, the sequences of siRNAs and their complementary

Noncoding RNAs: Molecular Biology and Molecular Medicine, edited by Jan Barciszewski and Volker A. Erdmann.

Table 1. Some microRNAs of plants and animals

MiRNAs	Organism	Method	Ref.
lin-4	*C. elegans*	genetic mapping, transformation rescue	6
let-7	*C. elegans*	genetic mapping, transformation rescue	14
let-7	*Drosophila*, human[1]	bioinformatics, northern analysis	16
miR-1 to miR-14	*Drosophila*	molecular cloning (embryos)	17
let-7, miR-15 to miR-33[2]	human	molecular cloning (HeLa cells)	17
miR-1, 2, miR-34 to miR-86[3]	*C. elegans*	molecular cloning	21
miR-1, 2, 42, 43, 52, 58, 60, 62, 72, 80, 81, miR-87 to miR-90	*C. elegans*	molecular cloning	22
miR-1	mouse, human, *Drosophila*	bioinformatics, northern analysis	22
miR-16,18,19a,19b,21,22,23,27 miR-91 to miR-121	human	immunoprecipitation, molecular cloning (HeLa cells)	23
let-7a-i, miR-1b-c, 9, 15a-b, 16, 18, 19b, 20, 22, 23a-b, 24, 26a-b, 27a-b, 29b/102, 29c, 30a-d, 30a-as, 99a-b, 101 miR-122 to 155	mouse	molecular cloning (various adult tissues)	24
MiR156 to miR171	*Arabidopsis*	molecular cloning (seedlings and flowers)	29
MiR159, 163, 167 MiR172 to miR179	*Arabidopsis*	molecular cloning (arial portions)	30
Small RNAs 1 to 125[4]	*Arabidopsis*	molecular cloning (infloresence tissue	28

1. let-7-like sequences were detected by hybridization in many diverse bilaterian animals, although their sequences were not determined.
2. miR-30 has been renamed miR-30a-as (antisense). Mouse miR-30a-s (sense) is the same as human miR-97.
3. miR-44 and miR-45 are identical but encoded at different chromosomal loci. Others have used a number suffix (e.g., *mir-6-1*) to indicate different chromosomal loci that encode identical miRNA sequences.
4. The small RNAs described by Llave et al.[32] were not assigned miRNA numbers when published; some are identical to miRNAs identified by Reinhart et al.[29] and Park et al.[30]

sites are essentially arbitrary. The best studied miRNAs have revealed that the biological roles of miRNAs are determined not only by their sequence, but also by their regulated expression.

Discovery and Isolation of miRNAs

Table 1 lists the first reports of over 300 miRNAs in animals and plants. The first two of these, *lin-4* and *let-7*, are the prototypes and about which most is known. *lin-4* was originally referred to as a small antisense RNA and a regulatory RNA. The discovery of *let-7* and its widespread conservation led to the coining of "small temporal RNA" (stRNA) to describe

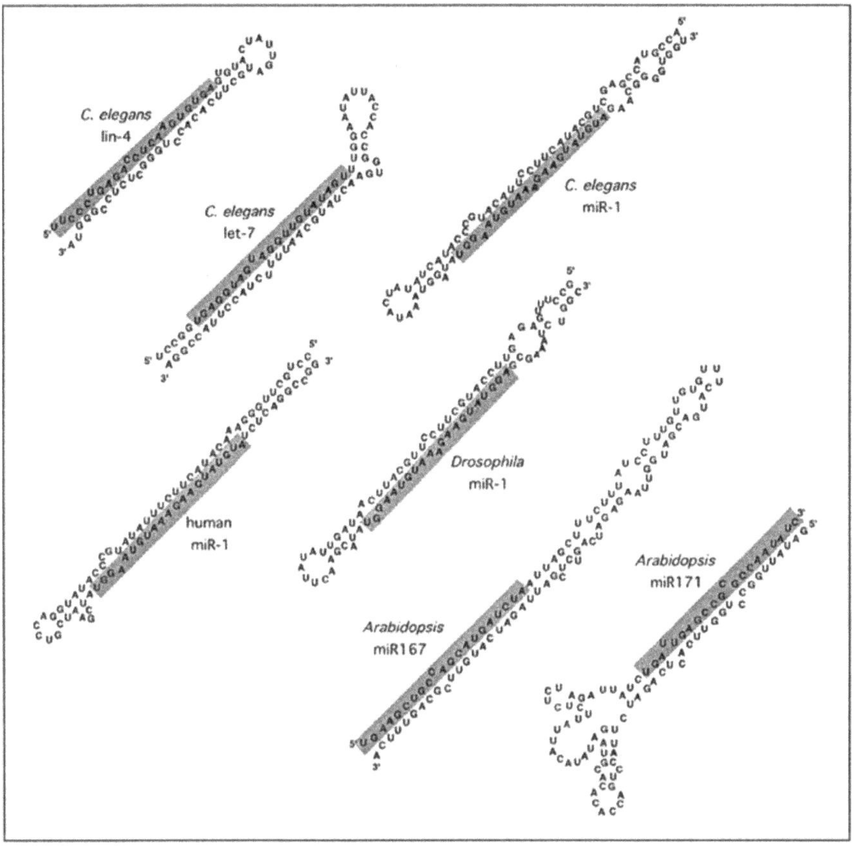

Figure 1. Some microRNAs and their precursors from animals and plants. miRNA precursors are extensively basepaired stem-loop structures with bulges and loops. The mature miRNAs (boxed regions) are cleaved by an RNase III activity to yield a single-stranded RNA of approximately 22nt. The miRNA may derive from either the 5' or 3' part of the stem. The miRNA precursors of plants appear to be longer and more complex than those of animals. All miRNA precursors may first be cleaved from longer short-lived transcripts. The *lin-4*L (long) RNA is the hairpin precursor of *lin-4*S (short), the mature form.

them, reflecting the fact that they act as temporally regulated developmental switches. The term "microRNA" (miRNA) now applies to these and the growing class of similar small RNAs.

lin-4 a Regulator of Gene Expression during Development

The first miRNA to be identified is *lin-4* of the nematode *Caenorhabditis elegans*. The *lin-4* gene was discovered by forward genetic analysis, in which a mutant that reiterated patterns of cell division and differentiation was found by chance.[5] When the gene was cloned extraordinary efforts were made to identify its product.[6] A small DNA fragment of about 700 base pairs could rescue the *lin-4* mutant phenotype, however, it contained no open reading frame conserved among homologous sequences from four species of *Caenorhabditis*, suggesting that it did not encode a protein. Two small RNAs complementary to one strand of the fragment were detected on northern blots: a 21 or 22nt RNA (*lin-4S*) and a less abundant 61nt RNA (*lin-4*L; (Fig. 1)). Nuclease-protection assays were performed to precisely map the ends of the RNAs, and the smaller lin-4S was found to constitute the 5' portion of lin-4L, suggesting that lin-4L may be a short-lived precursor. An inactivating point mutation in the *lin-4* gene was found within the lin-4S region. These data support the conclusion that the functional product of the *lin-4* gene is a very small RNA.

This conclusion provided a novel but plausible explanation for how the *lin-4* gene functions in development. Genetic analysis showed that *lin-4* negatively regulates expression of the gene *lin-14*, a protein-coding gene that is active early *C. elegans* larval development and repressed later.[7,8] The *lin-4* RNA is expressed stage-specifically corresponding to the time *lin-14* is repressed at the end of the first larval stage.[9,10] Two mutations of the *lin-14* gene delete a region of the 3' untranslated region (UTR) of the gene and result in continuous expression.[11] The developmental phenotype of these *lin-14* mutants is essentially identical to that of a *lin-4* mutant, suggesting that the *lin-4* product acts via the sequences within the region defined by the deletions to repress *lin-14* expression.[7] Seven sequence elements within the 3' UTR of *lin-14* display partial complementarity to the *lin-4* RNA, and these were predicted to be the sites through which it acts.[6,9] Mutation of all seven sites abrogates the *lin-4*-dependent regulation in reporter constructs.[12] A second gene regulated by *lin-4*, *lin-28*, has a single short sequence element in its 3' UTR with partially complementary to *lin-4*.[13] Deletion of this 15nt site causes misexpression of the *lin-28* gene, consistent with a failure of *lin-4* to repress its expression. Thus *lin-4* appears to basepair with specific sequences in the 3' UTRs of at least two target genes to repress their expression at a specific time in development. The mechanism by which *lin-4* regulates *lin-14* and *lin-28* is discussed below.

let-7 *and its Conservation Among Animals*

The gene *let-7* was also identified through mutant *C. elegans* animals that showed developmental defects.[14] Again, a DNA fragment that could rescue the mutant phenotype was found that contained no conserved open reading frame in the region of the mutations, and an abundant small RNA of 21nt, unrelated in sequence to *lin-4*, could be detected by northern blots and nuclease-protection experiments. Like *lin-4*, *let-7* is expressed stage-specifically, becoming most abundant during the last larval stage and continuing expression into adulthood. Genetic studies suggest that the gene *lin-41* is a target of *let-7* regulation.[14,15] Based on the analogy to *lin-4*, two sites with partial complementarity to *let-7* were identified in the 3' UTR of *lin-41*. This 3' UTR confers *let-7*-dependent regulation on reporter constructs and an 85nt deletion which removes the two predicted *let-7* complementary sites is not repressed, suggesting that *let-7* acts through these sites.[14,15]

Genomic sequences identical to the 21nt *let-7* sequence were found in the genomes of *Drosophila melanogaster* and humans.[16] In both species, RNAs of 21 to 22nt are detected on northern blots, suggesting the genomic sequences are expressed. Moreover, the expression is stage-specific in *Drosophila* and tissue-specific in humans. In *Drosophila*, *let-7* is abundant at late developmental stages that parallel its expression in *C. elegans*, suggesting for the first time that this 22nt RNA may be functionally conserved across great evolutionary distance.

By northern analysis, *let-7*-like sequences were detected in a diverse array of animals, including representatives of all three main clades of bilaterian animals.[16] However, conservation of *let-7* does not apparently extend to other metazoans, such as jellyfish or sponges, or to plants or unicellular organisms. The hybridization method used did not detect *lin-4*-like sequences, suggesting that only *let-7* had been conserved through evolution. However, it has been subsequently shown that other animals, including *Drosophila*, mouse and human, possess *lin-4*-like miRNAs, albeit with a few sequence differences.[17]

Although a precursor form of *let-7* of any species was initially not detected on northern blots, in all three species the genomic sequence of *let-7* has the potential to produce a longer RNA that could form partial duplexed hairpin structure like *lin-4L* (Fig. 1). The longer precursor

Table 2. Some miRNA homologues and paralogues

miRNA	Species	Sequence	Reference
lin-4	*C. elegans*	UCCCUGAGACCUCAAGUGUGA	6
miR-125	*Drosophila*	UCCCUGAGACCCUAACUUGUGA	24
miR-125a	Mouse	UCCCUGAGACCCUUUAACCUGUG	24
miR-125b	Mouse	UCCCUGAGACCCUAACUUGUGA	24
let-7	C. elegans	UGAGGUAGUAGGUUGUAUAGU	14
let-7	*Drosophila*	UGAGGUAGUAGGUUGUAUAGU	16
let-7a	mouse/human	UGAGGUAGUAGGUUGUAUAGU	16, 17
let-7b	mouse/human	UGAGGUAGUAGGUUGUGUGGUU	17, 24
let-7c	mouse/human	UGAGGUAGUAGGUUGUAUGGU	17, 24
let-7d	mouse/human	AGAGGUAGUAGGUUGCAUAGU	17, 24
let-7e	mouse/human	UGAGGUAGGAGGUGUAUAGU	17, 24
let-7f	mouse/human	UGAGGUAGUAGAUUGUAUAGUU	17, 24
let-7g	Mouse	UGAGGUAGUAGUUUGUACAGUA	24
let-7h	Mouse	UGAGGUAGUAGUGUGUACAGUU	24
let-7i	Mouse	UGAGGUAGUAGUUUGUGCU	24
miR-84	*C. elegans*	UGAGGUAGUAUGUAAUAUUGUA	21
miR-98	Human	UGAGGUAGUAAGUUGUAUUGUU	23
miR-1	*Drosophila*	UGGAAUGUAAAGAAGUAUGGAG	17
miR-1	*C. elegans*	UGGAAUGUAAAGAAGUAUGUA	21, 22
miR-1	Human	UGGAAUGUAAAGAAGUAUG	22
miR-1b	Mouse	UGGAAUGUAAAGAAGUAUGUAA	24
miR-1c	Mouse	UGGAAUGUAAAGAAGUAUGUAC	24
miR-1d	Mouse	UGGAAUGUAAAGAAGUAUGUAUU	24
miR170	*Arabidopsis*	UGAUUGAGCCGUGUCAAUAUC	29
miR171	*Arabidopsis*	UGAUUGAGCCGCGCCAAUAUC	28, 29
miR171	Rice	UGAUUGAGCCGCGCCAAUAUC	29

of approximately 70nt can be detected in *C. elegans*, *Drosophila*, and human cells when the Dicer nuclease responsible for generating the mature forms is not expressed.[18-20] Despite the conservation of the sequence of the mature *let-7* across metazoan phyla, the rest of the precursor sequence is not conserved.

Numerous miRNAs from Animals

To identify additional small RNAs like *lin-4* and *let-7* in *C. elegans*, both cDNA cloning and bioinformatic approaches have been taken, and the initial reports described 58 distinct miRNAs.[21,22] Contemporaneously, using a molecular approach 16 miRNAs were identified from *Drosophila* and 27 from human cells, including 6 different *let-7*-like RNAs (Table 2).[17] Subsequently, more miRNAs have been identified from human cells and from a variety of mouse tissues.[23,24] Collectively, these studies identified greater than 100 miRNAs and efforts to identify more miRNAs from these and other organisms have continued with the expectation that many more exist.

The sizes of the animal miRNAs vary from 21 to 25nt, with a few miRNAs occurring in multiple lengths. Some miRNAs are encoded at multiple genomic loci, but it is not known whether their expression is coordinately or differentially regulated. Some miRNAs from the same organism are very similar, differing by only one or a few nucleotides, but it is not known whether they function identically (Table 2). In addition to *let-7*, several others are conserved from *C. elegans* to mammals, such as miR-1. Because the molecular cloning methods have isolated some only once and bioinformatic searches alone are difficult, the current catalogue of miRNAs from these organisms must be far from complete. Nevertheless, these methods have proven to be highly effective in greatly expanding the membership in this class of noncoding RNAs.

cDNA Cloning

The procedure for cloning miRNAs was developed to isolate siRNAs from *Drosophila* extract.[4] Source material for miRNA isolation has been whole animals (*C. elegans* and *Drosophila*), cultured mammalian cells, and dissected adult tissues (mouse heart, liver, brain, etc.).[17,21-24] Essentially, total RNA is fractionated on a polyacrylamide gel, and RNA of the appropriate size range is purified, ligated to synthetic adaptors using T4 RNA ligase, reverse transcribed, amplified by PCR, cloned and sequenced. Dicer produces RNAs with a 5' phosphate and 3' hydroxyl group, characteristic of RNase III activity.[25] Therefore, one modification of this cloning procedure relied on the presence of a 5' phosphate and 3' hydroxyl group to capture small RNAs.[21] Given the effectiveness of this modified procedure, the RNAs captured must have had the expected chemical groups at their termini.

In some cases, certain miRNA sequences were identified from dozens of independent clones, others only once. A number of the small RNAs isolated by molecular cloning were identified as fragments rRNAs, mRNAs, tRNAs, or snRNAs, and were not been classified as miRNAs. It remains to be determined whether such fragments of other RNAs have a specific purpose.

Genomic Matches

The identification of miRNAs from cloned RNAs relies on matching a candidate to genomic sequence. For this reason miRNAs have been identified from organisms for which complete or nearly complete genomic sequence is available: *C. elegans*, *Drosophila*, mouse, human and *Arabidopsis*. Several miRNAs isolated from one species are predicted from genomic sequence to exist in another. For example, miRs-34, -60, -72, -79, and -84, which were identified in *C. elegans*, have exact genomic matches in *Drosophila*. Genomic sequences of other species, such as *Caenorhabditis briggsae* and the pufferfish Fugu, have been used to identify miRNAs encoded in them based on conservation. A number of cloned sequences that did not match the genome were not reported as miRNAs.

Because *lin-4* and *let-7* are derived from stem-loop precursors of about 70nt, the sequence surrounding each genomic match to a cloned miRNA sequence is examined for the potential to form such a structure containing the miRNA sequence. RNA secondary structures, helicies, loops and bulges, are predicted using mfold.[26] In the vast majority of cases miRNA sequences have the potential to form imperfect stem-loops with sequences a short distance away in the genome. These longer sequences are predicted to be the immediate precursors of the mature short forms of the RNAs (Fig. 1). Additional characteristics of these precursors are discussed below.

Bioinformatics

A small number of miRNAs have been identified directly from genomic sequences of *C. elegans* and *C. briggsae*.[22] These nematodes shared a common ancestor approximately 20 to 40 million years ago.[27] Highly conserved genomic regions not part of any known gene were searched for sequences with the potential to form RNA stem-loop structures of about 65nt. Candidate sequences were used to generate probes for northern analysis. Out of 40 candidates, 3 miRNAs were identified by the detection of the expression of approximately 22nt RNAs. One of these miRNAs was also identified by molecular cloning. Because mouse and human are separated by similar evolutionary distance, approximately 80 million years, a similar approach could be used to identify miRNA genes from mammals.

It has been possible to predict additional miRNAs where miRNAs match genomic sequence in clusters. miR-39, for example, was predicted from a conserved sequence in a cluster of miRNA loci highly conserved between *C. elegans* and *C. briggsae*, and was subsequently detected by northern analysis.[21]

Northern Analysis

Because molecular cloning and bioinformatic methods are not entirely reliable in identifying miRNAs or distinguishing them from artifacts, predicted miRNAs are tested for expression by northern analysis. Northern blots for small RNAs are performed using polyacrylamide gels, membranes such as GeneScreen Plus and Zeta-Probe GT, and oligonucleotide probes.[6,22] In most cases, an approximately 22nt species is the most abundant species detected and in some cases an approximately 70nt form can also be seen. In plants, longer forms are generally not detectable. Northern analysis also reveals temporal and tissue-specific patterns of expression. miR-1, for example, is abundant in adult human heart, but not detected in other tissues examined.[22]

let-7, which has the same sequence in nematodes, insects and mammals, has been detected by northern analysis in diverse species, including an annelid and a mollusc.[16] Several miRNAs isolated from human cells have been detected in adult zebrafish *Danio rerio*.[17] There is no reason to believe miRNAs could not be found in all multicellular organisms.

miRNAs of Plants

Over 100 miRNAs have been isolated by molecular cloning from tissues of the flowering plant *Arabidopsis thaliana*.[28-30] Some of these miRNAs have homologues in rice, tobacco and maize. Many of the general characteristics of miRNAs of animals are true of plant miRNAs, such as processing from partially duplexed precursors by a Dicer-like enzyme, named SIN1/CAF, and tissue-specific expression. None of the initial reports identified a miRNA conserved between plants and animals. Plant miRNA precursors are predicted to be longer and more complex than animal miRNA precursors (Fig. 1). An important apparent difference between plant and animal miRNAs is that multiple plant miRNAs are perfectly complementary to mRNAs of protein coding genes.[28,29,31,32]

Characteristics of miRNAs

All reported miRNAs are between 15 and 30nt in length and most are 21 to 24nt, which is characteristic of Dicer cleavage of siRNAs.[1,2] Although miRNAs have no specific sequence features common to them all, they show a bias for U and rarely G in the first position, and a general deficiency of C.[21] The majority of reported miRNAs have been detected on northern blots and are encoded at genomic loci that suggest they are expressed from transcription units in intronic regions or distant from other kinds of genes. In general, they do not appear to be derived from premRNAs, introns, mRNAs or other noncoding RNAs. Although processing by Dicer or SIN1/CAF has been established for only a few miRNAs, all are presumed to be generated by such a nuclease from a partially duplexed precursor yielding a stable single-stranded RNA with a 5' phosphate and a 3' hydroxyl group.

Precursors

Most reported miRNAs are predicted to be derived from precursor RNAs in which the miRNA is within a region of imperfect duplex containing bulges and/or loops (Fig. 1). In some cases in animals, approximately 70nt precursors can be detected on northern blots; no precursors of this size have been detected in plants although they are predicted from genomic sequences. The precursors themselves may be derived from even longer transcripts that are processed first in the nucleus to yield the precursor molecules.[33] This complex processing may explain why precursor forms of the expected size are not always detected. Phylogenetic evidence for the importance of the stem-loop precursor comes from a comparison of homologous miRNA genes. When homologous miRNAs of *C. elegans* and *C. briggsae* differ by a few nucleotides often compensatory changes exist in their opposite side of the precursor stems.[21] *Arabidopsis* miRNA precursors and sequences in the rice genome show significant conservation in secondary structure but not sequence of the precursors.[29] These observations suggest that the secondary structure is important for determining proper generation of the mature miRNA.

Mature miRNAs may be derived from either the 5' or 3' side of the stem-loop precursor (Fig. 1). In one assessment, about 25% of the miRNAs derive from the 5' arm of the precursor, the remaining coming from the 3' arm.[21] What determines the portion of a given precursor that will become the stable product is not known. In some cases stable miRNAs have been identified from both sides of a precursor, as is the case for miR-56 and miR-56* and miR-17 and miR-91, resembling siRNAs generated by Dicer cleavage.[21,23,29]

miRNA Genes

miRNAs are usually located in intergenic or intronic regions and therefore are not usually thought to be transcribed as part of a gene encoding another type of molecule. Some miRNAs are encoded in clusters in which the stem-loop precursors are separated by short sequences, suggesting that they may derive from a common transcript.[17,21] The miRNAs may be cleaved from the longer transcripts in a two step process that first generates the precursors and then the mature miRNAs.[33] Such clusters would allow coordinated regulation of miRNAs in time and space. Comparing a miRNA gene cluster between *C. elegans* and *C. briggsae* revealed that the cluster remained mostly the same, gaining or losing one miRNA and altering the sequences of others in the approximately 20 to 40 million years since the divergence of these species.[21]

Some miRNAs are encoded by multiple genes at different genomic locations. In plants, many of the miRNAs identified by molecular cloning match multiple genomic sequences, suggesting a high degree of redundancy among miRNA genes.[28] *let-7*a of humans was found to match four different chromosomal loci, two of which are exact duplications having the same precursor form.[17] In some cases distinct miRNA genes encoding closely related miRNAs appear to be coordinately expressed. Coordinate expression of disparate genes implies common regulatory elements in the genes' promoters. Promoters of miRNA genes have not been described. In the case of *lin-4* and *let-7*, relatively small fragments are capable of rescuing mutant phenotypes by transformation, suggesting that those fragments contain the promoters.

Expression Patterns

The hallmark of *lin-4* and *let-7* expression *C. elegans* is stage specificity: *lin-4* is expressed from the first larval stage to adulthood, and *let-7* from the fourth stage to adulthood, and these patterns of expression determine when their targets are repressed.[10,14] Several miRNAs isolated from *C. elegans* also show stage-specific patterns of expression.[21,22] The temporal pattern of *let-7* expression is conserved in *Drosophila* and other invertebrates.[16] In *Drosophila*, the expression of *let-7* has been shown to require the steroid hormone ecdysone,[34] suggesting that hormone signaling may play an important role in regulating temporal expression patterns of miRNAs.

Many miRNAs of unknown function also have tissue-specific expression patterns, supporting the idea that they are themselves regulators of gene expression.[17,22,29,30] For example, miR173 of *Arabidopsis* is abundant in inflorescences and leaves but low in roots.[30] One study isolated miRNAs from nine different adult mouse tissues, identifying several miRNAs that show a high degree of tissue specificity.[24] For example, in both mouse and human miR-1 is highly abundant in heart but low or undetectable in brain, liver and kidney. miR-122 is highly abundant in liver and not detected in the other tissues. Each of these miRNAs are the most abundant miRNAs in the tissues in which they are found. Other miRNAs dominate the populations of miRNAs in spleen, colon and brain. The developmental and tissue-specific expression profiles of miRNAs likely reflect their roles in regulating specific targets that have roles in development, cell fate determination, and differentiation.

Criteria for Defining a miRNA

One of the inherent problems in identifying miRNAs is distinguishing a small RNA that has an explicit purpose from a spurious sequence of similar size. In the three original reports of miRNAs isolated from *C. elegans*, *Drosophila*, and human cells, each group reports that many RNAs they cloned appeared to be fragments of mRNAs, rRNAs, tRNAs, and snRNAs. And although bioinformatic approaches have been successful, the sequence and structural diversity of miRNAs and their precursors makes such approaches unreliable at present. A few criteria have helped identify the likely functional miRNAs among candidates isolated by molecular cloning and bioinformatics.

1) An RNA of approximately 22nt is cloned or detected on northern blots (the 22nt form is most often the major species detected by northern analysis).
2) The short RNA sequence matches one or more genomic locus that does not encode the mature RNA of another gene, such as rRNA, mRNA, tRNA, or snRNA.
3) A longer precursor form is predicted from the genomic sequence which is approximately 70nt in animals and longer in plants and forms an extensively but incompletely basepaired stem loop, with the miRNA deriving from either the 5' or 3' side of the stem.
4) A similar or identical miRNA is encoded in the same species or another species.

Functional assays for many miRNAs may soon be available. In *C. elegans*, for example, systematic deletion of miRNA genes will help define their biological roles. In other organisms, misexpression of miRNAs from artificial gene constructs may cause phenotypes that will help assign biological functions to individual miRNAs. Once a candidate target gene is identified for an miRNA, direct effects on gene expression may be measured. Having a defined target for an miRNA will then allow detailed investigation of its mechanism of action.

The Mechanism of Action of *lin-4*

The mechanism of action of *lin-4* is the best-characterized of all miRNAs and some of its features are assumed to be true of other miRNAs, such as an ability to regulate specific genes post-transcriptionally. *lin-4* affects neither the abundance of the target mRNA nor its ability to initiate translation, instead it seems to function at the ribosome to prevent protein accumulation.[35,36] This phenomenon is not unique, although it is unusual among well-characterized post-transcriptional regulatory mechanisms. *lin-4* acts through elements of different sequences in the 3' UTRs of its targets, and it probably does not act alone.

The Heterochronic Genes

lin-4 and *let-7* genes of *C. elegans* belong to a class of genes called the heterochronic genes that are related by their developmental roles and encode diverse products.[37] The heterochronic genes are defined by mutations that cause transformations in cell fate such that development is either precocious, skipping stage-specific events, or retarded, reiterating stage-specific events. A hierarchy of heterochronic genes acts at each larval stage to coordinate successive changes in cell fate throughout the animal.

The known targets of *lin-4* and *let-7* are also heterochronic genes. The targets of *lin-4* are the genes *lin-14* and *lin-28*. The first evidence that *lin-4* represses *lin-14* and *lin-28* was based on classical genetic analysis.[38] The *lin-4* RNA is expressed late in the first larval stage and continuously thereafter.[10] The proteins encoded by *lin-14* and *lin-28* are expressed prior to *lin-4* and are down-regulated when *lin-4* is abundant.[9,13,36,39] A mutation in the *lin-4* gene causes constitutive expression of these proteins.

3' UTRs of lin-14 *and* lin-28

The 3' UTRs of *lin-14* and *lin-28* are 1.6kb and 530bp, respectively, and both show extensive conservation among other nematode species (Fig. 2).[9,13] Two gain-of-function mutations in the *lin-14* were identified by chance based on their mutant phenotypes. In each of these mutant *lin-14* genes a large fraction of the 3' UTR is deleted. Both cause continuous expression of the LIN-14 protein and a retarded developmental phenotype, essentially the same effect as a mutation in the *lin-4* gene.[40] Close examination of the conserved regions missing in these deletions identified seven sequence elements that have the potential to partially basepair with *lin-4* (Fig..2).[6,9] The wildtype *lin-14* 3' UTR confers *lin-4*-dependent regulation on a heterologous reporter, whereas a 3' UTR with a three nucleotide change in each of the seven potential complementary elements does not.[12]

In the 3' UTR of *lin-28*, a single potential *lin-4* complementary element was also identified among evolutionarily conserved regions (Fig. 2).[13] No naturally occurring mutations in the *lin-28* 3' UTR have been found. A mutant *lin-28* gene was constructed that lacked the 15nt *lin-4* complementary element and this mutation caused continued expression of *lin-28* into late larval stages along with a retarded development phenotype, indicating that the 15nt element is required for the *lin-4*-dependent repression of *lin-28*.[13]

A 3' UTR containing only multiple copies of a *lin-4* complementary element is sufficient to confer *lin-4* dependent regulation on a reporter, indicating that no other regions of the long 3' UTR of *lin-14* are involved in the regulation.[12] This is consistent with the recent result showing that miR-30 can post-transciptionally repress a reporter construct with a single synthetic complementary site modeled after predicted *lin-4* and *let-7* sites.[41] The *lin-4* complementary site multimerized in the reporter construct would cause a bulged C residue near the 5' end of *lin-4* when the two are duplexed. When a different site is multimerized, one that would not cause a bulged C, down-regulation in response to *lin-4* does not occur.[12] The bulged C duplex, therefore, might bind a specific protein factor essential to the regulation that the nonbulged duplex does not bind. Since three of the seven predicted *lin-4* complementary elements in *lin-14* would not cause a bulged C, it was surmised that these sites must have no activity on their own but cooperate with the other four sites, perhaps by acting as high-affinity sites for recruiting *lin-4* to the 3' UTR.[12]

However, the *lin-4* complementary element in *lin-28* would not cause a bulged C to form and this site is known to be essential for repression.[13] *lin-4* may act very differently on the *lin-14* and *lin-28* 3' UTRs. On the other hand, the context of the miRNA complementary sites may be critical. The sequences around the *lin-4* complementary sites in *lin-14* and *lin-28* are also highly conserved. Evidence for the importance of the context of the sites comes from the observation that *lin-4* is not sufficient to down-regulate the natural *lin-28* 3' UTR if a second, *lin-4* independent regulatory circuit is not functioning.[36] Thus, a natural miRNA target may contain a complex of regulatory sites, akin to promoters and enhancers that contain binding sites for multiple factors, and these sites may work differently together than when they are isolated.

mRNA-Specific Regulation at the Ribosome

Regulation through the 3' UTR of a gene implies a post-transcriptional mechanism. When the mRNAs of *lin-14* and *lin-28* were examined, their levels did not appreciably change over time, whereas their protein levels dropped substantially, at least 10-fold.[9,35,36] Translational regulation often involves a block to translation initiation and poly(A) tail length may be

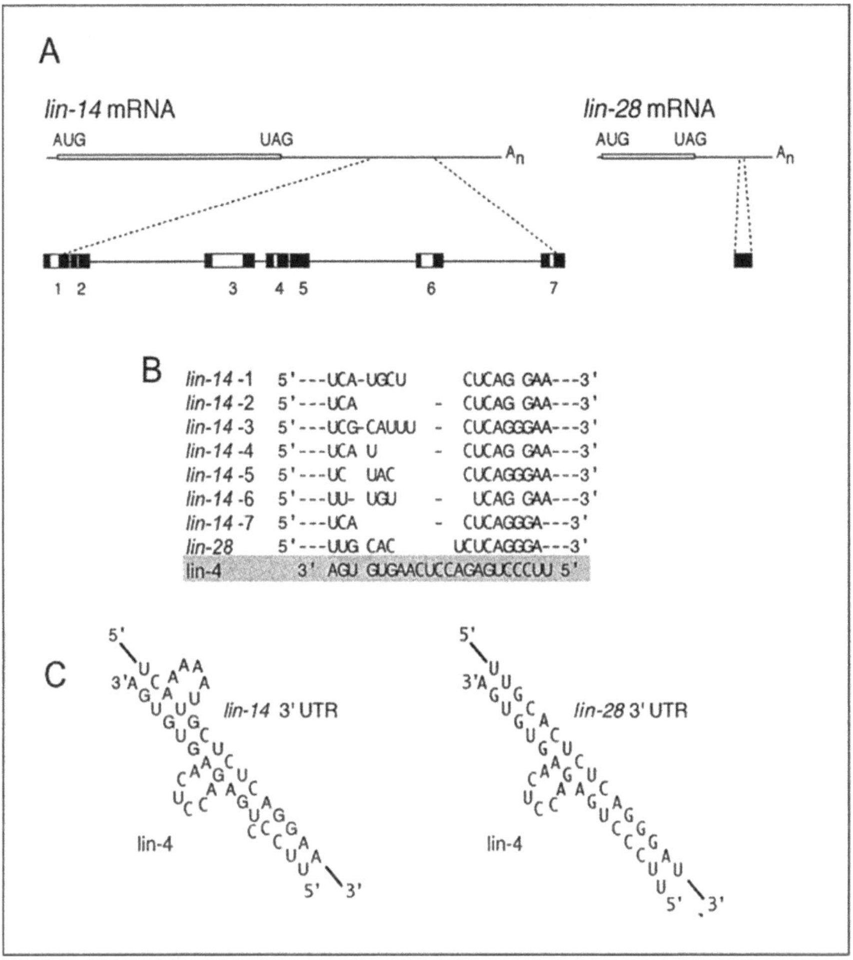

Figure 2. The microRNA *lin-4* of *C. elegans* is partially complementary to sites in the 3' UTRs of its target genes. A) The mRNAs of the heterochronic genes *lin-14* and *lin-28* drawn to the same scale. The 3' UTRs of *lin-14* and *lin-28* contain seven and one *lin-4*-complementary elements, respectively. The sites are indicated by black boxes within the expanded regions, and the white boxes indicate regions of loops in the predicted duplexes formed with *lin-4*. B) The *lin-4* complementary elements of *lin-14* and *lin-28* aligned to show potential basepairing with *lin-4*. The extent of complementarity, the positions and lengths of loops, and the presence of bulges varies among the sites. The alignments have not been confirmed experimentally and may be aligned in other variations. Dashes indicate the presence of nucleotides not shown. C. Two predicted secondary structures formed by *lin-4* and target sites. The *lin-14* site shown has a bulged C near the 5' end of *lin-4*, whereas the *lin-28* site does not. The any one of the three consecutive C residues may be drawn as bulged.

associated with that effect.[42] However, repression by *lin-4* causes no change in the *lin-14* mRNA level, its polyadenylation, nor its ability to initiate translation.[35] Ribosomes bound to repressed *lin-14* and *lin-28* mRNAs are capable of run-off translation in vitro.[36] These observations imply that *lin-4* affects one or more steps in gene expression concurrent with or subsequent to translation. Other examples of post-initiation regulation of protein expression are known, such as *Drosophila nanos*.[43] However, in these cases the role for miRNAs has not been reported.

A substantial fraction of *lin-4* is found in association with ribosomes.[35] Other miRNAs have also been found in polysomal fractions.[44] The miRNP which contains miRNAs[23] and RISC which mediates RNAi are similar in size and constitution.[45] Interestingly, RNAi also requires that mRNAs be actively translated, suggesting that both the miRNP and RISC function at the ribosome to affect the targets of small RNA.[46]

Proteins Involved in miRNA Biogenesis and Function

Several of the proteins involved in miRNA biogenesis and/or function have been characterized genetically, providing substantial insight in to the breadth of biological roles that miRNAs have. Other proteins have been identified biochemically, some of which have already been implicated in RNA metabolism or translational regulation. The fact that some of the proteins are also involved in the biogenesis and function of siRNAs complicates interpretations about whether the biological role of a particular protein is mediated by miRNAs or endogenously generated siRNAs.

Dicer and SIN1/CAF

Dicer of *Drosophila*, *C. elegans*, and mammals is required for the processing of long dsRNAs into siRNAs and for miRNA processing from precursors.[2,18-20,22,47] A homologous protein is encoded by *SHORT INTEGUMENTS1* (*SIN1*; also known as *SUSPENSOR1* [*SUS1*] and *CARPEL FACTORY* [*CAF*]) of the flowering plant *Arabidopsis thaliana* which is required for miRNA processing.[30,48,49] Dicer and SIN1/CAF possess a DEXD-box helicase domain, a so-called PAZ domain (shared with PPD proteins discussed below), two dsRNA binding domains, and two RNase III-type nuclease domains.[50] Dicer cleaves dsRNA to yield two basepaired approximately 22nt siRNAs that are offset by 2nt, whereas it cleaves an miRNA precursor most often yielding only a single-stranded RNA.[1,2,18] The difference in products may depend on the nature of associated proteins or on the precursors themselves. Dicer cleavage requires ATP, which may be hydrolyzed by the N-terminal helicase domain.[19,51] Dicer mutant animals and SIN1/CAF mutant plants have a broad array of severe developmental defects, indicating a fundamental role for small RNAs in organism development.[18,47-49]

PAZ-PIWI Domain Proteins

PAZ-PIWI domain (PPD) proteins, also called ARGONAUTE family proteins, either have been shown genetically to be involved in miRNA processing and/or function, or have been biochemically isolated in association with miRNAs. PPD proteins exist in animals, plants and the fission yeast *Schizosaccharomyces pombe*. Multicellular organisms encode a few to two dozen or so PPD proteins. The proteins contain an N-terminal PAZ domain (named for the PPD proteins PIWI, ARGONAUTE, and ZWILLE) and a C-terminal PIWI domain, but their biochemical functions are not yet known.

The *C. elegans* PPD proteins ALG-1 and ALG-2 are required for generation or stability of the mature forms of *lin-4* and *let-7* from their precursors, but do not affect the accumulation of siRNAs from exogenous dsRNA.[18] It has not been determined whether ALG-1/ALG-2 are required for the processing of other miRNAs in *C. elegans*, and because this animal possesses many PPD proteins, others may be involved in miRNA processing as well. *C. elegans* lacking ALG-2 also have defects in germline development,[52] which is not affected by either *lin-4* or *let-7*, suggesting this protein may act in the developing germline to process other miRNAs. The *C. elegans* PPD protein RDE-1, in contrast, is required for the generation of siRNAs, but not for miRNA processing. Thus, different members of the PPD family within one organism distinguish between miRNA and siRNAs. How this distinction is made is not known.

The PPD family member, eIF2C2 of humans, was found to be a component of an miRNA-protein complex called the miRNP.[23,45] Argonaute2 of *Drosophila* is a component of RISC, the complex that contains siRNAs and leads to cleavage of target RNAs.[53] In *Drosophila*, miRNAs have been found in a protein complex that contains Argonaute1 as well as other proteins found in RISC.[44] miRNAs have also been found in RISC or a similar complex, suggesting that siRNAs and miRNAs may act with many proteins in common.[44,45]

Other PPD family members of other species have been shown to be required for gene silencing, including ARGONAUTE of *Arabidopsis* and QDE1 of the mold *Neurospora crassa* (miRNAs have not been reported in fungi). And others have roles in development and differentiation suggesting the extent of miRNA function. For example, Aubergine of *Drosophila* is known to regulate events early in development.[54] It remains to be determined whether these important biological roles of PPD proteins are due to their involvement in the biogenesis of miRNAs, siRNAs, or another activity.

Gemin3 and Gemin4

A complex in human cells containing numerous miRNAs was found also to contain Gemin3, Gemin4, and the PPD protein eIF2C2, and has been termed the miRNP.[23] Gemin3 and Gemin4 are human proteins found in a complex with the Survival of Motor Neurons (SMN) protein, a protein missing or defective in the disease Spinal Muscular Atrophy.[55,56] The SMN protein complex is involved in the assembly and/or restructuring of diverse RNA-protein complexes (RNPs), however the miRNP is distinct from the SMN complex in that it does not contain SMN. Gemin3 is a DEAD-box RNA helicase, suggesting a role in facilitating RNA duplex unwinding. Other such helicases have been implicated in RNAi, including Dicer and Spindle-E, a *Drosophila* protein.[2,46] The function of Gemin4 is not known. The size of the miRNP is similar to that of RISC, the ribonuclease complex containing siRNAs that mediates target recognition – about 500kD.[51] In fact, both Gemin3 and eIF2C2 have been found in association with RISC, suggesting these small RNA-protein complexes may be the same.[45]

HEN1

HEN1 of *Arabidopsis* was identified genetically for its role in flower development.[57] The type and range of developmental phenotypes displayed by a *hen1* mutant resembles *sin1/caf* mutants. The proteins similar to HEN1 are encoded in the genomes of animals and fungi, but their function is not known.[30] Just like *sin1/caf* mutants, a *hen1* mutant fails to accumulate mature forms of miRNAs, suggesting that it has a role in the processing of miRNA precursors or the stability of mature miRNAs.[30]

VIG and FMR1

VIG1 and FMR1 of *Drosophila* are present in complexes with miRNAs.[44] VIG contains a putative RGG RNA binding motif, and it has homologues in other animals, plants and fungi. FMR1 is the *Drosophila* homologue of Fragile X mental retardation protein (FMRP), an RNA-binding protein containing two KH domains and an RGG box.[58-60] FMRP is known to be a regulator of translation and it has been speculated that it may do so in cooperation with miRNAs.[44] VIG1 and FMR1 are also in the RISC complex, suggesting that these proteins serve the same function for both siRNA- and miRNA-mediated gene regulation.[44]

Identifying miRNA Targets

In plants, some miRNAs appear to have complete complementarity with their targets and therefore may function like siRNAs.[28,29,31,32] For translational regulation it is assumed that incomplete complementarity is required. A direct interaction between the *C. elegans* miRNA *lin-4* and the 3'UTRs of *lin-14* and *lin-28* is strongly supported by genetic and molecular analysis. But the sites predicted within these 3' UTRs show significant variation such that the RNA duplexes predicted to form have few features in common (Fig. 2). *lin-4* binding sites are predicted to be complementary to 50 to 75% of the miRNA, with the 3' end of the site is usually entirely in Watson-Crick basepairs with the 5' end of the miRNA and the central region of the miRNA looped out. Bulges and loops of various sizes are predicted at several different positions. Although one study stressed the importance of a particular bulged C some *lin-4*-target site duplexes,[12] a systematic analysis of what range of complementary sites can respond to *lin-4* regulation has not been reported.

Based on our current understanding, any given miRNA may have dozens of complementary sites among protein coding genes. However, it seems unlikely that any miRNA-target interaction that shows a great range of structures could induce regulation, otherwise many spurious regulatory interactions might occur as gene sequences vary over time. Thus it seems extremely difficult with any certainty to predict the targets of a given miRNA based on partial complementarity alone given the available data.

The K box (CUGUGAUA) and Brd box (AGCUUUA) are each 3' UTR elements known to be important for the post-transcriptional regulation of developmentally important genes in *Drosophila*.[61,62] Lai has hypothesized that these sites, which are found in the 3' UTRs of genes from other animals as well, are in fact sites through which certain miRNAs act.[63] For example, miR-4 and miR-11 have the potential to form *lin-4*-like duplex structures with the K box and Brd box motifs in the 3' UTR of the developmental regulator Enhancer-of-split. Because these sites are already known to be involved in gene regulation, determining the role of the miRNAs in that regulation may be readily assessed by the misexpression of the miRNA or the identification of mutant alleles of their genes.

Artificial miRNAs

RNA-interference in mammalian cells can be induced by expression of short hairpin RNAs (shRNAs) containing a double-stranded stem with one strand that is complementary to a desired target.[64-69] The processed forms of these short hairpins are functionally siRNAs but may be called artificial miRNAs because like natural miRNAs they are explicitly encoded by genes that are transcribed into a hairpin precursor RNA. These artificial miRNAs are expressed as short transcripts from the pol III promoters of U6 and H1 RNAs. An artificially expressed miR-30, which has been shown to translationally regulate a synthetic target, was made in the context of its normal precursor form embedded in a protein coding gene.[41] In theory, any short sequence may be expressed from such constructs to allow constitutive or regulated expression of a natural or artificial miRNA for investigation of processing and function, and possibly for other experimental, genetic modification, and therapeutic purposes.

Summary and Conclusions

Naturally occurring miRNAs and the siRNAs that mediate gene silencing have significant similarities: both are approximately 21-25nt and cleaved from duplexed precursors by an RNase III activity of Dicer or a similar protein; both also require PPD family members for their accumulation; both exist in approximately 500kD and smaller cytoplasmic complexes that contain PPD proteins and other shared proteins; and both act on mRNA targets through complementary sequences.

But miRNAs and siRNA also have important differences. siRNA precursors are long dsRNAs and the products are two complementary RNAs in equal abundance. An miRNA is generated from a stem-loop that contains bulges and/or loops to yield a single-stranded RNA usually in vast excess over any complement. Distinct PPD proteins appear to be involved in the accumulation of the two kinds of RNAs. siRNAs are incorporated into RISC which then targets the siRNA to a perfectly complimentary site in an mRNA and cleaves it. miRNAs are incorporated into an miRNP, which shares at least some proteins with RISC and is a similar size. But in the best studied cases in animals, their target sites are not perfectly complementary and their target is not cleaved, but rather protein biogenesis is inhibited in some way.

A crucial distinction between miRNAs and siRNAs is that miRNAs are explicitly encoded in the genome. Other genes in the genome may give rise to siRNA, but the particular sequence and that of its complementary target are arbitrary – the function of the siRNA could be performed equally well by an RNA from a nearby sequence. Because miRNAs are so small and their target sites as small or smaller, the potential for rapid evolution of an miRNA and its target sites seems great. The complete conservation of *let-7* and other miRNAs widely in the animal kingdom suggests that its sequence is not arbitrary. Although, if an miRNA has many complementary target sites, then divergence might be very slow because of the number of sequences that must coevolve – a miRNA with one or few target sites might diverge quickly.

Is there is more significance to the sequences of a miRNAs and its target site than complementarity? Based on our present understanding of what constitutes an miRNA binding site, it seems that any given miRNA may have dozens of target sites in expressed sequences. Given the likely number and sequence diversity of all miRNAs in an organism, it would seem that no mRNA would escape binding by at least one miRNA, suggesting that there is more than complementarity to the regulatory interaction. However, the ability of an artificially expressed miR-30 to repress a synthetic binding site, as discussed above, is evidence to support the sufficiency of complementarity to induce regulation.[41]

Nevertheless, it is likely that basepairing between an miRNA and its target creates a binding site for specific proteins and that binding is dependent on the details of the RNA structure formed. A great number of RNA binding proteins exist in the genomes of plants and animals that have no known function. These proteins may bind miRNA-target structures and cause any number of regulatory consequences. Although no animal miRNA has been found to be perfectly complementary to any mRNA, endogenous *let-7* can function as a siRNA and lead to cleavage of an artificial target.[45] It has been speculated that translational control may allow more flexibility in regulation than target degradation.[24,45] So collectively, miRNAs may have diverse effects on their targets, including stimulating gene expression rather than repressing it. Given that miRNAs have long been overlooked, previously studied cases of post-transcriptional regulation may in fact involve miRNAs.

miRNAs have been called "the biological equivalent of dark matter – all around us but almost escaping detection".[70] The force exerted by these molecules on gene expression during an organisms development could be very great. All cells of a plant or animal probably express a subset of the organism's miRNAs. That subset may affect an even greater number of genes, if any one miRNA has more than one target, as does *lin-4*. This is a humbling thought since developmental biology has explored mechanisms of regulation in many contexts, some quite thoroughly, without the knowledge of miRNAs. Molecular biologists seeking to identify regulators of particular genes are compelled to consider whether miRNAs are involved.

References

1. Zamore PD, Tuschl T, Sharp PA et al. RNAi: double-stranded RNA directs the ATP-dependent cleavage of mRNA at 21 to 23 nucleotide intervals. Cell 2000; 101:25-33.
2. Bernstein E, Caudy AA, Hammond SM et al. Role for a bidentate ribonuclease in the initiation step of RNA interference. Nature 2001; 409(6818):363-366.
3. Hammond SM, Bernstein E, Beach D et al. An RNA-directed nuclease mediates post-transcriptional gene silencing in *Drosophila* cells. Nature 2000; 404(6775):293-296.
4. Elbashir SM, Lendeckel W, Tuschl T. RNA interference is mediated by 21- and 22-nucleotide RNAs. Genes Dev 2001; 15(2):188-200.

5. Chalfie M, Horvitz HR, Sulston JE. Mutations that lead to reiterations in the cell lineage of *C. elegans*. Cell 1981; 24:59-69.
6. Lee RC, Feinbaum RL, Ambros V. The *C. elegans* heterochronic gene *lin-4* encodes small RNAs with antisense complementarity to *lin-14*. Cell 1993; 75(5):843-854.
7. Ambros V, Horvitz HR. The *lin-14* locus of *Caenorhabditis elegans* controls the time of expression of specific postembryonic developmental events. Genes Dev 1987; 1:398-414.
8. Ruvkun G, Ambros V, Coulson A et al. Molecular genetics of the *Caenorhabditis elegans* heterochronic gene *lin-14*. Genetics 1989; 121(3):501-516.
9. Wightman B, Ha I, Ruvkun G. Posttranscriptional regulation of the heterochronic gene *lin-14* by *lin-4* mediates temporal pattern formation in *C. elegans*. Cell 1993; 75(5):855-862.
10. Feinbaum R, Ambros V. The timing of *lin-4* RNA accumulation controls the timing of postembryonic developmental events in *Caenorhabditis elegans*. Dev Biol 1999; 210(1):87-95.
11. Wightman B, Burglin TR, Gatto J et al. Negative regulatory sequences in the *lin-14* 3'-untranslated region are necessary to generate a temporal switch during *Caenorhabditis elegans* development. Genes Dev 1991; 5:1813-1824.
12. Ha I, Wightman B, Ruvkun G. A bulged *lin-4/lin-14* RNA duplex is sufficient for *Caenorhabditis elegans lin-14* temporal gradient formation. Genes Dev 1996; 10(23):3041-3050.
13. Moss EG, Lee RC, Ambros V. The cold shock domain protein LIN-28 controls developmental timing in *C. elegans* and is regulated by the *lin-4* RNA. Cell 1997; 88(5):637-646.
14. Reinhart BJ, Slack FJ, Basson M et al. The 21-nucleotide *let-7* RNA regulates developmental timing in *Caenorhabditis elegans*. Nature 2000; 403(6772):901-906.
15. Slack FJ, Basson M, Liu Z et al. The *lin-41* RBCC gene acts in the *C. elegans* heterochronic pathway between the *let-7* regulatory RNA and the LIN-29 transcription factor. Mol Cell 2000; 5:659-669.
16. Pasquinelli AE, Reinhart BJ, Slack F et al. Conservation of the sequence and temporal expression of *let-7* heterochronic regulatory RNA. Nature 2000; 408(6808):86-89.
17. Lagos-Quintana M, Rauhut R, Lendeckel W et al. Identification of novel genes coding for small expressed RNAs. Science 2001; 294(5543):853-858.
18. Grishok A, Pasquinelli AE, Conte D et al. Genes and mechanisms related to RNA interference regulate expression of the small temporal RNAs that control *C. elegans* developmental timing. Cell 2001; 106(1):23-34.
19. Hutvagner G, McLachlan J, Pasquinelli AE et al. A cellular function for the RNA-interference enzyme Dicer in the maturation of the *let-7* small temporal RNA. Science 2001; 293(5531):834-838.
20. Ketting RF, Fischer SE, Bernstein E et al. Dicer functions in RNA interference and in synthesis of small RNA involved in developmental timing in *C. elegans*. Genes Dev 2001; 15(20):2654-2659.
21. Lau NC, Lim LP, Weinstein EG et al. An abundant class of tiny RNAs with probable regulatory roles in *Caenorhabditis elegans*. Science 2001; 294(5543):858-862.
22. Lee RC, Ambros V. An extensive class of small RNAs in *Caenorhabditis elegans*. Science 2001; 294(5543):862-864.
23. Mourelatos Z, Dostie J, Paushkin S et al. miRNPs: a novel class of ribonucleoproteins containing numerous microRNAs. Genes Dev 2002; 16(6):720-728.
24. Lagos-Quintana M, Rauhut R, Yalcin A et al. Identification of tissue-specific microRNAs from mouse. Curr Biol 2002; 12(9):735-739.
25. Bass BL. Double-stranded RNA as a template for gene silencing. Cell 2000; 101(3):235-238.
26. Mathews DH, Sabina J, Zuker M et al. Expanded sequence dependence of thermodynamic parameters improves prediction of RNA secondary structure. J Mol Biol 1999; 288(5):911-940.
27. Kennedy BP, Aamodt EJ, Allen FL et al. The gut esterase gene (*ges-1*) from the nematodes *Caenorhabditis elegans* and *Caenorhabditis briggsae*. J Mol Biol 1993; 229(4):890-908.
28. Llave C, Kasschau KD, Rector MA et al. Endogenous and silencing-associated small RNAs in plants. Plant Cell 2002; 14(7):1605-1619.
29. Reinhart BJ, Weinstein EG, Rhoades MW et al. MicroRNAs in plants. Genes Dev 2002; 16(13):1616-1626.
30. Park W, Li J, Song R et al. CARPEL FACTORY, the Dicer homologue, and HEN1, a novel protein, act in microRNA metabolism in *Arabidopsis thaliana*. Curr Biol 2002; 12:1484-1495.

31. Rhoades MW, Reinhart BJ, Lim LP et al. Prediction of plant microRNA targets. Cell 2002; 110:513-520.
32. Llave C, Xie Z, Kasschau KD et al. Cleavage of scarecrow-like mRNA targets directed by a class of *Arabidopsis* miRNA. Science 2002; 297(5589):2053-2056.
33. Lee Y, Jeon K, Lee JT et al. MicroRNA maturation: stepwise processing and sub cellular localization. EMBO J 2002; 21(17):4663-4670.
34. Sempere LF, Dubrovsky EB, Dubrovskaya VA et al. The expression of the *let-7* small regulatory RNA is controlled by ecdysone during metamorphosis in *Drosophila melanogaster*. Dev Biol 2002; 244(1):170-179.
35. Olsen PH, Ambros V. The *lin-4* regulatory RNA controls developmental timing in *Caenorhabditis elegans* by blocking LIN-14 protein synthesis after the initiation of translation. Dev Biol 1999; 216(2):671-680.
36. Seggerson K, Tang L, Moss EG. Two genetic circuits repress the *C. elegans* heterochronic gene *lin-28* after translation initiation. Dev Biol 2002; 243(2):215-225.
37. Ambros V. Control of developmental timing in *Caenorhabditis elegans*. Curr Opin Genet Dev 2000; 10(4):428-433.
38. Ambros V. A hierarchy of regulatory genes controls a larva-to-adult developmental switch in *C. elegans*. Cell 1989; 57(1):49-57.
39. Arasu P, Wightman B, Ruvkun G. Temporal regulation of *lin-14* by the antagonistic action of two other heterochronic genes, *lin-4* and *lin-28*. Genes Dev 1991; 5(10):1825-1833.
40. Ambros V, Horvitz HR. Heterochronic mutants of the nematode *Caenorhabditis elegans*. Science 1984; 226:409-416.
41. Zeng Y, Wagner EJ, Cullen BR. Both natural and designed micro RNAs can inhibit the expression of cognate mRNAs when expressed in human cells. Mol Cell 2002; 9(6):1327-1333.
42. Mathews MB, Sonenberg N, Hershey JWB. Origins and targets of translational control. In: Hershey JWB, Mathews MB, Sonenberg N, eds,. Translational Control. Plainview, N.Y: Cold Spring Harbor Laboratory Press, 1996:1-29.
43. Clark IE, Wyckoff D, Gavis ER. Synthesis of the posterior determinant Nanos is spatially restricted by a novel cotranslational regulatory mechanism. Curr Biol 2000; 10(20):1311-1314.
44. Caudy AA, Myers M, Hannon GJ et al. Characterization of RNAi effector complexes: a possible role for RNAi in Fragile X syndrome. Genes Dev 2002; in press.
45. Hutvagner G, Zamore PD. A MicroRNA in a multiple-turnover RNAi enzyme complex. Science Aug 1 2002.
46. Kennerdell JR, Yamaguchi S, Carthew RW. RNAi is activated during *Drosophila* oocyte maturation in a manner dependent on *aubergine* and *spindle-E*. Genes Dev 2002; 16(15):1884-1889.
47. Knight SW, Bass BL. A role for the RNase III enzyme DCR-1 in RNA interference and germ line development in *Caenorhabditis elegans*. Science 2001; 293:2269-2271.
48. Ray A, Lang JD, Golden T et al. SHORT INTEGUMENT (SIN1), a gene required for ovule development in *Arabidopsis*, also controls flowering time. Development 1996; 122(9):2631-2638.
49. Jacobsen SE, Running MP, Meyerowitz EM. Disruption of an RNA helicase/RNAse III gene in *Arabidopsis* causes unregulated cell division in floral meristems. Development 1999; 126(23):5231-5243.
50. Cerutti L, Mian N, Bateman A. Domains in gene silencing and cell differentiation proteins: the novel PAZ domain and redefinition of the Piwi domain. Trends Biochem Sci 2000; 25(10):481-482.
51. Schwarz DS, Zamore PD. Why do miRNAs live in the miRNP?. Genes Dev 2002; 16(9):1025-1031.
52. Cikaluk DE, Tahbaz N, Hendricks LC et al. GERp95, a membrane-associated protein that belongs to a family of proteins involved in stem cell differentiation. Mol Biol Cell 1999; 10(10):3357-3372.
53. Hammond SM, Boettcher S, Caudy AA et al. Argonaute2, a link between genetic and biochemical analyses of RNAi. Science 2001; 293(5532):1146-1150.
54. Wilson JE, Connell JE, Macdonald PM. *aubergine* enhances oskar translation in the *Drosophila* ovary. Development 1996; 122(5):1631-1639.
55. Charroux B, Pellizzoni L, Perkinson RA et al. Gemin3: A novel DEAD box protein that interacts with SMN, the spinal muscular atrophy gene product, and is a component of gems. J Cell Biol 1999; 147(6):1181-1194.

56. Charroux B, Pellizzoni L, Perkinson RA et al. Gemin4. A novel component of the SMN complex that is found in both gems and nucleoli. J Cell Biol 2000; 148(6):1177-1186.
57. Chen X, Liu J, Cheng Y et al. HEN1 functions pleiotropically in *Arabidopsis* development and acts in C function in the flower. Development 2002; 129(5):1085-1094.
58. Ashley Jr CT, Wilkinson KD, Reines D et al. FMR1 protein: conserved RNP family domains and selective RNA binding. Science 1993; 262(5133):563-566.
59. Siomi H, Siomi MC, Nussbaum RL et al. The protein product of the fragile X gene, FMR1, has characteristics of an RNA-binding protein. Cell 1993; 74(2):291-298.
60. Wan L, Dockendorff TC, Jongens TA et al. Characterization of dFMR1, a *Drosophila melanogaster* homolog of the fragile X mental retardation protein. Mol Cell Biol 2000; 20(22):8536-8547.
61. Lai EC, Posakony JW. The Bearded box, a novel 3' UTR sequence motif, mediates negative post-transcriptional regulation of Bearded and Enhancer of split complex gene expression. Development 1997; 124(23):4847-4856.
62. Lai EC, Burks C, Posakony JW. The K box, a conserved 3' UTR sequence motif, negatively regulates accumulation of enhancer of split complex transcripts. Development 1998; 125(20):4077-4088.
63. Lai EC. Micro RNAs are complementary to 3' UTR sequence motifs that mediate negative post-transcriptional regulation. Nat Genet 2002; 30(4):363-364.
64. Brummelkamp TR, Bernards R, Agami R. A system for stable expression of short interfering RNAs in mammalian cells. Science 2002; 296(5567):550-553.
65. Lee NS, Dohjima T, Bauer G et al. Expression of small interfering RNAs targeted against HIV-1 rev transcripts in human cells. Nat Biotechnol 2002; 20(5):500-505.
66. Paddison PJ, Caudy AA, Bernstein E et al. Short hairpin RNAs (shRNAs) induce sequence-specific silencing in mammalian cells. Genes Dev 2002; 16(8):948-958.
67. Paul CP, Good PD, Winer I et al. Effective expression of small interfering RNA in human cells. Nat Biotechnol 2002; 20(5):505-508.
68. Sui G, Soohoo C, Affar el B et al. A DNA vector-based RNAi technology to suppress gene expression in mammalian cells. Proc Natl Acad Sci USA 2002; 99(8):5515-5520.
69. McManus MT, Petersen CP, Haines BB et al. Gene silencing using micro-RNA designed hairpins. RNA 2002; 8(6):842-850.
70. Ruvkun G. Molecular biology. Glimpses of a tiny RNA world. Science 2001; 294(5543):797-799.

CHAPTER 8

Short Interfering and MicroRNAs:

Tiny but Mighty

Martin Tabler, Alexandra Boutla, Kriton Kalantidis and Mina Tsagris

Abstract

Two functionally distinct classes of short noncoding RNAs consisting of ca. 20-25 nucleotides have been discovered recently. Both classes of RNAs are generated from longer single- or double-stranded RNA precursors by the same ribonuclease. Either of the two categories of short RNAs is involved in posttranscriptional regulation of target RNAs that are in part or fully complementary. Despite a great degree of similarity, the two classes of RNAs are generated in different pathways and have different functions as well as different effects on their target RNAs. MicroRNAs (miRNAs) modulate translation efficiency in a reversible interaction with their cognate target mRNA, while short interfering RNA (siRNA) direct the specific degradation of their matching RNA target in a process called RNA interference (RNAi) or posttranscriptional gene silencing (PTGS). The current knowledge about origin, functionality of these two categories of small noncoding RNAs is summarized and first attempts towards practical application in gene suppression are reviewed.

Peptides and RNA share some common features. Composed of amino acids or ribonucleotides they form polymeric macromolecules, that are usually linear, unbranched chains of variable size. The sequence of their units determines their overall structure and dynamics, a precondition for the functionality in a living cell. A further remarkable similarity between the two classes of biomolecules is the enormous size range. Peptides may be extremely large, like Titin, which is composed about 28,000 amino acids with a molecular weight of about 3 MD and a length of about 1 μm (reviewed in ref. 1). Quite remarkably, viral RNA has a similar size limit for its monomeric units. Coronaviruses contain the longest known RNA genome with a chain length of about 30,000 nucleotides,[2] but some mRNAs and their precursors can be considerably longer.

The other side of the size spectrum is formed by minimal peptides consisting of just two or three amino acids that are responsible for various bioactivities. For example the di- and tripeptides Pro-Gly (PG), Gly-Pro (GP), Pro-Gly-Pro (PGP) are members of the glyproline family that are involved—besides further activities - in suppression of some reactions of blood coagulation.[3] Similar to small peptides, there are at least two classes of petite noncoding RNAs: the 'short interfering RNA (siRNA)' and the 'microRNA (miRNA)'. Both types of RNAs have been discovered recently. They share common features, but exhibit also different function and are subject of this review.

Noncoding RNAs: Molecular Biology and Molecular Medicine, edited by Jan Barciszewski and Volker A. Erdmann.

MicroRNAs (miRNAs)

History of Regulatory Small RNAs

By chromosome walking Ambros and coworkers[4] isolated and cloned in 1993 the gene *lin-4* of *Caenorhabditis elegans*, which is essential for the normal temporal control of various post-embryonic developmental events by down-regulating another gene, *lin-14*, which had been observed already earlier.[5] However, *lin-4* was not a normal protein gene but turned out to be a small noncoding RNA. Mutagenesis studies confirmed that the noncoding RNA was responsible for the genetic function of *lin-4*. Two *lin-4* specific RNA transcripts could be detected of 22 and 61 nucleotides, respectively, which contained sequences complementary to a repeated sequence element in the 3' untranslated region (UTR) of *lin-14* mRNA, suggesting negative modulation of translation via an antisense RNA-RNA interaction. This assumption could be confirmed by transferring the sequence elements to a 3' UTR of marker genes.[6] It should be noted however, that the interaction did not result in the formation of a contiguous RNA duplex, an indication that the sequences had been selected by evolution for a reversible interaction. Seven years later, a second regulatory RNA, *let-7*, was also identified in *C. elegans*.[7] Also, *let-7* was hybridizing to the 3'UTR of target RNAs, in part the same target RNAs that were regulated by *lin-4*. In view of their small size and their temporary expression pattern these two RNAs were termed 'small temporal' (st) RNAs. Both RNAs were considered as rather exotic examples of RNA genes that regulate gene expression by an antisense mechanism in eukaryots. This view changed drastically when it was discovered last year that stRNAs are examples – and actually prototypes - of a much larger group of small RNAs. By cloning and sequencing naturally occurring small RNAs of about 22 nucleotides, three groups found independently that there is an abundant class of genome-derived small RNAs in such diverse species as *Drosophilae melanogaster*, *Caenorhabditis elegans* and humans.[8-10] Only a fraction of the short RNAs has been identified by cloning, which included, however, the already known stRNAs *lin-4* and *let-7* from *C. elegans*. RNAs related to *let-7* were found in *Drosophilae* and human.[8] The new class of RNAs was called 'microRNA' with the acronym 'miRNA'. Lately miRNAs were also identified in plants[11,12] indicating that they represent an evolutionary well conserved class of small noncoding RNAs. Collectively more than hundred miRNAs have been identified to date and it is estimated that there are more than hundred different miRNAs per species.

miRNA Genes

Already for *lin-4*, RNA transcripts of two different sizes were observed in Northern blots.[4] This suggested that the mature miRNA was formed from a precursor and then processed. It turned out that the processing is done by the ribonuclease III-type enzyme Dicer from *Drosophilae*,[13] which had been discovered to be responsible for the processing of double-stranded RNA (see chapter below). The same enzyme was shown responsible to process miRNAs from their longer precursor.[14-17] Precursors have a hairpin-type structure, however without a continuous double-strand. The RNA helix is interrupted by short bulges and interior loops. The mature miRNA can originate from the 5' arm of the hairpin as well as from the 3' arm of the hairpin, although the latter seems to be more frequent. At current it is not known how the processing sites are determined by Dicer. Most mature miRNA are in the range of 21-24 nucleotides. Precursors can encode a single miRNA gene (single locus), but there are also precursors known that contain several hairpin domains in a cluster (multiple locus), each domain encoding a miRNA.[10] There are also precursors that have a slightly deviating Y –type secondary structure (with two hairpins), including those that encode miRNAs of two different polarities.[11] Most precursor genes are located in intergenic regions, but some are found in introns. Several unique miRNA sequences originate from different precursors (different loci).

The polymerase responsible for the synthesis of the precursor RNA has not been identified. It could be RNA polymerase II or III. The observation that the expression of *let-7* in *Drosophilae melanogaster* is controlled by the steroid hormone ecdysone during metamorphosis[18] argues in favor of RNA polymerase II. However, since the *let-7* locus is downstream of a gene, but in the opposite polarity, it would require a separate transcription event. The ecdysone-dependence could be also explained if the hormone regulates transcription of a specific protein factor required for RNA polymerase III transcription. Hence, the ecydsone-specific expression of the miRNA would be an indirect effect.

Function of miRNAs

The cellular function of *lin-4* and *let-7* had been characterized before it was recognized that they are archetypes of a new class of RNAs. Therefore it is tempting to assume that also the residual miRNAs are negative regulators of translation. However, it is difficult to identify the target RNAs. Since *lin-4* and *let-7* bind to the 3' UTR of their target mRNAs in a nonperfect manner it is almost impossible to identify target motifs by bioinformatics, despite the availability of the complete genome sequence data for several organisms, where miRNAs have been characterized (*C. elegans, Drosophilae melanogaster, Arabidopsis thaliana* and humans). These difficulties are due to the flexible rule of forming RNA duplexes, which allow mismatches, bulge and interior loops of variable size, as well as G:U pairs. Any program that could predict a matching target site is likely to find too many false positives. Only in cases where there is a complete or almost complete match the interaction is evident. For example, Llave et al have identified in *Arabidopsis thaliana* several miRNAs that map to intergenic regions and also to protein-coding genes.[11] One RNA could even interact with three different protein genes, which had the same matching sequence. That would represent a genuine antisense regulation, which, however, may turn out to be more an exception than a rule for miRNA function.

Further evidence that miRNAs are involved in modulating translation comes from another observation. Lai[19] observed that the 5'-terminal 6-8 nucleotides of *Drosophilae* miRNAs *miR-2a,b, miR-6, miR-11* and *miR-13a,b* are related and complementary to the consensus sequence of the *K* box. This *K* box has been previously identified as a conserved motif in the 3' UTR sequence that negatively regulates *Drosophilae* genes of the *Enhancer split* complex.[20] The *GY* and *Brd* box are similar control elements in the 3' UTR of other genes.[21,22] As demonstrated by Lai,[19] for several of the genes with K, *GY* or *Brd* boxes, a miRNA can be identified that might base pair. This implies that the miRNA and their complementary target motifs are divided into two domains. The 5'-terminal nucleotides of the miRNAs represent the 'family' domain that matches to one of these boxes, while the residual domain resembles a 'forename' that matches specifically to a particular target. It should be added, however, that the interaction of target genes having *K* and *Brd* boxes with matching miRNAs is mechanistically slightly different than the *lin-4* mediated regulation. While the *lin-4* primarily down-regulates translation, in a step following initiation,[23] the *K* and *Brd* boxes also influence the stability and hence the concentration of the mRNA.[20,22] It remains to be seen whether the reduced mRNA stability is caused by the interaction with a matching miRNA.

At current it is too early to conclude that all miRNAs influence mRNA translation or stability by interaction with 3' UTRs. In our efforts to clone potential target genes that are regulated by miRNAs, we used a DNA oligonucleotide that was complementary to *miR-13a* as a primer for a polymerase chain reaction (PCR) with a pooled cDNA library from *Drosophilae* embryos and a second primer specific for the vector sequence. We expected that primarily target sites would match to the antisense primer, especially since its last six 3'-terminal nucleotides were expected to base-pair perfectly to the cDNA, since this domain corresponds to the 5'-terminal family domain of the miRNA. We could indeed identify some target genes that have in their 3' UTR a domain that allowed interaction with *miR-13*. Interestingly, however, we identified also other clones that had a binding site for *miR-13*, however, not in their 3'UTR. Especially, we identified potential target motifs in exon sequences close to exon/intron junctions as well as in an intron itself (such a sequence would not be expected in a cDNA library). Therefore it is at least conceivable that miRNAs have additional functions. By forming temporary RNA duplexes they could modulate additional RNA-related processes, such as for example splicing. If our observations are not purely accidental, one could envisage that miRNA influence pathways of alternative splicing, depending on whether a miRNA is bound or not. This would imply that at least a fraction of the miRNAs is located in the nucleus.

Mourelatos et al,[24] made another intriguing discovery. They identified a novel human ribonucleoprotein (RNP) that contained numerous miRNAs. This miRNP complex is large (~15S) and composed of the proteins Gemin3, Gemin4 and eukaryotic initiation factor 2C2 (eIF2C2). Gemin3 is a DEAD box RNA helicase and is known to bind to the Survival of Motor Neuron (SNM) protein. The latter protein forms with five other proteins, including Gemin3 and Gemin4 that are also part of the miRNP, the SNM complex. The loss of function of SNM protein results in a neurodegenerative disease called spinal muscular atrophy (SMA).[25] The SMN complex is known to be involved in the assembly and restructuring of diverse ribonucleoprotein complexes, including small nuclear RNPs (snRNP), small nucleolar RNPs (snoRNPs), hetergenous nuclear RNPs (hnRNPs) and transcriptosomes (for references see introduction of ref. 24). The third protein (eIF2C2) of the 15S miRNP is a protein of unknown function; however, it is member of the *Argonaute* family characterized by two conserved — yet biochemically not characterized — domains PAZ and PIWI, which are therefore also called PPD proteins. Originally the *argonaute* mutant had been isolated from *Arabidopsis thaliana*, where it caused pleiotropic effects on the general plant architecture.[26] Then, the *Drosophilae* homologue was isolated and its loss-of-function mutant showed developmental defects especially in the nervous system and finally it was shown that the *Argonaute*1 locus and its homologues in fungi and worms is required for the expression of stRNAs (which we would call now miRNA) as well as for posttranscriptional gene silencing (PTGS) in plants and RNA interference (RNAi) in animals.[14,27] A more detailed discussion on the PPD proteins and their cellular function, as well as the implication of the finding that eIF2C2 is PPD protein and a part of the miRNP can be found in reference 28.

Short Interfering RNA (siRNA)

Posttranscriptional gene silencing (PTGS) is a process that was originally discovered in plants. This term describes a cellular process where an RNA—typically a mRNA—is degraded in a sequence-specific manner. As a consequence, the steady-state concentration of the RNA is reduced, but the transcription proceeds — at least initially—at normal levels. The understanding of PTGS is complicated by its interrelation to transcriptional gene silencing where methylation and chromatin remodeling result in down-regulation of transcription. In the last couple of years the progress on PTGS in plants has been reviewed frequently[29-40] and the following text will just deal with the role of short RNAs during PTGS and RNA interference (RNAi).

Three years ago Hamilton and Baulcombe[41] observed that plants exhibiting PTGS produced a class of short RNAs specific for the silenced gene. The short RNAs consisted of both polarities: sense and antisense. Especially the occurrence of antisense polarity indicated that this class of RNAs was not a mere degradation products of target mRNA. Originally, the length of the RNAs was estimated to about 25 nucleotides and ensuing analysis has revealed that there is a population of RNAs slightly variable in size, which also depends on the species. Hamilton and Baulcombe[41] further observed that the small RNAs were generated regardless of the way PTGS had been initiated. Induction of PTGS by a transgene in the absence or presence of an endogenous gene resulted in the formation of target-specific small RNAs, as well as induction or by an RNA virus or by local infiltration using *Agrobacterium tumefacies* carrying a transformation vector. Therefore the occurrence of short RNAs is also a hallmark for induced PTGS and can also be used as diagnostic tool. In this way it could be shown that viroids, subviral plant RNA pathogens that replicate autonomously in the nucleus via double-stranded RNA intermediates, are most likely subject to a PTGS response.[42] In plants engineered to express virus-specific double-stranded RNA the occurrence of short RNAs can be taken as molecular marker. Those plants that generate small RNAs are likely to be resistant when challenged with the virus.[43]

RNA silencing is not restricted to plants. Highly related processes are found in a broad variety of different species. The most intensively studied species are *Neurospora crassa*,[44-46] where the process is called quelling, *Caenorhabditis elegans*[47-49] and *Drosophilae melanogaster*.[50,51] For the two latter invertebrates it could be shown that an artificially introduced double-stranded RNA induces a silencing process similar to that in plants for which the additional term 'RNA interference' for which the acronym 'RNAi' has been coined. The number of species where RNAi /PTGS operates was rapidly expanding from *E. coli*[52], and protozoa,[53,54] to vertebrates,[55,56] including mammals.[57,58] Taken together with the occurrence in plants, this indicates that the process is evolutionary conserved and most likely was developed already in the last common ancestor of plants and animals.

Similar to plants with activated PTGS, short RNAs are also found in animals. Using an in vitro system derived from *Drosophilae* cells, Zamore et al[59] could show that both strands of double-stranded RNA were cleaved in an ATP-dependent manner at intervals of 21 and 23 nucleotides, generating short double-stranded RNA fragments. The generation was independent of the presence of single-stranded target RNA. Yang et al[60] also observed the formation of short RNAs in *Drosophilae* embryos after injecting double-stranded RNA. The time of appearance and the persistence of the short RNA correlated with the inhibitory RNAi effect, indicating that the short RNAs play a functional role. Bernstein et al[13] identified the enzyme Dicer in *Drosophilae* that is responsible for converting double-stranded RNA into the characteristic short RNA fragments. Dicer is a member of the RNase III family that is specific for double-stranded RNA, however, without sequence-specificity. Dicer has besides a dual RNase III motif also a helicase domain and it shares also homology to the proteins of the *Argonaute* family (see above). Dicer has evolutionary conserved homologues, for example carpel factory (caf) in *Arabidopsis thaliana*,[61] which was originally identified as a gene that plays a role in floral meristem determinacy. But its involvement in cleavage of double-stranded RNA has also been demonstrated for *Caenorhabditis elegans*[14,17] and mammalian cells.[62]

A next important step in the understanding of RNAi and the functionality of the short RNA was the identification of a sequence specific-ribonuclease purified from cultured *Drosophilae* cells that had been transfected with double-stranded RNA.[63] Hammond et al could show that the enzyme contained an essential RNA component of the same size as the RNA products generated by Dicer. They called the multi-protein component enzyme 'RNA-induced silencing complex (RISC)'. An initial biochemical characterization showed that the active ribonucleoprotein is a complex of about 500 kD.[64] Hammond et al[63] concluded that the sequence-specificity was conferred to RISC by the incorporation of the short RNAs — the Dicer-generated processing products originating from double-stranded RNA — which would guide the ribonuclease to a matching single-stranded RNA target. RISC charged with the appropriate short RNA could then cleave a multiple of single-stranded cognate target RNAs. Besides Dicer and RISC, further proteins are required for the processing of double-stranded RNA to short RNA fragments, such as Argonaute2[64] in *Drosophilae*, another member of the *Argonaute* family, which has homologues in *C. elegans*[65,66] and *Neurospora crassa*.[67] These proteins may assist in bringing double-stranded RNA and Dicer in contact to each other. Very recently it was found that also single-stranded RNAs of the size of miRNA may become incorporated into RISC, provided they are 5' phosphorylated.[68,69] These data support the idea that the 15S miRNA (previous chapter) is identical to RISC.

The RNAi mechanism—as outlined so far - also explains why only substoichiometric amounts of double-stranded RNAs are sufficient to initiate a PTGS/RNAi response. There are two steps of amplification. First, a double-stranded RNA is cleaved into several short RNA fragments. The longer the trigger double-stranded RNA, the more short RNA fragments can be processed. This is consistent with the observation that longer double-stranded RNAs are better inducers of RNAi in *Drosophilae*.[60] Second, once RISC is loaded with the matching short RNA can—like any RNase—cleave a multiple of target molecules.

Besides their function to guide RISC to a matching target RNA, siRNAs have also a second role. It was shown that siRNA may act as primers to initiate the synthesis of double-stranded RNA from a single-stranded RNA template.[70,71] This reaction provides a further amplification step since it converts a single-stranded RNA into a double-stranded RNA from which further siRNAs can be made (secondary siRNAs). In case two RNA transcripts are coupled, induction of RNAi at the downstream gene may eventually induce silencing at the upstream gene, because the antisense strand of siRNA may direct the synthesis of an antisense RNA strand covering also the 5' located gene. This process is called 'transitive' RNAi[71], but spans only a few hundreds of nucleotides. The synthesis of double-stranded RNA is thought to be carried out by RNA-dependent RNA polymerase (RdRp). The first cellular RNA polymerase that was undoubtedly able to use an RNA template was discovered in tomato'.[72-74] Tomato RdRp, or its homologues, was shown to be involved in plant PTGS.[75-77] Equivalent enzymes have been found also in *Neurospora crassa*,[46] *Caenorhabditis elegans*,[78] *Chlamydomonas reinhardtii*[79] and *Dictyostelium discoideum*,[80] where they play a similar role in silencing. In some organisms, for example *Dictyostelium discoideum* and *C. elegans* and several RdRp genes are present that fulfil distinct functions during RNAi.[80,81] Surprisingly, no equivalent RdRp genes have been identified in the sequenced genomes of *Drosophilae melanogaster*. Therefore, it requires further experimental confirmation to prove that siRNAs can indeed act as primers for double-stranded RNA synthesis in *Drosophilae*. Also mammalian systems seem to be devoid of RdRP and introduction of the tomato gene into transgenic mice had no impact on silencing.[82]

The RNAi/PTGS mechanism described so far shows clearly that the short RNAs are not just simple degradation products of double-stranded RNA but play an essential role in the entire silencing mechanism. Since the short RNAs are generated by Dicer and related enzymes, they operate downstream of the double-stranded RNA trigger. Therefore it was an obvious idea to test whether also the short RNAs themselves would be able to initiate the silencing reaction. It was known that RNase III of *E. coli* releases cleavage products that are characterized by two nonpaired single-stranded nucleotides at the 3' terminus. Further, the cleavage of the phosphodiester bond occurs such that the RNA products contain a 5' monophosphate and 3' hydroxyl group.[83] It could be shown in various systems that duplexes of short RNAs, including chemically synthesized RNA cassettes, were able to trigger the RNAi reaction. The first indication came from application of short RNAs in an in vitro system derived from cultured *Drosophilae* cells.[84] In view of its potential to induce RNAi, Tuschl and coworkers proposed the term 'short interfering RNA' with the acronym 'siRNA'.[84] Like in the in vitro system, siRNA were also active in mammalian cell lines[85-87] and in whole organisms such as embryos of *Drosophilae*[88] and *Xenopus*.[89] They were equally useful to impair virus replication,[90,91] including human immunodeficiency virus (HIV).[92,93] The efficiency of inducing RNAi in *Drosophilae* embryos by synthetic RNA cassettes requires less amounts of RNA than induction by a long double-stranded RNA.[88] The induction of RNAi by short synthetic RNAs is not just an alternative to long double-stranded RNA but has a lot of implications. Especially, in mammalian systems the usage of short double-stranded RNA has principal advantages since it avoids the unspecific interferon response normally observed when a cell is exposed to long double-stranded

RNA, with the exception of embryonal cells.[62] Normally, long double-stranded RNA are not present in a mammalian cell, unless there is an infection with an RNA virus that replicates via double-stranded RNA intermediates. As an unspecific first line of defense against viruses, the mammalian cell activates interferon production, which then activates the transcription of at least 30 genes (reviewed in ref.[94]). The most prominent genes activated via this pathway are the 2'-5'-oligoadenylate synthetase and the 2'-5'-A-dependent RNase L, which cleaves single-stranded RNA in an unspecific manner, thus limiting the synthesis of viral RNA, as well as the interferon induced protein kinase (PKR), which gets activated by binding to double-stranded RNA, and then phosphorylates the translation initiation factors eIF2α, eventually impairing translation.[94] In particular the latter pathway mediates apoptosis as a response to double-stranded RNA, limiting its usage to trigger RNAi in mammalian cells. These problems are overcome by the application of siRNAs, which are too short to activate the interferon response. In view of this potential, some aspects of the application of siRNA and variations thereof will be discussed below.

Application of siRNAs to Trigger RNAi

Synthetic siRNA Cassettes

As mentioned in the previous chapter, synthetic RNAs could be used to induce RNAi in various systems, from cell lysates[84] to mammalian cell lines[85,86] to whole organisms[88,89] and the number of successful applications is rapidly expanding. Therefore it is important to know how a synthetic siRNA cassette should be designed. In general, the cassette works best if it resembles the authentic cleavage product released by Dicer. For example, it was shown that a 5' phosphorylated cassette is more efficient than the hydroxylated form.[88] The blunt ended short double-stranded RNA is a much less potent inducer of RNA than the authentic form with two 3' unpaired nucleotides.[88,95] Boutla et al reported that neither of the sense nor the antisense strand of the siRNA can be replaced by DNA,[88] but Elbashir et al[85,95] found that the two unpaired 3'-terminal nucleotides of the siRNA could be replaced by deoxy nucleotides, even nonmatching to the target RNA . In our own test system (*Drosophilae* embryos), we could see, however, a reduction of the silencing potential when we replaced ribo-nucleotides with deoxy nucleotides (Boutla et al, unpublished). Elbashir et al demonstrated that—at least in the *Drosophilae* lysate system—the optimal length for the siRNA is 21, which was slightly better than 22.[95] They could further show that the antisense strand of the siRNA determines the cleavage site. Cleavage will occur in the complementary target strand at position 10 and 11 nucleotides relative to the 5' end of the siRNA antisense strand.

Chemical modification may increase the stability of RNA and DNA and is widely used in antisense technology. It appears however, that the siRNA is very sensitive to chemical modifications. Substitution by 2' deoxy or 2' O -methyl residues in either strand abolished RNAi in the *Drosophilae* lysate test system,[95] but in *Caenorhabditis elegans* some chemical modifications were tolerated, primarily if they were located in the sense strand.[96] The latter RNAs were, however, small double-stranded RNAs, but not proper siRNAs with unpaired 3'-terminal nucleotides. Also the question of specificity seems to depend on the system tested. Boutla et al[88] introduced a single point mutation at a central position of the synthetic siRNA cassette in either the sense or the antisense strand or into both. None of these modifications had a dramatic effect on the silencing potential when injected into *Drosophilae* embryos. Similar results were obtained by Holen et al[97] in a mammalian tissue culture system. However, Elbashir et al[95] observed that a single mutation abolished RNAi in a *Drosophilae* lysate system.

For a systematic application it would be important to know whether each possible siRNA cassette will work equally well in inducing RNAi or whether there are preferred sites for target selection. One obvious test is to analyze whether the designed cassette has any sequence similarity to another gene. This is especially useful in organisms whose genome is fully sequenced. A BLAST analysis[98] will show whether the selected sequences share some undesired sequence similarity. Regardless of accidental sequence similarity, it appears that the nature of the selected sequence also plays a role. Holen et al[97] performed a systematic analysis synthesizing siRNAs targeting different sites on the same target mRNA (human Tissue Factor). They observed striking differences in silencing efficiency. Even a shift of as little as three nucleotides was able to influence the overall efficiency of the RNA cassette and in one case the authors observed over a shift of 3, 6 and 9 nucleotides a gradual conversion of siRNA activity from no mRNA depletion to 90% depletion. Surprisingly, the 'good' sites do not correlate with sites that are well accessible to cleavage by hammerhead ribozymes. This suggests that secondary structure of the target RNA plays a minor role and that other—so far unidentified—criteria determine an effective target site for siRNA.

Expression Cassettes for siRNA

The chemical synthesis of siRNA is relatively expensive (at current about 10 EURO per base). The cost may not be relevant for the development of drugs, but it excludes extensive usage of a great number of different siRNAs for scientific purposes. Moreover, the delivery of the synthetic RNA may be difficult. Further, the persistence of the RNAi effect after an application of dose of synthetic siRNA in mammalian systems is a matter of only a few days.[97] One alternative to overcome these problems is the expression of siRNA from a DNA template. This can be achieved by expressing both siRNA strands separately. A type III class promoter for DNA-dependent polymerase III, such as promoter for the U6 small nuclear RNA, is especially suitable for the expression of small RNAs because, unlike other Pol III promoters, no sequence elements downstream of the +1 position are required.[99] Also the termination signal is relatively simple, as just some consecutive U residues are required. Using such a Pol III system, the expression of siRNA could be accomplished,[100,101] but also a T7 RNA polymerase system has been used successfully.[102] An alternative to the separate expression of two short RNAs is the generation of a short hairpin construct, where the two strands are connected by a short loop, which will be cleaved in vivo. It was shown that such a construct delivers an RNA transcript that is able to effectively initiate RNAi in mammalian cells.[103-108] Surprisingly, such short hairpin RNAs designed for expression of short RNAs resemble an RNA that has been identified in silenced plants and which was able to induce efficiently silencing in *C. elegans*.[109] Straightforward expression systems making use of short hairpins may be used for functional genomics, i.e., for functional analysis of genes that have been identified in various sequencing projects but not yet characterized. Besides this scientific application there are numerous applications conceivable in biomedicine, such as target validation and development of anti-cancer and anti-virus strategies. In this sense the research on RNA silencing starting with analysis of some perplexing phenomenon on color formation in petunia[110,111] is an excellent example how basic research may eventually result in practical applications that could not be foreseen.

References

1. Trinick J, Tskhovrebova L. Titin: a molecular control freak. Trends Cell Biol 1999; 9(10):377-380.
2. Lai MM, Cavanagh D. The molecular biology of coronaviruses. Adv Virus Res 1997; 48:1-100.
3. Samonina G, Ashmarin I, Lyapina L. Glyproline peptide family: review on bioactivity and possible origins. Pathophysiology 2002; 8229-234.
4. Lee RC, Feinbaum RL, Ambros V. The *C. elegans* heterochronic gene lin-4 encodes small RNAs with antisense complementarity to lin-14. Cell 1993; 75(5):843-854.
5. Arasu P, Wightman B, Ruvkun G. Temporal regulation of lin-14 by the antagonistic action of two other heterochronic genes, lin-4 and lin-28. Genes Dev 1991; 5(10):1825-1833.
6. Wightman B, Ha I, Ruvkun G. Posttranscriptional regulation of the heterochronic gene lin-14 by lin- 4 mediates temporal pattern formation in *C. elegans*. Cell 1993; 75(5):855-862.

7. Reinhart BJ, Slack FJ, Basson M et al. The 21-nucleotide let-7 RNA regulates developmental timing in *Caenorhabditis elegans*. Nature 2000; 403(6772):901-906.
8. Lagos-Quintana M, Rauhut R, Lendeckel W et al. Identification of novel genes coding for small expressed RNAs. Science 2001; 294(5543):853-858.
9. Lee RC, Ambros V. An extensive class of small RNAs in *Caenorhabditis elegans*. Science 2001; 294(5543):862-864.
10. Lau NC, Lim LP, Weinstein EG et al. An abundant class of tiny RNAs with probable regulatory roles in *Caenorhabditis elegans*. Science 2001; 294(5543):858-862.
11. Llave C, Kasschau KD, Rector MA et al. Endogenous and silencing-associated small RNAs in plants. Plant Cell 2002; 14(7):1605-1619
12. Reinhart BJ, Weinstein EG, Rhoades MW et al. MicroRNAs in plants. Genes Dev 2002; 16(13):1616-1626.
13. Bernstein E, Caudy AA, Hammond SM et al. Role for a bidentate ribonuclease in the initiation step of RNA interference. Nature 2001; 409(6818):363-366.
14. Grishok A, Pasquinelli AE, Conte D et al. Genes and mechanisms related to RNA interference regulate expression of the small temporal RNAs that control *C. elegans* developmental timing. Cell 2001; 106(1):23-34.
15. Hutvagner G, McLachlan J, Pasquinelli AE et al. A cellular function for the RNA-interference enzyme Dicer in the maturation of the let-7 small temporal RNA. Science 2001; 293(5531):834-838.
16. Knight SW, Bass BL. A role for the RNase III enzyme DCR-1 in RNA interference and germ line development in *Caenorhabditis elegans*. Science 2001; 293(5538):2269-2271.
17. Ketting RF, Fischer SE, Bernstein E et al. Dicer functions in RNA interference and in synthesis of small RNA involved in developmental timing in *C. elegans*. Genes Dev 2001; 15(20):2654-2659.
18. Sempere LF, Dubrovsky EB, Dubrovskaya VA et al. The expression of the let-7 small regulatory RNA is controlled by ecdysone during metamorphosis in *Drosophilaee melanogaster*. Dev Biol 2002; 244(1):170-179.
19. Lai EC. MicroRNAs are complementary to 3' UTR sequence motifs that mediate negative post-transcriptional regulation. Nat Genet 2002; 30(4):363-364.
20. Lai EC, Burks C, Posakony JW. The K box, a conserved 3' UTR sequence motif, negatively regulates accumulation of enhancer of split complex transcripts. Development 1998; 125(20):4077-4088.
21. Leviten MW, Lai EC, Posakony JW. The *Drosophilaee* gene Bearded encodes a novel small protein and shares 3' UTR sequence motifs with multiple enhancer of split complex genes. Development 1997; 124(20):4039-4051.
22. Lai EC, Posakony JW. The Bearded box, a novel 3' UTR sequence motif, mediates negative post-transcriptional regulation of Bearded and Enhancer of split Complex gene expression. Development 1997; 124(23):4847-4856.
23. Olsen PH, Ambros V. The lin-4 regulatory RNA controls developmental timing in *Caenorhabditis elegans* by blocking LIN-14 protein synthesis after the initiation of translation. Dev Biol 1999; 216(2):671-680.
24. Mourelatos Z, Dostie J, Paushkin S et al. miRNPs: a novel class of ribonucleoproteins containing numerous microRNAs. Genes Dev 2002; 16(6):720-728.
25. Melki J. Spinal muscular atrophy. Curr Opin Neurol 1997; 10(5):381-385.
26. Bohmert K, Camus I, Bellini C et al. AGO1 defines a novel locus of *Arabidopsis* controlling leaf development. EMBO J 1998; 17(1):170-180.
27. Fagard M, Boutet S, Morel JB et al. AGO1, QDE-2, and RDE-1 are related proteins required for post- transcriptional gene silencing in plants, quelling in fungi, and RNA interference in animals. Proc Natl Acad Sci USA 2000; 97(21):11650-11654.
28. Schwarz DS, Zamore PD. Why do miRNAs live in the miRNP? Genes Dev 2002; 16(9):1025-1031.
29. Baulcombe DC. RNA as a target and an initiator of post-transcriptional gene silencing in transgenic plants. Plant Mol Biol 1996; 32(1-2):79-88.
30. Baulcombe D. Viruses and gene silencing in plants. Arch Virol Suppl 1999; 15189-201.
31. Meins Jr F. RNA degradation and models for post-transcriptional gene-silencing. Plant Mol Biol 2000; 43(2-3):261-273.
32. Sijen T, Kooter JM. Post-transcriptional gene-silencing: RNAs on the attack or on the defense? Bioessays 2000; 22(6):520-531.

33. Bender J. A vicious cycle: RNA silencing and DNA methylation in plants. Cell 2001; 106(2):129-132.
34. Finnegan EJ, Wang M, Waterhouse P. Gene silencing: fleshing out the bones. Curr Biol 2001; 11(3):R99-R102.
35. Matzke M, Matzke AJ, Kooter JM. RNA: guiding gene silencing. Science 2001; 293(5532):1080-1083.
36. Matzke MA, Matzke AJ, Pruss GJ et al. RNA-based silencing strategies in plants. Curr Opin Genet Dev 2001; 11(2):221-227.
37. Vance V, Vaucheret H. RNA silencing in plants—defense and counterdefense. Science 2001; 292(5525):2277-2280.
38. Vaucheret H, Fagard M. Transcriptional gene silencing in plants: targets, inducers and regulators. Trends Genet 2001; 17(1):29-35.
39. Voinnet O. RNA silencing as a plant immune system against viruses. Trends Genet 2001; 17449-459.
40. Waterhouse PM, Wang M, Lough T. Gene silencing as an adaptive defence against viruses. Nature 2001; 411834-842.
41. Hamilton AJ, Baulcombe DC. A species of small antisense RNA in posttranscriptional gene silencing in plants. Science 1999; 286(5441):950-952.
42. Papaefthimiou I, Hamilton A, Denti M et al. Replicating potato spindle tuber viroid RNA is accompanied by short RNA fragments that are characteristic of post-transcriptional gene silencing. Nucleic Acids Res 2001; 29(11):2395-2400.
43. Kalantidis K, Psaradakis S, Tabler M et al. The occurrence of CMV-specific short RNAs in transgenic tobacco expressing virus-derived double-stranded RNA is indicative of resistance to the virus. Mol Plant Microbe Interact 2002; 15(8):826-833.
44. Cogoni C, Macino G. Isolation of quelling-defective (qde) mutants impaired in posttranscriptional transgene-induced gene silencing in *Neurospora crassa*. Proc Natl Acad Sci USA 1997; 94(19):10233-10238.
45. Cogoni C, Macino G. Posttranscriptional gene silencing in *Neurospora* by a RecQ DNA helicase. Science 1999; 286(5448):2342-2344.
46. Cogoni C, Macino G. Gene silencing in *Neurospora crassa* requires a protein homologous to RNA-dependent RNA polymerase. Nature 1999; 399(6732):166-169.
47. Fire A, Xu S, Montgomery MK et al. Potent and specific genetic interference by double-stranded RNA in *Caenorhabditis elegans*. Nature 1998; 391(6669):806-811.
48. Montgomery MK, Xu S, Fire A. RNA as a target of double-stranded RNA-mediated genetic interference in *Caenorhabditis elegans*. Proc Natl Acad Sci USA 1998; 95(26):15502-15507.
49. Tavernarakis N, Wang SL, Dorovkov M et al. Heritable and inducible genetic interference by double-stranded RNA encoded by transgenes. Nat Genet 2000; 24(2):180-183.
50. Kennerdell JR, Carthew RW. Use of dsRNA-mediated genetic interference to demonstrate that frizzled and frizzled 2 act in the wingless pathway. Cell 1998; 95(7):1017-1026.
51. Misquitta L, Paterson BM. Targeted disruption of gene function in *Drosophilaee* by RNA interference (RNA-i): a role for nautilus in embryonic somatic muscle formation. Proc Natl Acad Sci USA 1999; 96(4):1451-1456.
52. Tchurikov NA, Chistyakova LG, Zavilgelsky GB et al. Gene-specific silencing by expression of parallel complementary RNA in *Escherichia coli*. J Biol Chem 2000; 275(34):26523-26529.
53. Ngo H, Tschudi C, Gull K et al. Double-stranded RNA induces mRNA degradation in *Trypanosoma brucei*. Proc Natl Acad Sci USA 1998; 95(25):14687-14692.
54. Shi H, Djikeng A, Mark T et al. Genetic interference in *Trypanosoma brucei* by heritable and inducible double-stranded RNA. RNA 2000; 6(7):1069-1076.
55. Wargelius A, Ellingsen S, Fjose A. Double-stranded RNA induces specific developmental defects in zebrafish embryos. Biochem Biophys Res Commun 1999; 263(1):156-161.
56. Li YX, Farrell MJ, Liu R et al. Double-stranded RNA injection produces null phenotypes in zebrafish. Dev Biol 2000; 217(2):394-405.
57. Wianny F, Zernicka-Goetz M. Specific interference with gene function by double-stranded RNA in early mouse development. Nat Cell Biol 2000; 2(2):70-75.
58. Svoboda P, Stein P, Hayashi H et al. Selective reduction of dormant maternal mRNAs in mouse oocytes by RNA interference. Development 2000; 127(19):4147-4156.

59. Zamore PD, Tuschl T, Sharp PA et al. RNAi: double-stranded RNA directs the ATP-dependent cleavage of mRNA at 21 to 23 nucleotide intervals. Cell 2000; 101(1):25-33.
60. Yang D, Lu H, Erickson JW. Evidence that processed small dsRNAs may mediate sequence-specific mRNA degradation during RNAi in *Drosophilaee* embryos. Curr Biol 2000; 10(19):1191-1200.
61. Jacobsen SE, Running MP, Meyerowitz EM. Disruption of an RNA helicase/RNAse III gene in *Arabidopsis* causes unregulated cell division in floral meristems. Development 1999; 126(23):5231-5243.
62. Billy E, Brondani V, Zhang H et al. Specific interference with gene expression induced by long, double- stranded RNA in mouse embryonal teratocarcinoma cell lines. Proc Natl Acad Sci USA 2001; 98(25):14428-14433.
63. Hammond SM, Bernstein E, Beach D et al. An RNA-directed nuclease mediates post-transcriptional gene silencing in *Drosophilaee* cells. Nature 2000; 404(6775):293-296.
64. Hammond SM, Boettcher S, Caudy AA et al. Argonaute2, a link between genetic and biochemical analyses of RNAi. Science 2001; 293(5532):1146-1150.
65. Tabara H, Sarkissian M, Kelly WG et al. The rde-1 gene, RNA interference, and transposon silencing in *C. elegans*. Cell 1999; 99(2):123-132.
66. Tabara H, Yigit E, Siomi H et al. The dsRNA binding protein RDE-4 interacts with RDE-1, DCR-1, and a DExH- box helicase to direct RNAi in *C. elegans*. Cell 2002; 109(7):861-871.
67. Catalanotto C, Azzalin G, Macino G et al. Gene silencing in worms and fungi. Nature 2000; 404(6775):245.
68. Martinez J, Patkaniowska A, Urlaub H et al. Single-stranded antisense siRNAs guide target RNA cleavage in RNAi. Cell 2002; 110(5):563.
69. Hutvagner G, Zamore PD. A MicroRNA in a Multiple-Turnover RNAi Enzyme Complex. Science 2002; 11.
70. Lipardi C, Wei Q, Paterson BM. RNAi as random degradative PCR. siRNA primers convert mRNA into dsRNAs that are degraded to generate new siRNAs. Cell 2001; 107(3):297-307.
71. Sijen T, Fleenor J, Simmer F et al. On the role of RNA amplification in dsRNA-triggered gene silencing. Cell 2001; 107(4):465-476.
72. Schiebel W, Haas B, Marinkovic S et al. RNA-directed RNA polymerase from tomato leaves. II. Catalytic in vitro properties. J Biol Chem 1993; 268(16):11858-11867.
73. Schiebel W, Haas B, Marinkovic S et al. RNA-directed RNA polymerase from tomato leaves. I. Purification and physical properties. J Biol Chem 1993; 268(16):11851-11857.
74. Schiebel W, Pelissier T, Riedel L et al. Isolation of an RNA-directed RNA polymerase-specific cDNA clone from tomato. Plant Cell 1998; 10(12):2087-2101.
75. Dalmay T, Hamilton A, Rudd S et al. An RNA-dependent RNA polymerase gene in *Arabidopsis* is required for posttranscriptional gene silencing mediated by a transgene but not by a virus. Cell 2000; 101(5):543-553.
76. Mourrain P, Beclin C, Elmayan T et al. *Arabidopsis* SGS2 and SGS3 genes are required for post-transcriptional gene silencing and natural virus resistance. Cell 2000; 101(5):533-542.
77. Xie Z, Fan B, Chen C et al. An important role of an inducible RNA-dependent RNA polymerase in plant antiviral defense. Proc Natl Acad Sci USA 2001; 98(11):6516-6521.
78. Smardon A, Spoerke JM, Stacey SC et al. EGO-1 is related to RNA-directed RNA polymerase and functions in germ- line development and RNA interference in *C. elegans*. Curr Biol 2000; 10(4):169-178.
79. Wu-Scharf D, Jeong B, Zhang C et al. Transgene and transposon silencing in *Chlamydomonas reinhardtii* by a DEAH-box RNA helicase. Science 2000; 290(5494):1159-1162.
80. Martens H, Novotny J, Oberstrass J et al. RNAi in *Dictyostelium*: the role of RNA-directed RNA polymerases and double-stranded RNase. Mol Biol Cell 2002; 13(2):445-453.
81. Simmer F, Tijsterman M, Parrish S et al. Loss of the putative RNA-directed RNA polymerase RRF-3 makes *C. elegans* hypersensitive to RNAi. Curr Biol 2002; 12(15):1317.
82. de Wit T, Grosveld F, Drabek D. The tomato RNA-directed RNA polymerase has no effect on gene silencing by RNA interference in transgenic mice. Transgenic Res 2002; 11(3):305-310.
83. Nicholson RH, Nicholson AW. Molecular characterization of a mouse cDNA encoding Dicer, a ribonuclease III ortholog involved in RNA interference. Mamm Genome 2002; 13(2):67-73.
84. Elbashir SM, Lendeckel W, Tuschl T. RNA interference is mediated by 21- and 22-nucleotide RNAs. Genes Dev 2001; 15(2):188-200.

85. Elbashir SM, Harborth J, Lendeckel W et al. Duplexes of 21-nucleotide RNAs mediate RNA interference in cultured mammalian cells. Nature 2001; 411(6836):494-498.
86. Caplen NJ, Parrish S, Imani F et al. Specific inhibition of gene expression by small double-stranded RNAs in invertebrate and vertebrate systems. Proc Natl Acad Sci USA 2001; 98(17):9742-9747.
87. Leirdal M, Sioud M. Gene silencing in mammalian cells by preformed small RNA duplexes. Biochem Biophys Res Commun 2002; 295(3):744-748
88. Boutla A, Delidakis C, Livadaras I et al. Short 5'-phosphorylated double-stranded RNAs induce RNA interference in *Drosophilae*. Curr Biol 2001; 11(22):1776-1780.
89. Zhou Y, Ching YP, Kok KH et al. Post-transcriptional suppression of gene expression in *Xenopus* embryos by small interfering RNA. Nucleic Acids Res 2002; 30(7):1664-1669.
90. Gitlin L, Karelsky S, Andino R. Short interfering RNA confers intracellular antiviral immunity in human cells. Nature 2002; 418(6896):430-434.
91. Hu W, Myers C, Kilzer J et al. Inhibition of retroviral pathogenesis by RNA interference. Curr Biol 2002; 12(15):1301.
92. Jacque JM, Triques K, Stevenson M. Modulation of HIV-1 replication by RNA interference. Nature 2002; 418(6896):435-438.
93. Novina CD, Murray MF, Dykxhoorn DM et al. siRNA-directed inhibition of HIV-1 infection. Nat Med 2002; 8(7):681-686.
94. Kaufman RJ. Double-stranded RNA-activated protein kinase mediates virus-induced apoptosis: a new role for an old actor. Proc Natl Acad Sci USA 1999; 96(21):11693-11695.
95. Elbashir SM, Martinez J, Patkaniowska A et al. Functional anatomy of siRNAs for mediating efficient RNAi in *Drosophilae* melanogaser embryo lysate. EMBO J 2001; 20(23):6877-6888.
96. Parrish S, Fleenor J, Xu S et al. Functional anatomy of a dsRNA trigger. Differential requirement for the two trigger strands in RNA interference. Mol Cell 2000; 6(5):1077-1087.
97. Holen T, Amarzguioui M, Wiiger MT et al. Positional effects of short interfering RNAs targeting the human coagulation trigger Tissue Factor. Nucleic Acids Res 2002; 30(8):1757-1766.
98. Altschul SF, Madden TL, Schaffer AA et al. Gapped BLAST and PSI-BLAST: a new generation of protein database search programs. Nucleic Acids Res 1997; 25(17):3389-3402.
99. Paule MR, White RJ. Survey and summary: transcription by RNA polymerases I and III. Nucleic Acids Res 2000; 28(6):1283-1298.
100. Lee NS, Dohjima T, Bauer G et al. Expression of small interfering RNAs targeted against HIV-1 rev transcripts in human cells. Nat Biotechnol 2002; 20(5):500-505.
101. Miyagishi M, Taira K. U6 promoter-driven siRNAs with four uridine 3' overhangs efficiently suppress targeted gene expression in mammalian cells. Nat Biotechnol 2002; 20(5):497-500.
102. Donze O, Picard D. RNA interference in mammalian cells using siRNAs synthesized with T7 RNA polymerase. Nucleic Acids Res 2002; 30(10):e46.
103. Brummelkamp TR, Bernards R, Agami R. A system for stable expression of short interfering RNAs in mammalian cells. Science 2002; 296(5567):550-553.
104. McManus MT, Petersen CP, Haines BB et al. Gene silencing using micro-RNA designed hairpins. RNA 2002; 8(6):842-850.
105. Paddison PJ, Caudy AA, Bernstein E et al. Short hairpin RNAs (shRNAs) induce sequence-specific silencing in mammalian cells. Genes Dev 2002; 16(8):948-958.
106. Paul CP, Good PD, Winer I et al. Effective expression of small interfering RNA in human cells. Nat Biotechnol 2002; 20(5):505-508.
107. Sui G, Soohoo C, Affar el B et al. A DNA vector-based RNAi technology to suppress gene expression in mammalian cells. Proc Natl Acad Sci USA 2002; 99(8):5515-5520.
108. Yu JY, DeRuiter SL, Turner DL. RNA interference by expression of short-interfering RNAs and hairpin RNAs in mammalian cells. Proc Natl Acad Sci USA 2002; 99(9):6047-6052.
109. Boutla A, Kalantidis K, Tavernarakis N et al. Induction of RNA interference in *Caenorhabditis elegans* by RNAs derived from plants exhibiting post-transcriptional gene silencing. Nucleic Acids Res 2002; 30(7):1688-1694.
110. Napoli C, Lemieux C, Jorgensen R. Introduction of a chimeric chalcone synthase gene into petunia results in reversible cosuppression of homologous genes in trans. Plant Cell 1990; 7599-609.
111. van der Krol AR, Mur LA, Beld M et al. Flavonoid genes in petunia: addition of a limited number of gene copies may lead to a suppression of gene expression. Plant Cell 1990; 2(4):291-299.

CHAPTER 9

Post-Transcriptional Gene Silencing in Plants

Matthew A. Escobar and Abhaya M. Dandekar

Abstract

Accumulating genetic and biochemical evidence suggests that antisense-mediated gene silencing, cosuppression, RNA interference and virus-induced gene silencing are all unique inputs into a common RNA silencing pathway triggered by double stranded RNA. This pathway, termed post-transcriptional gene silencing (PTGS) is characterized by accumulation of 21-25 nt small-interfering RNAs, sequence-specific degradation of target mRNAs, and methylation of target gene sequences. PTGS appears to be ancient and highly conserved, as several groups of homologous genes required for silencing in plants, animals, and fungi have been identified. Though biochemical dissection of PTGS is still in its infancy, several key activities have been identified, such as Dicer, the endonuclease responsible for synthesis of short-interfering RNAs, and RISC, the nucleoprotein complex which mediates mRNA degradation. Several lines of evidence suggest that PTGS plays a key role in viral defense in plants, but further study is required to investigate the intriguing possibility that PTGS can act as an endogenous gene control mechanism.

Introduction

The characterization of antisense RNA-mediated controls of plasmid replication and maintenance in prokaryotes prompted several studies of mRNA silencing in eukaryotic genes in the mid-1980s.[1,2] Rothstein et al (1987) first demonstrated an "antisense effect" in intact plants by silencing an integrated nopaline synthase (*nos*) transgene through expression of antisense *nos* RNA in tobacco.[3] Interestingly, this eukaryotic antisense silencing appeared to result from increased turnover of the targeted *nos* mRNA, in contrast to the inhibitory effects on translation[4] or DNA synthesis[5] commonly associated with antisense RNA in bacteria. A similar homology-dependent RNA degradation phenomenon, resulting from the over expression of an endogenous gene in transgenic plants, was discovered soon thereafter.[6,7] These phenomena, termed antisense-mediated gene silencing and cosuppression, became powerful techniques for plant improvement and functional gene analysis, but relatively little was understood about their underlying mechanisms.[8]

The seminal discoveries that double stranded RNA (dsRNA) and plant viruses are potent initiators of gene silencing has recently invigorated basic research on silencing mechanisms.[9,10] It has become apparent that the diverse threads of antisense-mediated gene silencing, cosuppression, dsRNA-mediated gene silencing (RNA interference), and virus-induced gene silencing converge with the synthesis of dsRNA.[11] This dsRNA is digested to 21-25 nt small-interfering RNAs (siRNAs) which play an integral role in homology-dependent RNA

Noncoding RNAs: Molecular Biology and Molecular Medicine, edited by Jan Barciszewski and Volker A. Erdmann.

turnover and methylation of homologous DNA coding sequences in plants.[12] This dsRNA-triggered sequence-specific RNA degradation pathway has been termed post-transcriptional gene silencing (PTGS). Studies of PTGS mutants and cell-free extracts capable of in vitro PTGS have demonstrated that PTGS is genetically and mechanistically conserved among plants, animals, and fungi, suggesting an ancient origin prior to the divergence of these kingdoms.[13] In plants, PTGS appears to be a significant component of viral defense and possesses functional and mechanistic similarities to other epigenetic gene control mechanisms, such as paramutation and transcriptional gene silencing.[14] In the following review, we discuss the current understanding of natural and artificial RNA inducers of PTGS in plants, models of the RNA degradation mechanism, and potential roles of PTGS in plants.

Initiation of PTGS in Plants Many Roads to dsRNA

Antisense Mediated Gene Silencing

Expression of antisense RNA from integrated transgenes was the first, and until recently the most widespread, method of initiating PTGS in plants.[8] Antisense constructs, usually consisting of an inverted gene coding sequence driven by a strong constitutive promoter, are moderately efficient inducers of PTGS, with ~5-20% of transformed individuals displaying a reduction in target mRNA accumulation.[15,16] The extent of mRNA suppression is quite variable from individual to individual, ranging from no detectable effect to >99% reduction in steady state RNA levels.[8] In addition, silencing is highly dependent upon antisense RNA:mRNA homology, with decreasing nucleotide sequence identity (to ~70%) minimizing the extent of mRNA suppression.[17] Nuclear run-on assays have demonstrated that antisense-silenced lines are generally not deficient in transcription of the target mRNA, indicating that silencing is a post-transcriptional effect.[8]

The presence of abundant double stranded siRNA in antisense-silenced plants strongly supports the long-standing hypothesis that homologous pairing of sense and antisense RNAs produces dsRNA in vivo, and that this dsRNA is the primary initiator of silencing.[18,19] Although sense:antisense RNA duplex formation is likely the primary mechanism of dsRNA production in antisensed plants, Stam et al (2000) have reported that dsRNA can also arise from the read through transcription of antisense transgene constructs integrated as inverted repeats, resulting in production of self-complementary RNA molecules.[20] Presumably the frequency and extent of PTGS in antisense RNA-expressing plant lines is dependent upon the efficiency of RNA:RNA hybridization, which may be affected at the individual gene level by secondary mRNA structure and at the individual transformant level by the abundance of antisense RNA.[18]

RNA Interference (RNAi)

In 1998, Fire et al demonstrated that dsRNA is a substantially more efficient inducer of PTGS in *Caenorhabditis elegans* than sense or antisense RNA alone.[9] Microinjection of substoichiometric levels of dsRNA into the gut of *C. elegans* triggered highly-specific silencing of the targeted endogenous genes in 100% of tested individuals. Later studies determined that dsRNA-induced silencing occurs post-transcriptionally, through degradation of homologous mRNA.[21] This highly efficient induction of PTGS by direct production of dsRNA, termed RNAi, was subsequently demonstrated in plants through expression of inverted-repeat transgenes or through simultaneous expression of sense and antisense transgenes.[22] Several groups have now reported near-absolute suppression of various endogenous and transgene mRNA species in >90% of transformants by the expression of complementary gene sequences separated by a nonhomologous linker region or a spliceable intron.[23,24] The stem-loop or hairpin RNAs resulting from transcription of these constructs must possess a dsRNA region of at least ~100 nt to efficiently induce PTGS in plants, which likely explains why the limited secondary structure of endogenous mRNAs does not trigger silencing.[16] Predictably, abundant siRNA is associated with RNAi in both plants and animals.[13,16] The exceptional potency and efficiency of dsRNA as an inducer of PTGS was recently exemplified by the systematic functional analysis of *C. elegans* chromosomes I and III through feeding studies with dsRNA-expressing *E. coli* bacteria.[25,26]

Cosuppression

Cosuppression was first described in two similar studies of the chalcone synthase (*chs*) gene in *Petunia*.[6,7] Attempts to increase gene expression by introduction of an additional copy of *chs* fused to a strong, constitutive promoter unexpectedly produced a significant number of transgenic plants displaying decreased abundance of both transgenic and endogenous *chs* mRNA, as manifested by wholly or partially white flower petals. This silencing caused by over expression of an endogenous gene was termed cosense-suppression, or cosuppression. Like antisense silencing and RNAi, cosuppression is a post-transcriptional, homology-dependent process associated with siRNA accumulation.[15,19] As with antisense RNA, sense RNA transgene constructs are only moderately efficient inducers of PTGS (silencing in ~5-20% of transformants).[15,16] However, unlike antisense RNA and RNAi, the high-level expression of sense RNA provides no apparent mechanism for production of a dsRNA trigger of PTGS. Although read through transcription of inverted transgene inserts or spurious transcription of antisense RNAs from cryptic transcription start sites could potentially account for cosuppression in some cases,[27] genetic evidence suggests that cosuppression is mechanistically separable from antisense silencing and RNAi.

Analysis of *Arabidopsis thaliana* mutant lines which are partially deficient in PTGS have revealed several components of a catalytic pathway that could produce dsRNA from overabundant sense RNA. Mutagenesis of plant lines displaying stable cosuppression of the GUS or GFP marker genes resulted in characterization of four unique genes that were required for silencing.[28-31] The *sde1/sgs2* gene is highly homologous to plant RNA-dependent RNA polymerases (RdRP), and similar genes are required for silencing in *Neurospora (qde1)* and *C. elegans (ego1, rrf1)*. Similarly, the putative helicase gene *sde3*, has homologs in *Chlamydomonas (mut6)* and *C. elegans (smg2)* that are also required for silencing. A putative translation initiation factor *(ago1)* with homology to the silencing effectors *qde2 (Neurospora)* and *rde1 (C. elegans)* was also necessary for silencing, as was *sgs3*, a gene with no known homologs which may be unique to silencing in plants. The cross-kingdom conservation in sequence and function of these genes strongly indicates that they are ancient components of the gene silencing pathway. However, these genes appear to uniquely affect PTGS initiated by sense RNA, as recent studies have demonstrated that RNAi and virus-induced gene silencing (see below) is unaffected in *sde1/sgs2, sde3, ago1*, and *sgs3* mutants.[32] These results suggest that a unique catalytic mechanism, likely involving *sde1/sgs2*-mediated synthesis of antisense RNA from a sense mRNA template, mediates production of dsRNA in the cosuppression pathway. How specific sense mRNA species are targeted for this catalytic process is unknown, though various models have invoked the presence of an mRNA accumulation threshold or the production of ill-defined "aberrant RNA" molecules from highly expressed transgenes.[15]

In addition to the unique catalytic elements related to initiation of PTGS, cosuppressed plant lines also display a striking systemic spread of PTGS termed systemic acquired silencing.[33] Voinnet and Baulcombe (1997) showed that transient ectopic expression of a sense GFP transgene in a single leaf of GFP-expressing *Nicotiana benthamiana* plants caused a local initiation of PTGS which gradually spread throughout the plant, resulting in almost complete silencing.[34] The spread of PTGS mirrored the movement of cytoplasmic RNA viruses, with cell-to-cell transmission presumably through plasmodesmata and long-distance transmission through phloem. Similarly, several groups have demonstrated that cosuppression is graft-transmissible to nonsilenced plants.[33,35,36] The systemic silencing signal is apparently stable and pervasive, as PTGS was transmitted from a GUS-cosuppressed rootstock to a GUS-expressing scion through 30 cm of wild-type (nonGUS-expressing) stem.[33] Plants displaying antisense-mediated PTGS and RNAi appear incapable of systemic silencing (Escobar and Dandekar, unpublished results).[36] This suggests that a signal required for systemic silencing is generated during the catalytic reactions specific to dsRNA production in the cosuppression pathway. The nature of this systemic silencing signal is unknown, but the sequence-specific nature of the signal suggests the existence of a nucleic acid component.

Virus-Induced Gene Silencing (VIGS)

Nepoviruses, caulimoviruses, and several other types of RNA virus induce an unusual pattern of symptom development, called recovery, on some host plants.[10] Infected plants initially display severe localized symptom development and high viral load, but tissues which emerge subsequent to the initial infection are a symptomatic and show very low virus accumulation. These "recovered" tissues are resistant to subsequent infection by the inducing virus as well as other closely related viruses. Ratcliff et al (1997) demonstrated that viral recovery is mediated by a PTGS-like RNA degradation mechanism which is post-transcriptional and homology dependent.[10,37] The finding that tobacco plants infected with potato virus X (PVX) accumulate PVX-homologous siRNA conclusively demonstrated that plant viruses can be potent initiators and targets of PTGS.[12] Subsequently, several laboratories have developed recombinant viral vectors designed to initiate PTGS in plants.[37,38] Silencing of transgenes and endogenous genes can be efficiently induced by the integration of as little as 23 nt of target gene sequence into these viral vectors.[39]

The utilization of PTGS-inducing recombinant viral vectors has allowed the dissection of several unique characteristics of VIGS. Like cosuppression, VIGS is systemic, as demonstrated by the spread of silencing from a single leaf infiltrated with a coat protein mutant (cell autonomous) recombinant PVX vector.[40] The same study showed that the viral protein p25 is capable of suppressing the systemic silencing effect caused by either cosuppression or VIGS. Importantly, p25 had a differential effect on the local induction of silencing in the infiltrated leaf depending on whether the inducer of PTGS was sense RNA or virus. The high-efficiency induction of local silencing by over expression of sense mRNA was completely blocked by p25, but induction of PTGS in tissues expressing the viral vector was unaffected.[40] In combination with mutant studies showing that VIGS can operate independent of several plant-encoded enzymes which are required for cosuppression (see above), these results suggest that viruses may produce dsRNA via two independent pathways.[40] The direct production of dsRNA viral replication intermediates by the virus-encoded RdRP likely provides the *sde1/sgs2*-independent, p25-resistant inducer of PTGS. In addition, abundant single stranded viral genomic RNA and mRNA could independently trigger dsRNA synthesis through the plant-encoded, cosuppression-type pathway, thus leading to production of the systemic silencing signal.

A General Model for dsRNA Production in Plants

Homology dependent RNA degradation and siRNA, the key elements that unify antisense-mediated gene silencing, cosuppression, RNAi, and VIGS, are thought to act downstream of dsRNA production in the PTGS pathway.[32,41] The evidence above suggests that the various initiators of PTGS in plants differ primarily in their mechanism(s) of dsRNA production.[40,42] Cosuppression, and to some extent VIGS, appear to produce dsRNA from sense RNA via a catalytic mechanism requiring the plant genes *sde1/sgs2*, *sde3*, *ago1*, and *sgs3*. This "indirect" pathway is wholly inhibited by the p25 viral protein and generates an as yet uncharacterized systemic silencing signal. In contrast, antisense-mediated silencing, RNAi, and VIGS produce dsRNA through homologous pairing or through catalytic reactions independent of

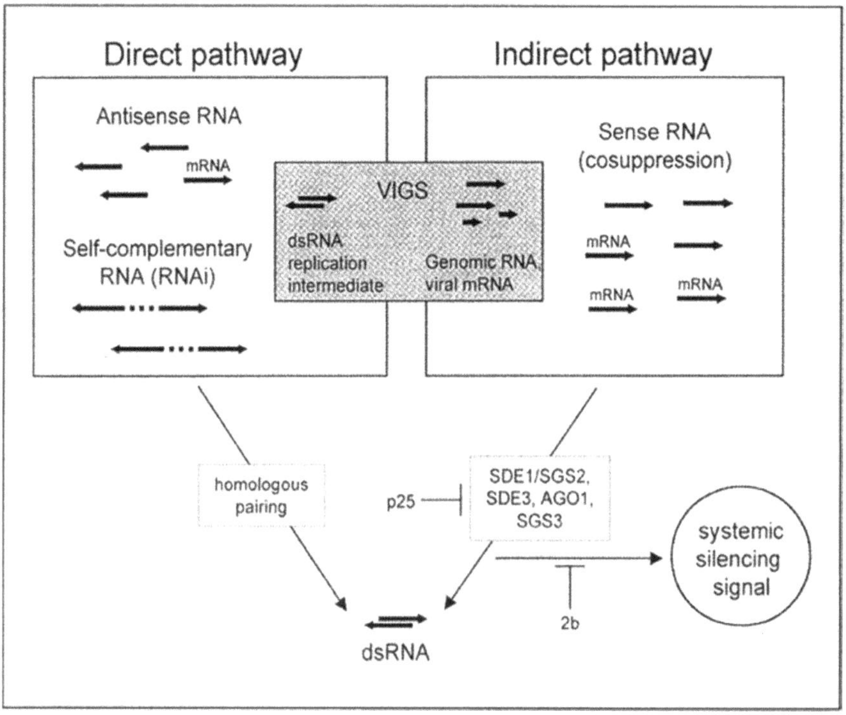

Figure 1. Mechanisms of dsRNA production in plants. Analyses of PTGS mutants and viral suppressors of PTGS suggest that two primary pathways of dsRNA production exist in plants. In the direct pathway, dsRNA is generated through pairing of homologous RNA molecules (antisense, RNAi), or through synthesis of dsRNA viral replication intermediates by virus-encoded RdRPs (VIGS). Alternatively, production of dsRNA by the indirect pathway is reliant upon recognition of an abundant sense RNA species and subsequent synthesis of a complementary antisense strand by the plant encoded genes *sde1/sgs2, sde3, ago1,* and *sgs3* (cosuppression, VIGS). The indirect pathway generates a systemic silencing signal and is blocked by the p25 viral suppressor. For further details, see accompanying text.

the plant-encoded genes described above. This "direct" pathway is p25-resistant and does not generate systemic silencing. A simplified binary model of dsRNA production in plants is presented in (Fig. 1)

From dsRNA to Silencing Models of the RNA Degradation Pathway

The majority of significant insights regarding the RNA degradation mechanisms of PTGS have come from animal models, especially *C. elegans* and RNAi-competent cell-free *Drosophilae* extracts. Although PTGS appears to be highly conserved across kingdoms (see Table 1), it is possible that the specific mechanisms of PTGS in plants could differ significantly from the models described below and in (Fig. 2)

Dicer and Primary siRNA

Cell-free extracts from *Drosophilae* embryos or S2 cells perform rapid, ATP-dependent cleavage of exogenously supplied dsRNA into discrete 21-22 nt siRNA molecules.[43,44] Structurally, these siRNAs consist of a 19-20 nt double stranded region with 3' terminal single-stranded

Table 1. Common elements of PTGS in animals, fungi, and plants

	Caenorhabditis elegans	*Neurospora crassa*	*Arabidopsis thaliana*
Initiators:			
sense RNA	+	+	+
antisense RNA	+	?	+
double-stranded RNA	+	?	+
viruses	?	?	+
Mechanism components:			
siRNA	+	?	+
RdRP	EGO1, RRF1	QDE1	SDE1/SGS2
RNA helicase	SMG2	?	SDE3
RNAse III	DCR	?	CAF?, T25K16.47
Argonaute gene family	AGO2, RDE1	QDE2	AGO1

tails of 2 nt.[45] This structure is characteristic of the products of RNAse III endonuclease digestion of dsRNA templates.[45]

Recently, Bernstein et al (2001) identified an RNAse III nuclease in *Drosophilae* which specifically cleaves dsRNA into ~22 nt fragments in vitro.[46] This ATP-dependent nuclease, called Dicer, possesses an amino terminal helicase domain, a PAZ domain, two RNAse III motifs, and a dsRNA binding motif. The function of glutamine-rich PAZ domains has not been experimentally determined, but this domain is shared with several members of the Argonaut gene family that are required for PTGS in various organisms.[46] Dicer displays decreased activity on dsRNA substrates of less than 200 nt, which correlates with dsRNA size requirements for RNAi in *Drosophilae*. In addition, a 6-7 fold decrease in Dicer activity in vivo substantially reduces RNAi competence. These results suggest that Dicer is a key element in the synthesis of siRNA from dsRNA.[46] Bioinformatic analyses have identified three unique genes encoding proteins with domain architecture identical to *Drosophilae* Dicer-1: *C. elegans* K12H4.8, *Arabidopsis* T25K16.4, and *Arabidopsis carpel factory (caf)* (Escobar and Dandekar, unpublished results).[46] K12H4.8 was recently characterized as a functional ortholog of Dicer which is required for RNAi in *C. elegans*.[47] The putative *Arabidopsis* Dicer orthologs T25K16.4 and CAF are 97% identical at the amino acid level, and previous developmental studies of *caf* mutants have suggested that this protein suppresses cell division in floral meristems.[48] The functional importance of these proteins for PTGS in *Arabidopsis* is currently being investigated.

The RNA-Induced Silencing Complex (RISC)

Dicer is biochemically separable from the sequence-specific RNA degradation activity associated with PTGS, suggesting that the processes of siRNA production and mRNA turnover are performed by unique groups of enzymes.[44] Hammond et al (2000) purified a ~500 kD nucleoprotein complex, termed RISC, that mediates sequence-specific RNA degradation in vitro.[44] The partial characterization of RISC has revealed a 22-25 nt RNA component (presumably siRNA) and a 130 kD protein component (Agronaute2), both of which are required for sequence-specific RNA degradation.[44,49] Though the identity of the nuclease component has not yet been determined, the catalytic activity of RISC has been described in some detail. Exogenous application of synthetic 21-22 nt dsRNA molecules with structural characteristics identical to siRNA (see above) is sufficient to trigger silencing of homologous mRNA molecules in the in vitro *Drosophilae* system.[45] By introducing a single synthetic siRNA species,

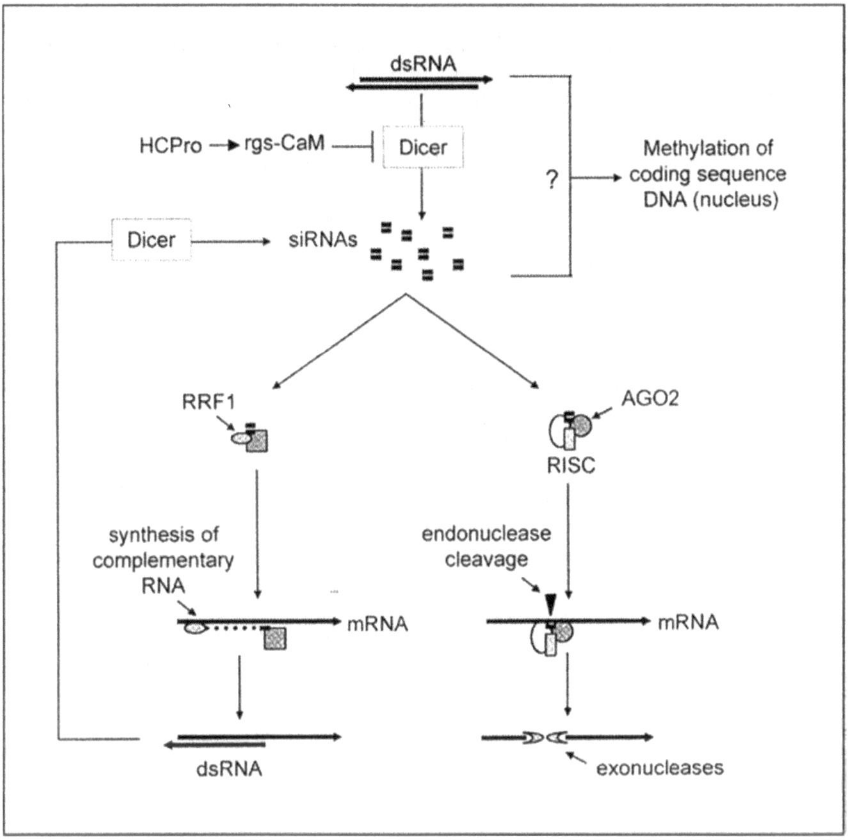

Figure 2. A model for the mode of action of PTGS. Regions of dsRNA of >100 nt are digested by the RNAse III-type endonuclease Dicer, generating 21-25 nt siRNAs. siRNAs can be integrated into the nucleoprotein RNA-induced silencing complex (RISC), providing sequence specificity to the mRNA degradation activity. siRNAs can also cooperate with RRF-1, hybridizing with mRNA molecules to prime synthesis of an antisense RNA strand. Resultant secondary dsRNA can be digested by Dicer, amplifying the population of siRNAs. Many examples of PTGS are characterized by an RNA-directed DNA methylation component, but no specific mechanism has been determined for this activity. For further details, see accompanying text.

Elbashir et al (2001) were able to precisely map the resultant cleavage site on homolgous mRNA molecules.[45] Digestion of the mRNA occurred at a single site corresponding to the center of the siRNA molecule (11-12 nt from the 3' end of the siRNA). Presumably, the antisense strand of the siRNA component of RISC hybridizes with complementary mRNA, forming a short RNA duplex which is subsequently cleaved.[45] This endonuclease reaction would generate substrates for nonspecific cellular exonucleases which could fully digest the mRNA.

Secondary siRNA

Several aspects of PTGS cannot be readily explained by a simple model which posits that the catalysis of dsRNA produces a population of primary siRNAs which act as the sole determinants of RISC specificity. For example, it is difficult to explain the massive RNA degradation response triggered in *C. elegans* by microinjection of tiny amounts of short (400-500 bp) dsRNA without the existence of some mechanism of dsRNA amplification.[9] Further, several studies have shown that PTGS can target RNA sequences outside the original dsRNA inducer molecule. Sijen et al (2001) reported that a transcriptional fusion of the endogenous gene *unc-22* and GFP (*unc-22*::GFP) could be silenced in *C. elegans* by microinjection of GFP dsRNA.[50] This result was not unexpected, as the GFP sequences of the *unc-22*::GFP mRNA molecule should be degraded, destabilizing the entire transcript. However, the endogenous *unc-22* gene (which possesses no homology to GFP) was also silenced, suggesting that target sequences for silencing were somehow expanded through some interaction with the *unc-22*::GFP transcript. This phenomenon was termed transitive silencing. Subsequent RNAse protection experiments showed accumulation of siRNAs homologous to *unc-22*, with a higher abundance of *unc-22* sequences which lay closer to the *unc-22*/GFP junction in the fusion transcript.[50] Mutant analyses demonstrated that the RNA-dependent RNA polymerase rrf-1 was required for production of these "secondary" siRNAs which could not have arisen directly from digestion of the introduced dsRNA molecule. In addition, rrf-1 mutants displayed a large decrease in total siRNA accumulation and were incapable of RNAi in somatic tissues.

Based upon this data, a model was proposed in which a relatively small population of primary siRNAs is derived from direct digestion of the introduced dsRNA molecules.[50] These primary siRNAs can pair with homologous mRNA and directly or indirectly prime extension of an antisense RNA strand in the 5' to 3' direction by rrf-1. This catalytic activity would produce more dsRNA, ultimately generating a large population of secondary siRNAs, some of which would extend beyond the boundaries of the original dsRNA trigger. This amplification effect would appear to be required for PTGS, at least in the somatic tissues of *C. elegans*. However, the fact that transitive silencing has been described in *Nicotiana benthamiana*[51] suggests that a similar mechanism operates in plants, though it may play a less vital role in plants constitutively producing large amounts of dsRNA from integrated transgenes. It is also possible that systemic silencing in plants is caused by a related amplification effect in which a mobile signal molecule produced at the local PTGS initiation site primes *de novo* dsRNA synthesis from homologous mRNA templates in distant tissues.

DNA Methylation Induced by PTGS

In addition to a cytoplasmic, sequence-specific mRNA degradation activity, PTGS is also commonly associated with an increase in methylation of the target gene coding sequence in plants. Methylation of integrated transgenes has been correlated with antisense-silencing, cosuppression, RNAi, and VIGS, though endogenous genes may not be subject to this silencing-induced methylation.[52,53] The fact that cytoplasmic RNA viral vectors can cause methylation of target genes in the nucleus suggests that methylation is mediated by a mobile, RNA-containing effecter, as has previously been hypothesized for viroid-induced methylation.[54]

The importance of methylation for initiation and maintenance of PTGS in plants remains unclear. Mutations in the *Arabidopsis* genes *ddm1*, a chromatin remodeling factor, or *met1*, a DNA methyltransferase, cause significant genome-wide reductions in DNA methylation. When *ddm1*-mutant *Arabidopsis* lines are crossed to lines displaying cosuppression of an integrated GUS transgene, a small fraction of the resultant progeny show a complete loss of silencing.[55] Alternatively, a relatively high percentage of progeny from a cross between *met1*-mutant and GUS-silenced lines showed localized loss of PTGS in developing tissues at some time during development.[55] Thus it would appear that methylation plays some role in the maintenance of PTGS in plants. However, this hypothesis is not supported by the results of Wang and Waterhouse (2000), who showed that chemical demethylation of DNA by 5-azacytidine treatment did not release PTGS of a GUS transgene in rice calli.[52] Clearly, further studies will be necessary to determine the mechanism and functional importance of methylation induced by PTGS.

Roles of PTGS in Plants

Defense Against Viruses and Viroids

Several lines of evidence suggest that PTGS is an important defense mechanism against invasive nucleic acid parasites. DNA viruses, RNA viruses, and viroids are all initiators of PTGS in plants.[38,56] Accordingly, PTGS mediates viral recovery and cross protection, whereby plants display a reduction in disease symptoms and develop *de novo* resistance to infection by viruses closely related to the primary inoculum.[10] Further, loss of PTGS function in *sde1/sgs2, sde3, ago1*, and *sgs3* mutants results in hypersensitivity to infection by certain viruses.[32] However, the strongest evidence for an antiviral role of PTGS is the existence of viral proteins which suppress specific PTGS mechanisms in plants.

Although suppression of PTGS appears to be a widespread strategy among plant viruses,[57] only three viral PTGS suppressors have been studied in detail: p25, 2b, and HC-Pro. As previously described, the potato virus X movement protein p25 presumably suppresses some aspect of the "indirect" dsRNA synthesis pathway (Fig. 1), blocking systemic silencing and local silencing induced by cosuppression.[40] Alternatively, the cucumoviral 2b protein has no discernible effect on local silencing, but prevents establishment of PTGS in tissues which emerge subsequent to viral infection.[58] Thus, 2b most likely prevents synthesis or translocation of the systemic silencing signal and/or prevents the signal from initiating PTGS in distant tissues.[59] Perhaps most interesting is the potyviral HC-Pro protein, which abolishes local silencing and siRNA accumulation, but does not block PTGS-associated methylation or systemic transmission of silencing to tissues which are not expressing HC-Pro.[58] Yeast two-hybrid studies have shown that HC-Pro physically interacts with the calmodulin-related plant protein *rgs-CaM*.[60] HC-Pro upregulates expression of *rgs-CaM* in vivo, and *rgs-CaM* alone has PTGS suppressor activity, so HC-Pro may operate primarily through manipulation of endogenous silencing-suppression pathways.[60]

Genome Stability

Indirect evidence suggests that PTGS may help to maintain genome integrity by suppressing the activity of transposable elements. Transposons in plant genomes are generally located in heterochromatic regions that are highly methylated and transcriptionally inactive. There is a correlation between decreased genome methylation and increased transposon activity in the *ddm1* and *met1* mutants of *Arabidopsis*, and these mutants also display defects in maintenance of PTGS, as described above.[55] Similarly, the PTGS-deficient mutants *mut7, rde2*, and *rde3* in *C. elegans* and mut6 in *Chlamydomonas* display transposon activation.[61] In addition, the cloning and sequencing of siRNAs from tobacco has revealed a population of siRNAs homologous to integrated retrotransposon sequences.[61] As both PTGS and transcriptional gene silencing produce siRNAs and induce DNA methylation, it is likely that one or both of these pathways is involved in suppression of transposon activity in plants.

Regulation of Endogenous Gene Expression

It is tempting to speculate that an ancient and highly conserved pathway such as PTGS could play a broader role in the control of endogenous gene expression. At present there is no clear data to verify this hypothesis in plants, but several lines of evidence support the possibility. Mutants of some known and putative PTGS pathway components, notably *ago1* and *caf*, display highly abnormal development, as do transgenic plants over expressing the endogenous PTGS suppressor *rgs-CaM*.[30,48,60] Thus, it is possible that PTGS plays a role in development, but it is equally possible that *ago1, caf*, and *rgs-CaM* have separate developmental and PTGS-related activities. Naturally-occurring antisense RNA transcripts have been described for several endogenous genes in plants, but the functional importance of these antisense RNAs has not been determined.[8] Currently, the only example of PTGS-like control of endogenous gene expression is the small temporal RNA (stRNA) system in *C. elegans*.[47] The *let-7* and *lin-4* genes of *C. elegans* produce short, self-complementary mRNAs which form stem-loop structures in vivo. The duplex RNA sequence is digested into ~21 nt fragments by Dicer, and these stRNAs suppress expression of various homologous mRNA targets which are involved in developmental timing. However, stRNA is thought to operate by hindering translation of mRNA rather than by the RNA degradation mechanism associated with siRNA and PTGS.[47] Thus, several qualitatively different epigenetic control pathways may intersect with PTGS to coordinate endogenous gene regulation.

Conclusions

More has been learned about PTGS in the last four years than in the preceding fourteen years, dating back to the first demonstration of antisense-silencing in eukaryotes (Izant and Weintraub, 1984). In the near future, biochemical approaches using cell-free PTGS systems should allow further characterization of the nucleoprotein complexes mediating RNA degradation, DNA methylation, and secondary amplification of dsRNA, which should facilitate corresponding discoveries in plants. Still, many mysteries remain. Key questions of importance for students of PTGS in plants will be (1) What is the nature of the systemic silencing signal? (2) What is the functional relationship between methylation and PTGS? and (3) Does PTGS play a role in endogenous gene regulation? Further elucidation of the basic mechanisms of PTGS in the future should allow a continued refinement of gene silencing as a tool for functional genomics,[26] plant biotechnology,[62] and even human medicine.[63]

References

1. Izant JG, Weintraub H. Inhibition of thymidine kinase gene expression by antisense RNA: a molecular approach to genetic analyses. Cell 1984; 36:1007-1015.
2. Ecker J, Davis R. Inhibition of gene expression in plant cells by expression of antisense RNA. Proc Natl Acad Sci USA 1986; 83:5372-5376.
3. Rothstein SJ, DiMaio J, Strand M et al. Stable and heritable inhibition of the expression of nopaline synthase in tobacco expressing anti sense RNA. Proc Natl Acad Sci USA 1986; 84:8439-8443.
4. Simons RW, Kleckner N. Translational control of IS10 transposition. Cell 1983; 34:683-691.
5. Tomizawa J, Itoh T. The importance of RNA secondary structure in ColE1 primer formation. Cell 1982; 31:575-583.
6. Napoli C, Lemieux C, Jorgensen R. Introduction of a chimeric chalcone synthase gene into petunia results in reversible cosuppression of homologous genes in trans. Plant Cell 1990; 2:279-289.
7. Van der Krol AR, Mur LA, Beld M et al. Flavonoid genes in petunia: addition of a limited number of gene copies may lead to a suppression of gene expression. Plant Cell 1990; 291-299.
8. Bourque JE. Antisense strategies for genetic manipulation in plants. Plant Sci 1995; 105:125-149.
9. Fire A, Xu S, Montgomery MK et al. Potent and specific genetic interference by double-stranded RNA in *Caenorhabditis elegans*. Nature 1998; 391:806-811.
10. Ratcliff F, Harrison BD, Baulcombe DC. A similarity between viral defense and gene silencing in plants. Science 1997; 276:1558-1560.
11. Vance V, Vaucheret H. RNA silencing in plants- defense and counter defense. Science 2001; 292:2277-2280.
12. Hamilton AJ, Baulcombe DC. A species of small antisense RNA in posttranscriptional gene silencing in plants. Science 1999; 286:950-952.
13. Sharp PA. RNA interference- 2001. Genes Devel 2001; 15:485-490.
14. Mette MF, Aufsatz W, van der Winden J et al. Transcriptional silencing and promoter methylation triggered by double-stranded RNA. EMBO J 2000; 19:5194-5201.
15. Baulcombe DC. RNA as an initiator of post-transcriptional gene silencing in transgenic plants. Plant Mol Biol 1996; 32:79-88.

16. Wesley SV, Helliwell CA, Smith NA et al. Construct design for efficient, effective, and high-throughput gene silencing in plants. Plant J 2001; 27:581-590.
17. Elomaa P, Helariutta Y, Kotilainen M et al. Transformation of antisense constructs of the chalcone synthase gene superfamily into *Gerbera hybrida*: differential effect on the expression of family members. Mol Breed 1996; 2:41-50.
18. Nellen W, Lichtenstein C. What makes mRNA anti-sensitive? Trends Biochem Sci 1993; 18:419-423.
19. Di Serio F, Schob H, Iglesias A et al. Sense and antisense-mediated gene silencing is inhibited by the same viral suppressors and is associated with accumulation of small RNAs. Proc Natl Acad Sci USA 2001; 98:6506-6510.
20. Stam M, deBruin R, van Blokland R et al. Distinct features of post-transcriptional gene silencing by antisense transgenes in single copy and inverted T-DNA repeat loci Plant J 2000; 21:27-42.
21. Mongomery MK, Xu S, Fire A. et al. RNA as a target of double-stranded RNA-mediated genetic interference in *Caenorhabditis elegans*. Proc Natl Acad Sci USA 1998; 95:15502-15507.
22. Waterhouse PM, Graham MW, Wang M. Virus resistance and gene silencing in plants can be induced by simultaneous expression of sense and antisense RNA. Proc Natl Acad Sci USA 1998; 95:13959-13964.
23. Chuang C, Meyerowitz EM. Specific and heritable genetic interference by double-stranded RNA in *Arabidopsis thaliana*. Proc Natl Acad Sci USA 2000; 97:4985-4990.
24. Smith NA, Singh SP, Wang M et al. Total silencing by intron-spliced hairpin RNAs. Nature 2000; 407:319-320.
25. Gonczy P, Echeverri C, Oegema K et al. Functional genomic analysis of cell division in *C. elegans* using RNAi of genes on chromosome III. Nature 2000; 408:331-336.
26. Fraser AG, Kamath RS, Zipperlen P et al. Functional genomic analysis of *C. elegans* chromosome I by systematic RNA interference. Nature 2000; 408:325-330.
27. Montgomery MK, Fire A. Double stranded RNA as a mediator in sequence specific genetic silencing and cosuppression. Trends Genet 1998; 14:255-258.
28. Mourrain P, Beclin C, Elmayan T et al. *Arabidopsis* SGS2 and SGS3 genes are required for post-transcriptional gene silencing and natural virus resistance. Cell 2000; 101:|533-542.
29. Dalmay T, Hamilton A, Rudd S et al. An RNA-dependent RNA polymerase gene in *Arabidopsis* is required for posttranscriptional gene silencing mediated by a transgene but not by a virus. Cell 2000; 101:543-553.
30. Fagard M, Boutet S, Morel J-B et al. AGO1, QDE-2 and RDE-1 are related proteins required for post-transcriptional gene silencing in plants, quelling in fungi, and RNA interference in animals. Proc Natl Acad Sci USA 2000; 97:11650-11654.
31. Dalmay T, Horsefield R, Brautenstein TH et al. SDE3 encodes an RNA helicase required for post-transcriptional gene silencing in *Arabidopsis*. EMBO J 2001; 20:2067-2077.
32. Vaucheret H, Beclin C, Fagard M. Post-transcriptional gene silencing in plants. J Cell Sci 2001; 114:3083-3091.
33. Palauqui J-C, Elmayan T, Pollien J-M et al. Systemic acquired silencing: transgene specific post-transcriptional gene silencing is transmitted by grafting from silenced stocks to nonsilenced scions. EMBO J 1997; 16:4738-4745.
34. Voinnet O, Baulcombe DC. Systemic signaling in gene silencing. Nature 1997; 389:553.
35. Sonoda S, Nishiguchi M. Graft transmission of post-transcriptional gene silencing: target specificity for RNA degradation is transmissible between silenced and nonsilenced plants, but not between silenced plants. Plant J 2000; 21:1-8.
36. Crete P, Leuenberger S, Iglesias VA et al. Graft transmission of induced and spontaneous post-transcriptional silencing of chitinase genes. Plant J 2001; 28:493-501.
37. Ruiz MT, Voinnet O, Baulcombe DC. Initiation and maintenance of virus-induced gene silencing. Plant Cell 1998; 10:937-946.
38. Peele C, Jordan CV, Muangsan N et al. Silencing of a meristematic gene using geminivirus-derived vectors. Plant J 2001; 27:357-366.
39. Thomas CL, Jones L, Baulcombe DC et al. Size constraints for targeting post-transcriptional gene silencing and for RNA-directed methylation in *Nicotiana benthamiana* using a potato virus X vector. Plant J 2001; 25:417-425.

40. Voinnet O, Lederer C, Baulcombe DC. A viral movement protein prevents spread of the gene silencing signal in *Nicotiana benthamiana*. Cell 2000; 103:157-167.
41. Finnegan EJ, Wang M-B, Waterhouse P. Gene silencing: fleshing out the bones. Curr Biol 2001; 11:R99-R102.
42. Johansen LK, Carrington JC. Silencing on the spot. Induction and suppression of RNA silencing in the *Agrobacterium*-mediated transient expression system. Plant Physiol 2001; 126:930-938.
43. Zamore PD, Tuschl T, Sharp PA et al. RNAi: double stranded RNA directs the ATP-dependent cleavage of mRNA at 21 to 23 nucleotide intervals. Cell 2000; 101:25-33.
44. Hammond SM, Bernstein E, Beach D et al. An RNA-directed nuclease mediates post-transcriptional gene silencing in *Drosophilae* cells. Nature 2000; 404:293-296.
45. Elbashir SM, Lendeckel W, Tuschl T. RNA interference is mediated by 21- and 22-nucleotide RNAs. Genes Devel 2001; 15:188-200.
46. Bernstein E, Caudy AA, Hammond SM et al. Role for a bidentate ribonuclease in the initiation step of RNA interference. Nature 2001; 409:363-366.
47. Ketting RF, Fischer SEJ, Berstein E et al. Dicer functions in RNA interference and in synthesis of small RNA involved in developmental timing in *C. elegans*. Genes Devel 2001; 15:2654-2659.
48. Jacobsen SE, Running MP, Meyerowitz EM. Disruption of an RNA helicase/RNAse III gene in *Arabidopsis* causes unregulated cell division in floral meristems. Development 1999; 126:5231-5243.
49. Hammond SM, Boettcher S, Caudy AA et al. Argonaute 2, a link between genetic and biochemical analyses of RNAi. Science 2001; 293:1146-1150.
50. Sijen T, Fleenor J, Simmer F et al. On the role of RNA amplification in dsRNA-triggered gene silencing. Cell 2001; 107:465-476.
51. Voinnet O, Vain P, Angell S et al. Systemic spread of sequence-specific transgene RNA degradation in plant is inhibited by localized introduction of ectopic promoterless DNA. Cell 1998; 95:177-187.
52. Wang M-B, Waterhouse PM. High-efficiency silencing of a _-glucuronidase gene in rice is correlated with repetitive transgene structure but is independent of DNA methylation. Plant Mol Biol 2000; 43:67-82.
53. Jones L, Hamilton AJ, Voinnet O et al. RNA-DNA interactions and DNA methylation in post-transcriptional gene silencing. Plant Cell 1999; 11:2291-2301.
54. Wassenegger. M. RNA-directed DNA methylation. Plant Mol Biol 2000; 43:203-220.
55. Morel J-B, Mourrain P, Beclin C et al. DNA methylation and chromatin structure affect transcriptional and post-transcriptional transgene silencing in *Arabidopsis*. Curr Biol 2000; 10:1591-1594.
56. Papaefthimiou I, Hamilton AJ, Denti MA et al. Replicating potato spindle viroid RNA is accompanied by short RNA fragments that are characteristic of post-transcriptional gene silencing. Nucleic Acids Res 2001; 29:2395-2400.
57. Voinnet O, Pinto YM, Baulcombe DC. Suppression of gene silencing: a general strategy used by diverse DNA and RNA viruses of plants. Proc Natl Acad Sci USA 1999; 96:14147-14152.
58. Brigneti G, Voinnet O, Li W-X et al. Viral pathogenicity determinants are suppressors of transgene silencing in *Nicotiana benthamiana*. EMBO J 1998; 17:6739-6746.
59. Li W-X, Ding S-W. Viral suppressors of RNA silencing. Curr Opin Biotech 2001; 12:150-154.
60. Anandalakshmi R, Marathe R, Ge X et al. A calmodulin-related protein that suppresses post-transcriptional gene silencing in plants. Science 2000; 290:142-144.
61. Okamoto H, Hirochika H. Silencing of transposable elements in plants. Trends Plant Sci 2001; 6:527-534.
62. Escobar MA, Civerolo EL, Summerfelt KR et al. RNAi-mediated oncogene silencing confers resistance to crown gall tumorigenesis. Proc Natl Acad Sci USA 2001; 98:13437-13442.
63. Caplen NJ, Taylor JP, Statham VS et al. Rescue of polyglutamine-mediated cytotoxicity by double-stranded RNA-mediated RNA interference. Hum Mol Genet 2002; 11:175-184.

CHAPTER 10

RNA-Directed DNA Methylation and Chromatin Modifications

Marjori A. Matzke, M. Florian Mette, Tatsuo Kanno, István Papp, Werner Aufsatz and Antonius J.M. Matzke

Abstract

Discovered nearly ten years ago in viroid-infected transgenic plants, RNA-directed DNA methylation (RdDM) provided the first example of an RNA-mediated epigenetic alteration of homologous nuclear DNA. Until recently, research on RdDM was conducted exclusively by plant biologists, who viewed it as part of a larger phenomenon that they termed 'homology-dependent gene silencing'. Interest in RdDM among nonplant scientists intensified during the last several years following confirmation that RdDM requires a double stranded RNA that is processed to short RNAs 21-24 nucleotides in length. This finding strengthened the link to RNA interference (RNAi) and related post-transcriptional gene silencing processes, such as PTGS in plants and quelling in fungi, that are similarly triggered by short RNAs derived from double stranded RNA. The field of RNA-guided genome alterations has recently exploded, with parallel studies from diverse organisms demonstrating that components of the RNAi machinery are needed for targeting diverse chromatin modifications to specific regions of the genome. RdDM thus appears to be one of several genome-level manifestations of an elaborate RNA-guided system that can silence gene expression at the transcriptional and post-transcriptional levels.

Introduction

Enzyme complexes that degrade or modify nucleic acids can act with extraordinary specificity by incorporating short nucleic acid guides that are complementary to target DNA or RNA sequences. RNA guides that specify regions to be modified in eukaryotic ribosomal RNAs have been known for some time.[1] Nevertheless, it was impossible to anticipate the deluge of recent findings revealing a vast new world of small guide RNAs that appear to carry out diverse genetic regulatory functions. Small temporal RNAs,[2] microRNAs,[3-5] short interfering RNAs,[6] and heterochromatic short interfering RNAs[7] are all names given to various classes of tiny RNAs (~ 20-26 nucleotides in length) that appear to act at different levels of gene expression in both the cytoplasm and the nucleus to silence eukaryotic genes (Fig. 1). In this chapter, we concentrate on emerging evidence indicating that short RNAs can target DNA methylation and chromatin modifications to specific chromosomal regions. It is becoming apparent that RNA not only issues from DNA, but can also feed back on the genome, providing essential sequence-specific information for establishing the epigenetic architecture of eukaryotic nuclei.

Noncoding RNAs: Molecular Biology and Molecular Medicine, edited by Jan Barciszewski and Volker A. Erdmann.

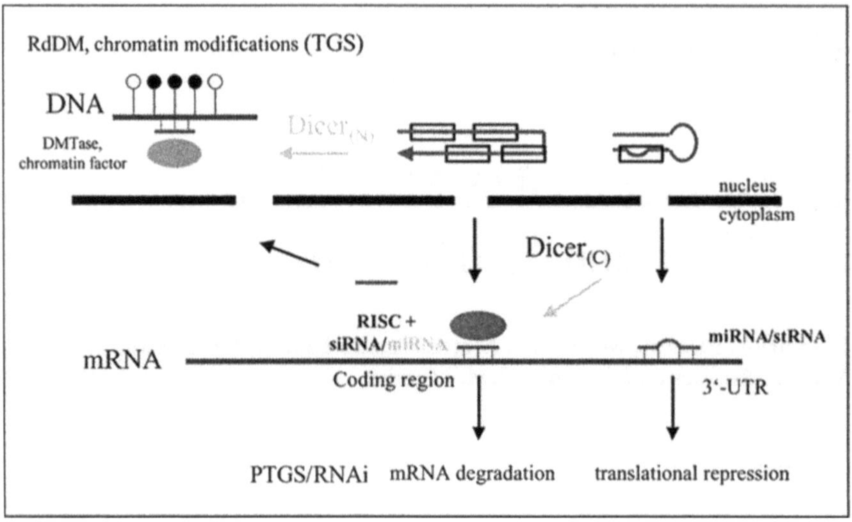

Figure 1. RNA silencing phenomena. RNA silencing refers to epigenetic gene silencing induced by double stranded RNA that is processed to short RNAs by an RNaseIII-like activity termed Dicer. Plants have multiple Dicer-like proteins, some of which contain putative nuclear localization signals (N). Humans have only one located in the cytoplasm (C). RNA silencing phenomena include: PTGS/RNAi—targeted, post-transcriptional degradation of mRNA (or replicating virus RNA) via an RNase complex termed RISC that incorporates a short RNA guide. PTGS/RNAi occurs predominantly in cytoplasm. RNA-directed DNA methylation—de novo methylation (filled circles) of almost all cytosine residues independent of sequence context within a region of sequence identity between the triggering RNA and target DNA. Chromatin factors (light grey oval) can be targeted by RNA generated by the RNAi machinery. Translation repression—a block in translation induced by imperfect base-pairing between a micro (mi) RNA (or small temporal (st) RNA) and the 3'-UTR of the target mRNA. Short RNAs produced by Dicer cleavage of dsRNA are ~ 20-26 nucleotides in length. They were first discovered in plants.[95] Functionally distinct types of short RNA are produced from different types of dsRNA precursors: siRNA—short interfering RNAs trigger PTGS/ RNAi of perfectly complementary target mRNAs. They are derived from all regions of long perfect duplex RNAs encoded by transgenes or produced during RNA virus replication, but only the antisense orientation induces PTGS/RNAi. If siRNAs of both polarities are present, suggesting that RdDM is occurring (at least in plants) via base pairing to dsDNA. Other chromatin modifications leading to transcriptional gene silencing (TGS) might also be guided by short RNAs. stRNA—small temporal RNAs resulting from Dicer cleavage of imperfect RNA duplexes; usually only the antisense polarity accumulates. The first discovered stRNAs (e.g.,*lin-4* and *let-7*) regulate developmental timing in *C. elegans* by incomplete base pairing to 3'-UTR of the target mRNA and blocking translation (Ruvkun, 2001). More recently, additional short RNAs dubbed 'microRNAs' (miRNA) have been discovered in animals and plants. miRNAs are derived from imperfect RNA duplexes and usually accumulate from only one side of one region of the precursor dsRNA. miRNAs in plants differ from known ones in animals by being perfectly homologous to a portion of the coding region the target mRNA, and hence act like siRNAs in triggering RNase degradation of the cognate mRNA. Light colored arrows and letters indicate plant-specific enzymes or functions.

Homology-Dependent Gene Silencing

Although most biologists first became aware of unusual gene silencing effects involving RNA following the discovery of RNA interference (RNAi) in *Caenorhabditis elegans* in 1998,[8] plant scientists have been grappling since the late 1980's with a variety of gene silencing phenomena that appeared to be triggered by homologous DNA and/or RNA sequence interactions (for a review see ref. 9). Homology-dependent gene silencing (HDGS) phenomena included posttranscriptional gene silencing (PTGS; also called cosuppression), an mRNA degradation process thought originally to be induced by complementary 'aberrant' RNA. Other types of HDGS included a form of transcriptional gene silencing (TGS) that possibly involved DNA-DNA pairing, and RNA-directed DNA methylation (RdDM). To plant scientists, the discovery of RNAi provided support for their findings on homology-dependent gene silencing. From their perspective, the novelty of RNAi concerned the focus on double stranded RNA as an elicitor of silencing.

Once the nature of the triggering RNA was known, PTGS, similarly to RNAi, could be reproduced at will in plants by using various transgene constructs to synthesize dsRNA.[10,11] Indeed, PTGS is now considered the plant equivalent of RNAi, and both phenomena are related to quelling, which was discovered independently in the filamentous fungus *Neurospora crassa*.[12] Recent genetic and biochemical studies have demonstrated that RNAi, PTGS and quelling share an evolutionarily conserved mechanism: Briefly, short interfering (si) RNAs 21-24 nucleotides in length, which are derived from processing of dsRNA by an RNase III-like activity termed Dicer, serve as guides for targeted cleavage of cognate mRNAs by an RNase complex called RISC (RNA-induced silencing complex) (Fig. 1). Details of the RNAi/PTGS/quelling mechanism are presented in a number of recent reviews.[13-16]

While the universality of the RNAi/PTGS mechanism was rapidly accepted by most biologists, the other types of homology-dependent gene silencing observed in plants (DNA-DNA pairing, RdDM) received less attention from nonplant scientists. Interest in homologous DNA interactions also waned somewhat among plant researchers,[17] but RdDM continued to be investigated by several plant groups. As we will discuss in this chapter, new findings on RdDM in plants, including a requirement for dsRNA and a link to TGS via dsRNAs containing promoter sequences, have helped to unite mechanistically RdDM - and very recently, chromatin modifications in other organisms - with RNAi/PTGS. In view of these mechanistic overlaps and a common molecular trigger, the term 'RNA silencing' has been introduced to refer to all types of epigenetic gene silencing triggered by dsRNA[13] (Fig. 1).

Discovery and Characteristics of RNA-Directed DNA Methylation

Discovered in 1994, RNA-directed DNA methylation (RdDM) in plants furnished the first example of an RNA-mediated epigenetic alteration of homologous nuclear DNA.[18] RdDM was initially detected in viroid-infected transgenic plants. Viroids are plant pathogens consisting solely of noncoding, closed circular RNAs several hundred bases in length (for a review see ref. 19). In plants inoculated with replication-competent viroids, cDNA copies of the viroid that had been integrated as transgenes into plant chromosomes became methylated de novo.[20] The replicating viroid RNA was thus able to induce methylation of homologous DNA sequences in the plant genome. RdDM was subsequently observed in a number of nonpathogenic, transgenic plant systems.[21-26]

RdDM produces a signature pattern of DNA methylation. While most methylation in eukaroytic genomes is present in symmetrical CGs (and in plants, CNGs), bisulfite sequencing has revealed that RdDM leads to dense de novo methylation of almost all Cs, independent of sequence context, specifically within the region of sequence identity between the triggering RNA and the target DNA.[26,27] The confinement of methylation to the region of sequence homology implies strongly that RNA-DNA base pairing interactions delimit the extent of DNA methylation.

Although viroids replicate via a dsRNA intermediate, it was not evident from the original study[20] that dsRNA is needed for RdDM. This requirement was demonstrated most convincingly in a promoter RNA-directed TGS system in tobacco.[22] Cre/*lox*-mediated recombination was used to convert a transcribed direct repeat comprising two copies of the nopaline synthase promoter (NOSpro) into an inverted repeat, a step that initiated transcription of a NOSpro hairpin RNA. Only the RNA hairpin, not the corresponding single stranded NOSpro RNA, was capable of eliciting methylation of target NOSpro DNA sequences in trans.[22] The NOSpro dsRNA active in the RdDM pathway was processed into short RNAs~21-24 nts in length,[22] similarly to the dsRNA involved in RNAi/PTGS. Processed dsRNA thus emerged as a common molecular trigger in PTGS and RdDM, which could lead to TGS of unlinked target promoters if dsRNAs contained promoter sequences. Given that RdDM was originally discovered with a viroid system,[20] it is noteworthy that short viroid RNAs of both orientations have been detected recently in viroid-infected plants.[28,29]

Although an exhaustive survey has not yet been performed, it is assumed that any DNA sequence can become methylated in the presence of homologous dsRNA/short RNAs. However, susceptibility to RdDM might be enhanced if the target DNA sequence contains reiterated elements.[26,30] Some degree of DNA repetitiveness may be necessary to attract and sequester homologous RNAs. Further experiments are needed to demonstrate that completely single copy DNA sequences can become methylated via RdDM.

DNA methylation exerts its strongest influence on gene expression when promoter sequences become modified, which usually results in transcriptional silencing of the affected gene. Methylation of a particular promoter using dsRNA can be induced effectively by transcribing an IR of the desired promoter sequence, with the two halves of the IR separated by a spacer, thereby producing a promoter RNA with a stem-loop structure. This method has been used successfully to methylate the promoters of transgenes[22,26] and endogenous genes[25] in plants. A second way to produce promoter dsRNAs is to engineer a plant RNA virus genome with sequences homologous to the promoter of a nuclear gene. During virus replication, which occurs exclusively in the cytoplasm, an RNA species is produced that can find its way into the nucleus and methylate identical promoter sequences, leading to their transcriptional inactivation.[23,24]

Protein coding regions of genes silenced by PTGS in plants also frequently acquire methylation,[31] presumably via RdDM. However, the consequences of methylation in coding regions are not yet clear. Although such methylation does not usually disrupt transcription elongation in plants, it might play a role in maintaining PTGS in an as yet unknown way.[32,33] Because DNA methylation is immediately repressive when present in promoter regions, the following discussion on the mechanism of RdDM concerns primarily promoter methylation and TGS induced by dsRNA.

Mechanism of RdDM

The mechanism of RdDM is beginning to be dissected. Genetic analyses are being carried out in *Arabidopsis thaliana* on a NOSpro-mediated TGS system[26] and a second TGS system based on a seed-specific promoter (T. Kanno and A. Matzke, unpublished results). Biochemical studies will also be important for reconstructing the mechanism of RdDM. In addition to an RNA signal, the minimal requirements for RdDM are presumably one or more DNA methyltransferases (DMTase) that can establish and maintain the unusual methylation pattern that is the hallmark of RdDM. It is likely that DNA and/or RNA helicases are required to unwind dsRNAs/short RNAs and the target DNA, allowing them to interact by base pairing. Given the known connections between DNA methylation and chromatin remodelling,[34-36] and DNA methylation and histone modifications, such as deacetylation[37] and methylation,[38,39] enzymes catalyzing these reactions probably contribute at some point to the RNA-mediated TGS pathway.

It is important to determine not only the molecular machinery required for RdDM but also the order of events linking DNA methylation to modulations in chromatin structure. DNA methylation is often considered a secondary consequence of chromatin alterations; i.e., methylation occurs downstream of initial histone modifications and serves mainly a reinforcing function to lock in silencing.[40] However, this might not be invariably true and in this respect, RdDM provides an interesting case for study. The restriction of methylation to the region of RNA-DNA sequence identity suggests that the primary event in assembling a transcriptionally inactive state is recognition of the heteroduplex by a de novo DMTase(s). Whether this occurs in a chromatin context is unclear. We will return to this point in Section V.C.

RNA Species Required for RdDM

As discussed above, RdDM requires a dsRNA that is processed to short RNAs. Whether the short RNAs or the dsRNA directly guide homologous DNA methylation has not yet been firmly established. However, the minimum DNA target size of RdDM (~30 bp)[41] is consistent with short RNA involvement. Intuitively, short RNAs would seem to have easier access to DNA than larger dsRNAs. Support for short RNA involvement in RdDM has come from a recent study that used plant virus proteins that are able to differentially prevent short RNA accumulation.[42]

In plants, a number of proteins encoded by RNA and DNA viruses have been shown to suppress PTGS.[43] This reflects a counter-defensive strategy employed by viruses to overcome the natural antiviral function of PTGS.[44,45] Viral proteins that suppress PTGS provide unique tools for studying RNA silencing mechanisms in plants. Hamilton and coworkers found that different viral proteins prevent the accumulation of distinct size classes of short RNA. A 'longer' class (24-26 nt) was shown to be responsible for RdDM and a 'shorter' class (21-22 nt) for inducing PTGS.[42] The generality of these results for other RdDM systems remains to be verified. However, given the potential diversity of Dicer-like proteins in *A. thaliana* (discussed in Section IV.A.2) it is entirely possible that at least in plants, functionally distinct short RNA sub-populations are generated and are active in different cellular compartments. Further support for the involvement of short RNAs in epigenetic modifications of the genome comes from studies showing impaired formation of centromeric heterochromatin in *Schizosaccharomyces pombe* mutants defective in Dicer (Section V.A).

Complex Short RNA Populations

The nature of short RNAs can provide information about their function and mode of action. For example, it might be anticipated that single stranded short RNAs will be rapidly degraded unless they can bind to a target nucleic acid; thus, detecting particular short RNAs implies that they have a function. Data so far support this notion. In *A. thaliana*, sense and antisense short RNAs that are associated with RdDM of a target promoter are derived from all portions of the precursor dsRNAs (Fig. 1) and they delimit the extent of DNA methylation (I. Papp, M.F. Mette and A. Matzke unpublished results). The coexistence of both sense and antisense polarities is consistent with their stabilization by base pairing with both strands of homologous DNA. This would permit de novo methylation of both strands of nonreplicating DNA (Fig. 2)

The accumulation of both sense and antisense short RNAs derived from the entire dsRNA molecule involved in triggering RdDM is in contrast to micro (mi) RNAs, which act to silence at the mRNA level. Most known plant miRNAs accumulate from only one side of the RNA duplex (antisense to the target mRNA) and from only one ~20-24 bp region, not the entire dsRNA[46-48] (Fig. 1). Stabilization of only the antisense polarity is consistent with the role of miRNAs in base pairing to mRNAs to either block translation or induce mRNA degradation (Fig. 1).

In addition to the differential stability of short RNAs depending on the availability of a homologous or complementary target nucleic acid, other variables potentially affecting the characteristics of short RNA populations in plants include: (1) the identity of the Dicer-like enzyme executing dsRNA cleavage; (2) the subcellular location of dsRNA processing (nucleus

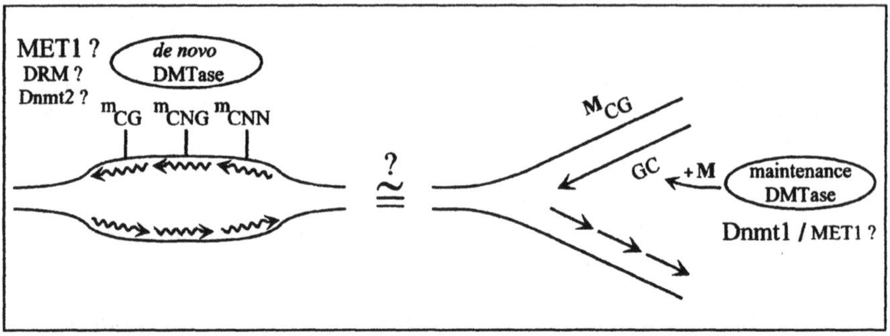

Figure 2. Hypothetical role for MET1, the CG maintenance DMTase in plants, in the de novo methylation step of RdDM. Hybrids (left) between short RNAs (wavy lines) and DNA might resemble a DNA replication fork (right), a region where Dnmt1, the mammalian homolog of MET1, is concentrated to carry out CG maintenance methylation (+M) on the newly synthesized DNA strand. Other candidates for the de novo methylation step of RdDM in plants are members of the DRM family, which are the traditional de novo DMTases, and the plant homolog of Dnmt2, which in other organisms can catalyze nonCG methylation (text Section IV.D) and which might be involved in RNA-guided centromere formation (text Section V.C).

versus cytoplasm); and (3) the nature of the dsRNA precursor (a shorter imperfect duplex versus a long, perfect duplex) (Fig. 1).

dsRNA Processing Dicer-Like Proteins in *Arabidopsis*

In addition to using plant viral suppressors of PTGS, a second approach for understanding the functions of various short RNAs is to recover mutations in the enzyme(s) responsible for dsRNA processing, thus preventing the production of short RNAs. This experiment is complicated in plants owing to the presence of multiple Dicer-like genes, all of which potentially encode functional proteins.

The *A. thaliana* genome comprises four Dicer-like (*DCL*) genes and five additional ones with at least one RNaseIII-like domain.[47] Two of the four DCL proteins contain one or more putative nuclear localization signals (NLSs), hinting that they localize to the nucleus and function in that compartment. Indeed, work with HC-Pro (helper component proteinase), a plant viral suppressor of PTGS that blocks the accumulation of short RNAs in an unknown way,[49,50] suggested that nonpolyadenylated dsRNAs synthesized in the nucleus might be processed and sequestered in that compartment.[51] The fact that the resulting short RNAs could induce RdDM of homologous DNA sequences but were unable to induce appreciable degradation (i.e., PTGS) of the corresponding single stranded, polyadenylated RNA supports the existence of separately compartmentalized, functionally distinct short RNA populations in plants (Fig. 1).

The possibility that dsRNA processing can occur throughout plant cells, including the nucleus, contrasts to the situation in mammals (Fig. 1). Humans have only one Dicer-like protein that is located exclusively in the cytoplasm[52] and RNAi is restricted to this cellular compartment.[53] The presence of multiple Dicer-like proteins in plants but not mammals points toward different ways of dealing with nuclear dsRNAs in each group of organisms. Indeed, in mammals, long RNA duplexes in the nucleus are not processed to short RNAs but instead undergo adenosine (A) to inosine (I) editing by enzymes termed ADARs (adenosine deaminases that act on RNA).[54] ADARs do not appear to be encoded in known plant genomes. The apparent lack of a nuclear Dicer activity in mammals does not exclude the possibility that RdDM could occur in mammals (see Section IV.D), but short RNAs produced via Dicer cleavage of dsRNA in the cytoplasm would have to relocate to the nucleus (Fig. 1). The likelihood that this can occur has been demonstrated by experiments in plants showing that cytoplasmically replicating RNA viruses can induce RdDM of homologous nuclear transgenes, presumably by migration of an RNA species into the nucleus.[23,24]

The only plant DCL protein for which any information is available so far is DCL1 [formerly called variously ABNORMAL SUSPENSOR1 (SUS1), SHORT INTEGUMENT1 (SIN1) and CARPEL FACTORY (CAF)[55]]. Mutations in the *DCL1* gene produce obvious developmental defects in *A. thaliana*,[55,56] suggesting a general requirement for dsRNA processing in plant development. Moreover, DCL1 was recently shown to be required for producing several plant miRNAs,[57,58] providing additional evidence that short RNAs help to regulate *A. thaliana* development. However, DCL1 is not the only Dicer-like activity involved in producing functional short RNAs in *A. thaliana*. For example, *dcl1* mutants do not appreciably impair processing of short RNAs from nonpolyadenylated (and presumably nuclear) dsRNAs that trigger RdDM of homologous promoters (M.F. Mette, W. Aufsatz, M. Matzke, unpublished results). This result was initially surprising given that DCL1 contains two putative NLSs and is predicted to be a nuclear protein. The differential effects of the *dcl1^{caf}* mutation on miRNA formation versus promoter short RNA formation might be due to the nature of the dsRNA precursors, which consist of imperfect or perfect RNA duplexes, respectively. Perhaps individual DCL proteins recognize different types of dsRNA substrate. In any case, the disparate effects of *dcl1* mutations on short RNA production in different systems are consistent with the notion of functionally distinct sub-populations produced by multiple Dicer-like proteins. Additional work testing mutants defective in other DCL proteins is needed to fully understand dsRNA processing in plants and the functions of different classes of short RNAs.

DNA MTase(s) Required for RdDM

RdDM must be considered with respect to both de novo and maintenance methylation steps. It is assumed that de novo methylation at Cs in all sequence contexts—the hallmark of RdDM—will be catalyzed continuously as long as the triggering RNA and the appropriate de novo DMTase(s) are present. If the RNA signal is withdrawn, only methylation that can be maintained by maintenance DMTase activities will be retained. The pattern of methylation resulting from RdDM at any given time during development is therefore dependent on the availability of a suitable RNA signal and the required DMTases.

In *A. thaliana*, there are four known or putative classes of DMTase.[59] The 'domains rearranged' (DRM) class, which is homologous to the mammalian Dnmt3 family, appears to be the major de novo DMTase.[60] Two classes of DMTase provide primarily maintenance functions. MET1, which is homologous to Dnmt1 in mammals, maintains methylation in CG dinucleotides.[61] CMT3, a member of the chromomethylase family that is unique to plants,[62] maintains methylation primarily in CNG trinucleotides.[63-65] The Dnmt2 class is an enigmatic family of putative DMTase that is found in vertebrates, plants, and two organisms not normally thought to methylate their DNA: *Drosophilae melanogaster*[66] and *S. pombe*.[67] Although the type of DNA methylation, if any, catalyzed by Dnmt2 is uncertain,[68] nonCGs are possibly preferred target sites.[67,69]

With respect to DMTases and RdDM, the most information available so far concerns MET1. In *Nicotiana benthamiana* plants in which RdDM was triggered by a cytoplasmically replicating RNA virus containing a sequence homologous to a promoter of a nuclear transgene, MET1 was necessary for maintaining specifically CG methylation, even in progeny plants that were virus-free.[24] In these experiments, the MET1 gene was not mutated but was silenced by PTGS. Somewhat different results were obtained in a study using *met1* genetic mutants in *A. thaliana*. In a *met1* mutant background, considerable loss of methylation at both CGs and nonCGs was observed in a promoter RNA-mediated TGS system.[26] This unusual result, which runs counter to the traditional view that *MET1* serves to maintain primarily CG methylation, requires further examination. It is conceivable that in addition to its conventional role, *MET1* can also act as a de novo MTase, particularly in the presence of RNA signals (Fig. 3). Indeed, the mammalian homolog of MET, Dnmt1, has weak de novo DMTase activity.[68] An intriguing possibility is that hybrids between short RNAs and DNA resemble a DNA replication fork (Fig. 2), where mammalian Dnmt1 has been shown to localize during S phase.[70]

CMT3 was originally suggested as a promising candidate for RdDM, owing to the presence of a chromodomain, which can potentially act as an RNA-protein interaction module.[71] However, recent studies on the NOSpro RNA-mediated TGS system[26] suggests that the *cmt3* mutation has little or no effect on silencing (W. Aufsatz, M.F. Mette and A. Matzke, unpublished results). Experiments are in progress using the NOSpro RNA-mediated TGS system to test mutations in DRM1 and DRM2, the traditional de novo DMTase activities.[60]

It is important to keep in mind that more than one de novo DMTase activity might be involved in RdDM (Figs. 2 and 3). Moreover, requirements for maintenance DMTase activities might become apparent only after the RNA signal (and the possibility of continuous de novo methylation) is removed.[24,26] Analyses of double mutants and experimental systems employing Cre/*lox*-mediated recombination to abolish dsRNA synthesis[26] will help to resolve these issues.

RdDM and Chromatin Modifications

Little is known about chromatin modifications that are associated with establishing and/or maintaining RNA-mediated promoter methylation and TGS. A particular chromatin modifying enzyme could either be specific for the RdDM pathway, or catalyze chromatin changes generically with no preference for RNA-targeted sequences. An ongoing genetic analysis using the NOSpro RNA-mediated TGS system in *A. thaliana* has suggested that perhaps at least one chromatin factor, an Rpd3-like histone deacetylase HDA6, might be specialized for the RNA-guided pathway of genome modifications (W. Aufsatz et al, submitted for publication). Moreover, HDA6 appears to be dispensable for the de novo methylation step of RdDM but be required primarily for maintaining CG methylation induced by RNA. In an interesting parallel, an Rpd3-like histone deacetylase (HDAC) was found to be required for maintenance but not establishment of transcriptional repression in *D. melanogaster*.[72] The degree to which HDACs carry out specialized functions was revealed in a recent study in the yeast *Saccharomyces cereviseae*, where the five different types of HDAC were found to target specific sequences by mechanisms that remain to be discovered.[73] It is possible that Rpd3-like HDACs are recruited to specific promoters in diverse organisms, including yeast, by guide RNAs.

The NOSpro RNA-mediated TGS system has also been tested in *A. thaliana* mutants defective in *ddm1* (decrease in DNA methylation), a putative chromatin remodeling complex.[34] While a decrease in TGS and NOSpro methylation was observed in *ddm1* mutants that continued to make wild type levels of NOSpro dsRNA/short RNAs, the effect of the mutation appeared to be indirect since the release of TGS and loss of methylation was nonuniform in the population and could be delayed over several generations.[26] It might be anticipated that chromatin remodeling would be needed to provide triggering RNAs access to the target DNA, particularly in view of the possibility that direct RNA-DNA interactions outside of a chromatin context are required for RdDM (Section V.C). Further genetic screens will potentially recover mutations in chromatin remodeling proteins that, unlike DDM1, play a direct role in the RdDM pathway.

RdDM in Other Organisms

RdDM has not yet been reported for organisms other than plants, but this apparent deficiency might simply reflect the absence of the proper DMTase and/or developmental variations

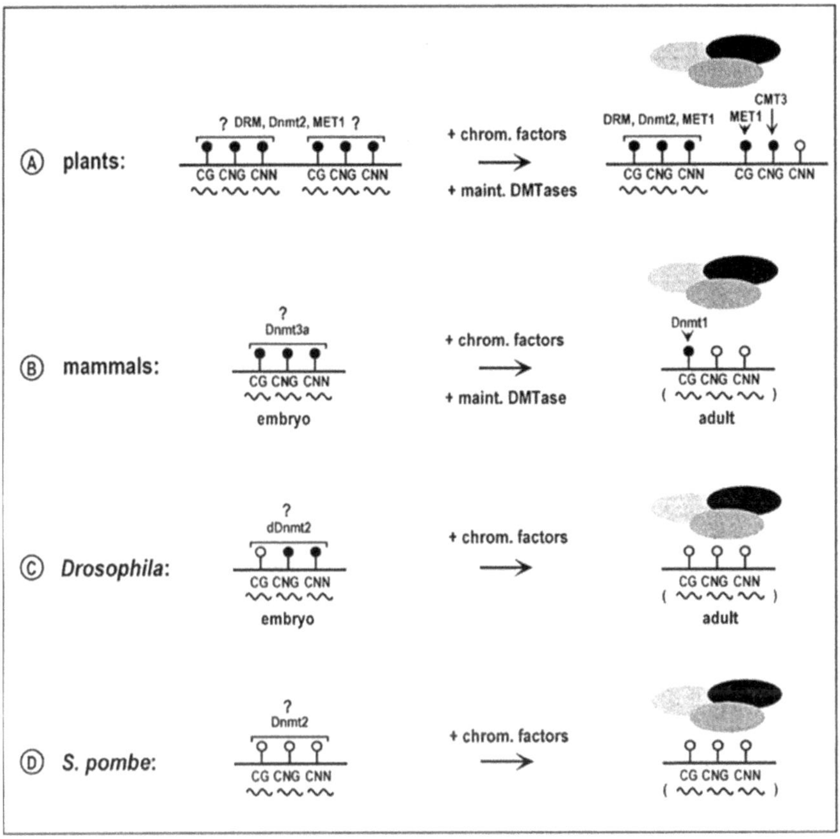

Figure 3. Hypothetical scenarios for RNA-guided establishment and maintenance of transcriptionally silent genomic regions in different organisms. Known or putative de novo DMTases and the type of DNA methylation (filled circles) they catalyze (CG, CNG, CNN) in response to short guide RNAs (wavy lines) are indicated in the left column. After the initial pattern of methylation is laid down according to RNA signals, maintenance DMTases in plants and mammals, and chromatin factors (gray and black ovals), such as histone modifying enzymes and polycomb proteins, in all organisms are recruited to stabilize and strengthen the silent state (right column). (A) In plants, maintenance DMTases, MET1 and CMT3, can maintain CG and CNG methylation, respectively, even if the RNA signals are removed. De novo methylation can continue if suitable RNA signals are present owing to the continued activity in adult plants of de novo DMTase(s). (B) In mammals, Dnmt1 will maintain CG methylation whether RNA signals remain or not (wavy lines in parentheses), but de novo methylation effectively ceases owing to decreased activity of the required de novo DMTase (Dnmt3a) during development. (C) NonCG methylation is present early in *Drosophilae* development, but then is lost in adult cells, possibly because of reduced activity of dDnmt2 and lack of an appropriate maintenance DMTase. (D) In *S. pombe*, the mutated Dnmt2 cannot catalyze DNA methylation, but it possibly recognizes an RNA-DNA hybrid and subsequently recruits chromatin factors. This scenario accounts for the distribution of cytosine methylation (or lack thereof) in these organisms. DNA methylation, which can be retained at symmetrical CGs and CNGs during DNA replication (Fig. 2), might be preserved in cells of adult plants and animals owing to a greater need to ensure heritability of transcriptionally silent states during mitoses.

in the incidence of RdDM in diverse species. As discussed in Section III, methylation of Cs in any sequence context is a hallmark of RdDM. Therefore, one hint that RdDM might be occurring is the presence of nonCG methylation.

Most methylation in animal genomes has been thought to be concentrated in CG dinucleotides. However, recent work in mammals and *D. melanogaster* - which has, surprisingly, putative homologs of vertebrate Dnmt2 and a methyl-binding protein[66] - indicates that substantial nonCG methylation might indeed be present early in animal development.[66,74] The de novo MTase catalyzing the nonCG and CG methylation in mammals could be Dnmt3a, which is active early in mammalian development.[74] The putative DMTase encoded in the *D. melanogaster* genome, dDnmt2, is active mainly in embryos where most of the DNA methylation in flies is observed.[66] Both Dnmt3a and dDnmt2 have been implicated in catalyzing nonCG methylation.[66,74] Thus, the de novo DMTases required for RdDM may be active only at the initial stages of animal development. Accordingly, if RdDM occurs in animals, it would be restricted to this early period. This proposal is illustrated in (Fig. 3), which also shows further hypothetical events leading to the distribution of methylation observed in adult cells of mammals, flies and plants.

According to this scheme, an initial interval of RdDM would establish sequence-specific epigenetic marks. Chromatin factors would subsequently take their cue from the RNA-guided de novo methylation pattern and catalyze chromatin modifications that contribute to stabilizing the silent state. Once RdDM ceases, through loss of either the appropriate DMTase and/or RNA signals, maintenance DMTase activities, if present, would determine methylation patterns in adult somatic cells. Since in mammals the sole maintenance DMTase is believed to be Dnmt1, only methylation in CG dinucleotides will be maintained, hence giving rise to the familiar pattern of predominantly CG methylation in adult animal cells (Fig. 3). In *D. melanogaster*, the activity of dDnmt2 subsides during development.[66] In the absence of a maintenance DMTase activity, de novo methylation initially triggered by RdDM in embryos will be lost and stable silencing in adult cells would rely on chromatin modifications imposed according to the original methylation pattern (Fig. 3).

In contrast to the developmentally variable distributions of DNA methylation and the methylation machinery in animals, the three major classes of DMTase in plants (*MET1*, DRM and CMT3) remain active throughout growth and development. This means that continuous de novo methylation can occur (via DRM activity) and methylation in CG and CNG nucleotides groups will be maintained (via MET1 and CMT3, respectively) (Fig. 3). Persistent activity of all three classes of DMTase probably accounts for the easy detection of RdDM in adult plants and the high degree of symmetrical and nonsymmetrical C methylation in the genomes of somatic plant cells.

RNA Guiding Chromatin Modifications

While RdDM has remained largely in the plant domain since its discovery eight years ago, exciting new results are extending the range of RNA-guided genome modifications, even to organisms that do not methylate their DNA. RdDM might thus be only one type of RNA-guided genome modification that happens to be prominent in plants.

RNAi Mutants Defective in Chromatin Modifications

Genetic analyses performed with *D. melanogaster, S. pombe, and C. elegans* have revealed that chromatin modifications require components of the RNAi machinery. Both RNAi and polycomb-dependent transcriptional silencing were impaired in *D. melanogaster* mutants defective in the Piwi protein, which is a member of the Argonaute family required for RNAi.[75] In *S. pombe*, mutants defective in homologs of three proteins required for RNAi (Dicer, a putative RNA-dependent RNA polymerase and Argonaute) released transcriptional silencing of a transgene in centromeric heterochromatin and prevented both histone methylation and recruitment of Swi6 (an *S. pombe* homolog of animal heterochromatin protein, HP1).[76] Short RNAs of both polarities were detected from the *S. pombe* centromeric repeats,[7] which is consistent with RNA-guided formation of centromeric heterochromatin. The centromeric short RNAs were termed 'heterochromatic short interfering RNAs' to distinguish them from regular siRNAs that target mRNAs[7] (Fig. 1). In *C. elegans*, an RNAi-mediated genetic screen to find genes required for RNAi uncovered several ORFs encoding putative chromatin proteins.[77] It is thus becoming clear that the dual nature of RNA silencing mechanisms first described in plants (RNA-guided mRNA degradation in the cytoplasm and RNA-guided genome modifications in the nucleus) is common to many eukaryotes.

RNAi and DNA Methylation Machineries Frequently Coexist

When contemplating shared components of the various RNA silencing pathways (Fig. 1), one is struck by a correlation between the presence of the machineries for both RNAi and DNA methylation. For example, plants and mammals have both; *S. cerevisiae* has neither. (An exception is *C. elegans*, which performs RNAi but not DNA methylation). Particularly striking, however, are *S. pombe* and *D. melanogaster*, both of which carry out RNAi and both of which were thought until recently to lack DNA methylation. This perception changed when the genome sequences of both organisms revealed putative DMTase genes.

As discussed in Section IV.D, the *D. melanogaster* genome encodes homologs to mammalian Dnmt2 and a methyl-binding protein. The *S. pombe* genome encodes a mutated DMTase that is similar to mammalian Dnmt2. The sparse methylation of the *D. melanogaster genome* in embryos indicates that the dDnmt2 is a functional enzyme.[66] Interestingly, even though *S. pombe* DNA is unmethylated, the Dnmt2 gene in this organism is expressed.[78] The retention and expression in *S. pombe* of the mutated Dnmt2, which is unable to catalyze cytosine methylation,[67] hints that it might play an alternative, perhaps structural role in an RNA-guided pathway of chromatin modifications. There is evidence that DMTases, particularly members of the Dnmt3 family of de novo DMTases in mammals, can provide structural functions to assemble condensed chromatin in a manner independent of their traditional catalytic activity.[79] Dnmt2 provides an example that is relevant to the present discussion of RNA-guided genome modifications.

Model for Chromatin Condensation Initiated by Guide RNAs and Proteins of the DNA Methylation Machinery

In addition to a possible correlation between the coexistence of the RNAi and DNA methylation machineries, a similar correlation has been proposed between centromere structure and the presence of Dnmt2. Bestor has invoked the unusual phylogenetic distribution of Dnmt2-like proteins to argue that they might play a role in the formation of centromeres, whose structure and function are relatively conserved among organisms that possess Dnmt2 homologs (e.g.,plants, vertebrates, *D. melanogaster* and *S. pombe*). In contrast, species without Dnmt2 either lack discrete centromeres (*C. elegans*) or have atypical centromeres (*S. cerevisiae*).[68] If Dnmt2 indeed plays a role in centromere formation, then it follows that even Dnmt2 proteins that lack conventional DMTase activity (e.g., the Dnmt2 homolog in *S. pombe*) contribute to centromere formation, presumably by serving a structural function and/or by recruiting other chromatin factors.

A possible role for Dnmt2 in centromere formation can be considered along with the recent report mentioned above that connects components of the RNAi machinery with histone methylation at centromeric repeats in *S. pombe*.[76] Furthermore, a link between histone methylation and DNA methylation has emerged from studies showing a requirement for histone methyltransferases to maintain full cytosine methylation in *N. crassa*[38] and *A. thaliana*.[39]

With all of this information in mind, it becomes possible to envision a hypothetical pathway for chromatin condensation in which a protein(s) of the DNA methylation machinery initially recognizes an RNA-DNA hybrid - a step which may or may not actually lead to cytosine methylation - followed by recruitment of chromatin factors and maintenance DMTases, if they exist in a given organism, to stabilize the silent state (Fig. 3). While this hypothesis overturns models proposing that DNA methylation occurs after histone modifications,[40] it is more consistent with the pattern of DNA methylation triggered by RdDM.

It is difficult to imagine how cytosine methylation resulting from RdDM could be targeted so precisely to a specific DNA region, unless a DMTase initially senses and modifies the RNA-DNA heteroduplex. The resolution of RdDM can be regulated virtually down to the single nucleotide level[26] over a DNA target region as short as 30 bp[41]. This contrasts to histone modifications, which are restricted to a nucleosomal unit (146 bp of DNA wrapped around an octamer of four core histones). The facts that the entire DNA target region spanned by the triggering RNAs appears to be equally accessible to DMTases, and that methylation does not spread considerably beyond the region of RNA-DNA sequence identity,[26] argue for RdDM requiring direct DNA-RNA interactions taking place outside of a complex chromatin context. The identification of a HDAC that was required mainly for maintenance of CG methylation but not the de novo methylation step of RdDM in *A. thaliana* supports a primary role for RNA signals and components of the DNA methylation machinery in assembly of repressed chromatin (W. Aufsatz et al, submitted for publication). In the case of *S. pombe*, which does not methylate its genome, the recognition and binding of Dnmt2 to the RNA-DNA hybrid alone could be sufficient for recruitment of chromatin proteins and formation of centromeres, which possibly does not require strict boundaries delimited at the single nucleotide level (Fig. 3).

RNA and Epigenetic Modifications Guide vs. Scaffold Function

In this chapter, we have been concerned with short RNAs derived from dsRNA cleavage and their role in guiding DNA methylation and/or other types of chromatin modifications to specific regions of the genome via sequence homology. The action of these short guide RNAs can be local and affect specific promoters. Moreover, short guide RNAs can act in trans to transcriptionally silence and methylate genes on separate chromosomes, perhaps because their small size facilitates movement through the nucleoplasm. This type of RNA-mediated epigenetic change might be distinct from the action of RNAs involved in dosage compensation, such as *Xist* RNA, and various antisense RNAs, such as *H19* and *Air*,[80] involved in genomic imprinting in mammals.

Xist and imprinting RNAs are long (sometimes many kilobases[80]) and there is no evidence that they undergo processing to short RNAs, even though they can contain potential regions of secondary structure that resemble miRNA precursors.[81] In addition, these RNAs are active in *cis* only and they silence and modify whole chromosomes or large chromosomal domains. Finally, it is not clear whether *Xist* and imprinted RNAs interact with their target DNA through sequence homology. Long, *cis*-acting RNAs might serve primarily as scaffolds for assembling heterochromatin proteins and they might have evolved specifically to operate in cases of dosage compensation that require heterochromatinization of large genomic regions. It will be important in the future to distinguish between informational (i.e., guide) and structural (i.e., scaffold) roles for various RNAs involved in epigenetic silencing phenomena.

Evolution of RNA Silencing Mechanisms

DNA methylation and RNAi/PTGS probably originated as host defenses to parasitic sequences.[82] The fundamental feature activating host defenses is DNA or RNA sequence reiteration, which seems to be perceived by eukaryotic cells as 'foreign'. This is consistent with the nature of transposable elements (TEs) and viruses, which can be present in multiple copies in

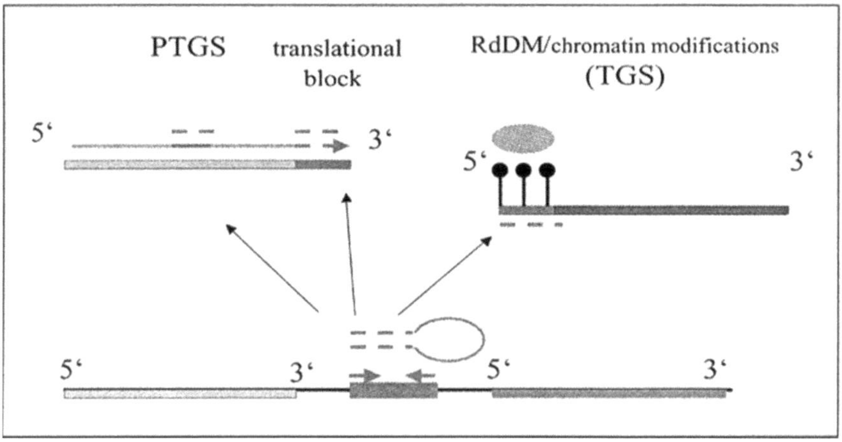

Figure 4. Scenario for evolution of RNA silencing mechanisms from transposable elements (TEs). Short intergenic TEs several hundred base pairs in length (bottom, short bar) containing terminal inverted repeats 20-26 bp (TIRs; short arrows) are transcribed to produce dsRNA that is processed into short RNAs (dashes) at one or both TIRs. The short RNAs can target complementary mRNAs or homologous DNA regions that contain remnants of the TE family (top, short bars). Depending on the location of the TE sequence, RNA silencing can involve PTGS (coding region), translational block (3'-UTR) or TGS via RNA-directed methylation (filled circles) and/or chromatin modifications (grey oval) of promoter sequences. A type of TE similar to the one proposed here is 'nemis', which are found in several *Neisseria* species (Section VI).

host genomes and/or produce dsRNA during their replication cycles; consequently, these invasive entities can be controlled by silencing mechanisms that rely on recognition of nucleic acid sequence homology. The host defense function of homology-dependent gene silencing mechanisms is best evidenced by the mobilization of TEs in *C. elegans* mutants defective in RNAi,[83,84] and in plant mutants impaired in PTGS[85] and DNA methylation.[63,86-89] In addition, some PTGS-defective mutants in plants are more susceptible to RNA virus infection.[90]

Despite the existence of host defenses, TEs and viral sequences have accumulated to vast numbers in most plant and animal genomes over evolutionary time and they have invaded host genes. This relentless buildup of intrusive DNA has probably been due to several inherent weaknesses of host defense systems.[82] Although TE insertions can debilitate host genes, the possibility that 'foreign' sequences have contributed beneficial traits to the host must also be considered. One possible positive outcome of invasive DNA accumulation is the evolution of epigenetic regulatory mechanisms required to silence host genes in cells where their protein products are not required. The basic idea is that host genes containing TE or viral sequences might themselves appear foreign to defense machinery, thus imposing the type of epigenetic regulation associated with host defense on host genes.[91]

When thinking about how specifically RNA silencing mechanisms (Fig. 1) could have evolved from primordial host defenses, the following scenario comes to mind. What is required are relatively short (~100-300 bp) intergenic TEs containing terminal inverted repeats (TIRs) that are transcribed to produce a dsRNA that is processed by Dicer to short RNAs. The short RNAs can then interact with homologous or complementary target sequences that have become integral parts of host genes as a result of insertions by other members of the TE family. Depending on the position of the target sequence in the host gene (coding region, 3'-UTR, promoter), silencing could be initiated, respectively, at different levels: mRNA degradation (PTGS/RNAi); translational block; or TGS accompanied by RdDM of promoter sequences (Fig. 4).

Although TEs possessing the required characteristics have not been among the traditional ones identified in higher eukaryotes, there is a type of mobile element in several species of the bacterial genus *Neisseria* that is remarkably similar to miRNA 'genes'.[48] So called 'nemis' (neisseria miniature insertion sequences) are around 150 bp; they feature TIRs of ~26 bp; and they are transcribed to produce dsRNA that is processed at one or both TIRs by RNaseIII (Dicer-like) activity. Moreover, nemis integrate close to host genes, bringing these genes under the regulation of RNase III.[92,93] In a similar manner, ancient TEs resembling nemis might have provided sources of miRNAs and their target sequences to plant genes.[48]

A number of plant miRNAs are homologous to DNA sequences at multiple intergenic sites in the *A. thaliana* genome.[46,47,57,58] Often these regions have the potential to form RNAs with stable secondary structures, and it is this feature, not the overall DNA sequence that appears to be conserved, suggesting selection for the formation of functional short RNAs.[48] Although it remains to be demonstrated that the sequences ultimately giving rise to short RNAs in eukaryotes are indeed derived from TEs resembling bacterial 'nemis', it is difficult to envision how these sequences have become amplified in plant genomes without resorting to explanations based on transposition.

To determine whether endogenous genes can be silenced transcriptionally by short RNAs that target DNA methylation and/or chromatin modifications, it will be necessary to obtain a full accounting of the short RNA populations in plants. Although promoter regions are not normally thought to be transcribed, the fact that many miRNAs in *A. thaliana* are derived from intergenic regions makes one optimistic that short RNAs homologous to endogenous promoters will be identified. Indeed, seven *A. thaliana* miRNAs identified in a recent study showed similarity or complementary in at least 15 nucleotides to potential promoter sequences (arbitrarily set at 0.1-1 kb upstream of the ATG start codon).[58] These miRNAs provide the first candidates for RNA guides that trigger TGS and epigenetic modifications of endogenous promoters.

Outlook

"Tiny RNA genes may be the biological equivalent of dark matter–all around us but almost escaping detection."[94]

This quote reflects the profound and immediate impact that the discovery of short RNAs has had on eukaryotic biology. Indeed, it is difficult to remember a time when we labored without thinking about short RNAs and the various modes of RNA silencing that they can induce. Although it would be unwise to rule out additional surprises, the basic outline of RNA silencing networks, which are proving to be essential for eukaryotic development and chromosome structure, appears to have been drawn. Now that RNA-guided genome modifications have been detected not only in plants but also in other genetically tractable organisms, the combined efforts of fly, yeast, worm and plant biologists should soon fill the major gaps in our understanding of RNA silencing mechanisms in the nucleus.

Acknowledgements

Work in our lab is supported by grants from the Austrian Fonds zur Förderung der wissenschaftlichen Forschung (Z21-MED and P15611) and the European Union (Contract QLRT-2000-00078).

References

1. Lafontaine D, Tollervey D. Birth of the snoRNPs: the evolution of the modification-guide snoRNAs. Trends Biochem Sci 1988; 23:383-388.
2. Ambros V. Dicing up RNAs. Science 2001; 293:811-813.

3. Lau N, Lim L, Weinstein E et al. An abundant class of tiny RNAs with probable regulatory roles in *Caenorhabditis elegans*. Science 2001; 294:858-862.
4. Lee R, Ambros V. An extensive class of small RNAs in *Caenorhabditis elegans*. Science 2001; 294:862-864.
5. Lagos-Quintana M, Rauhut R, Lendeckel W et al. Identification of novel genes coding for small expressed RNAs. Science 2001; 294:853-858.
6. Elbashir S, Lendeckel W, Tuschl T. RNA interference is mediated by 21- and 22-nucleotide RNAs. Genes Dev 2001; 15:188-200.
7. Reinhart B, Bartel D. Small RNAs correspond to centromere heterochromatic repeats. Science 2002; 297:1831.
8. Fire A, Xu S, Montgomery M et al. Potent and specific genetic interference by double-stranded RNA in *Caenorhabditis elegans*. Nature 1998; 391:806-811.
9. Matzke M, Aufsatz W, Kanno T et al. Homology-dependent gene silencing and host defense in plants. Adv Genet 2002; 46:235-275.
10. Waterhouse P, Graham M, Wang M.-B. Virus resistance and gene silencing in plants can be induced by simultaneous expression of sense and antisense RNA. Proc Natl Acad Sci USA 1998; 95:13959-13964.
11. Smith N, Singh S, Wang M.-B et al. Total silencing by intron-spliced hairpin RNAs. Nature 2000; 407:319-320.
12. Romano N, Macino G. Quelling: transient inactivation of gene expression in *Neurospora crassa* by transformation with homologous sequences. Mol Microbiol 1992; 6:3343-3353.
13. Matzke M, Matzke A, Kooter J. RNA: guiding gene silencing. Science 2001; 293:1080-1084.
14. Cogoni C. Homology-dependent gene silencing mechanisms in fungi. Annu Rev Microbiol 2001; 55:381-406.
15. Zamore P. Ancient pathways programmed by small RNAs. Science 2002; 296:1265-1269.
16. Hannon G. RNA interference. Nature 2002; 418:244-251.
17. Matzke M, Mette MF, Jakowitsch J. et al. A test for transvection in plants: DNA pairing may lead to trans-activation of silencing of complex heteroalleles in tobacco. Genetics 2001; 158:451-461.
18. Wassenegger M. RNA-directed DNA methylation. Plant Mol Biol 2000; 43:203-220.
19. Flores R. A naked plant-specific RNA ten-fold smaller than the smallest known viral RNA: the viroid. CR Acad Sci Paris 2001; 324:943-952.
20. Wassenegger M, Heimes S, Riedel L et al. RNA-directed de novo methylation of genomic sequences in plants. Cell 1994; 76:567-576.
21. Mette MF, van der Winden J, Matzke M et al. Production of aberrant promoter transcripts contributes to methylation and silencing of unlinked homologous promoters in trans. EMBO J 1999; 18:241-248.
22. Mette MF, Aufsatz W, van der Winden J et al. Transcriptional silencing and promoter methylation triggered by double-stranded RNA. EMBO J 2000; 19:5194-5201.
23. Jones L, Hamilton A, Voinnet O et al. RNA-DNA interactions and DNA methylation in post-transcriptional gene silencing. Plant Cell 1999; 11:2291-2301.
24. Jones L, Ratcliff F, Baulcombe D. RNA-directed transcriptional gene silencing in plants can be inherited independently of the RNA trigger and requires Met1 for maintenance. Curr Biol 2001; 11:747-757.
25. Sijen T, Vijn I, Rebocho A et al. Transcriptional and posttranscriptional gene silencing are mechanistically related. Curr Biol 2001; 11:436-440.
26. Aufsatz W, Mette MF, van der Winden J et al. RNA-directed DNA methylation in *Arabidopsis*. Proc Natl Acad Sci USA 2002; 99Suppl4:16499-16500.
27. Pélissier T, Thalmeir S, Kempe D et al. Heavy de novo methylation at symmetrical and nonsymmetrical sites is a hallmark of RNA-directed DNA methylation. Nucl Acids Res 1999; 27:1625-1634.
28. Papaefthimiou I, Hamilton A, Denti M. et al. Replicating potato spindle tuber viroid RNA is accompanied by short RNA fragments that are characteristic of post-transcriptional gene silencing. Nucl Acids Res 2001; 29:2395-2400.
29. Itaya A, Folimonov A, Matsuda Y et al. Potato spindle tuber viroid as inducer of RNA silencing in infected tomato. Mol Plant-Microb Inter 2001; 11:1332-1334.

30. Jakowitsch J, Papp I, Moscone E et al. Molecular and cytogenetic characterization of a transgene locus that induces silencing and methylation of homologous promoters in trans. Plant J 1999; 17:131-140.
31. Ingelbrecht, Van Houdt H, Van Montagu et al. Posttranscriptional silencing of reporter transgenes in tobacco correlates with DNA methylation. Proc Natl Acad Sci USA 1994; 91:10502-10506.
32. Kovarik A, Van Houdt H, Holy A et al. Drug-induced hypomethylation of a posttranscriptionally silenced transgene locus of tobacco leads to partial release of silencing. FEBS Lett 2000; 467:47-51.
33. Morel J-B., Mourrain P, Béclin C et al. DNA methylation and chromatine structure affect transcriptional and post-transcriptional transgene silencing in *Arabidopsis*. Curr Biol 2000; 10:1591-1594.
34. Jedelloh J, Stokes T, Richards E. Maintenance of genomic methylation requires a SWI/SNF2-like protein. Nat Genet 1999; 22:94-97.
35. Gibbons R, McDowell T, Raman S et al. Mutations in *ATRX*, encoding a SWI/SNF-like protein cause diverse changes in the pattern of DNA methylation. Nat Genet 2000; 24:368-371.
36. Dennis K, Fan T, Geiman T et al. Lsh, a member of the SNF2 family, is required for genome-wide methylation. Genes Dev 2001; 15:2940-2944.
37. Dobosy J, Selker E. Emerging connections between DNA methylation and histone acetylation. Cell Mol Life Sci 2001; 58:721-727.
38. Tamaru H, Selker E. A histone H3 methyltransferase controls DNA methylation in *Neurospora crassa*. Nature 2001; 414:277-283.
39. Jackson J, Lindroth A, Cao X. et al. Control of CpNpG DNA methylation by the KRYPTONITE histone H3 methyltransferase. Nature 2002; 416:556-560.
40. Jenuwein T. An RNA-guided pathway for the epigenome. Science 2002 in press.
41. Pélissier T, Wassenegger M. A DNA target of 30 bp is sufficient for RNA-directed DNA methylation. RNA 2000; 6:55-65.
42. Hamilton A, Voinnet O, Chappell L et al. (2002) Two classes of short interfering RNA in RNA silencing. EMBO J 2002; 21:4671-4679.
43. Voinnet O, Pinto Y, Baulcombe D. Suppression of gene silencing: a general strategy used by diverse DNA and RNA viruses of plants. Proc Natl Acad Sci USA 1999; 96:14147-14152.
44. Vance V, Vaucheret H. RNA silencing in plants – defense and counter defense. Science 2001; 292:2277-2280.
45. Voinnet O. RNA silencing as a plant immune system against viruses. Trends Genet 2001; 17:449-459.
46. Rhoades M, Reinhart B, Lim L et al. Prediction of plant microRNA targets. Cell 2002; 110:513-515.
47. Llave C, Kasschau K, Rector M et al. Endogenous and silencing-associated small RNAs in plants. Plant Cell 2002; 14:1605-1619.
48. Mette MF, van der Winden J, Matzke M. et al. Short RNAs can identify new candidate transposable element families in *Arabidopsis*. Plant Physiol 2002; in press.
49. Llave C, Kasschau K, Carrington J. Virus-encoded suppressor of posttranscriptional gene silencing targets a maintenance step in the silencing pathway. Proc Natl Acad Sci USA 2000; 97:13401-13406.
50. Mallory A, Ely L, Smith T et al. HC-Pro suppression of transgene silencing eliminates the small RNAs but not transgene methylation of the mobile signal. Plant Cell 2001; 13:571-583.
51. Mette MF, Matzke A, Matzke M. Resistance of RNA-mediated TGS to HC-Pro, a viral suppressor of PTGS, suggests alternative pathways for dsRNA processing. Curr Biol 2001; 11:1119-1123.
52. Billy E, Brondani V, Zhang H et al. Specific interference with gene expression induced by long, double-stranded RNA in mouse embryonal teratocarcinoma cell lines. Proc Natl Acad Sci USA 2001; 98:14428-14433.
53. Zeng Y, Cullen B. RNA interference in human cells is restricted to the cytoplasm. RNA 2002; 8:855-860.
54. Bass B. RNA editing by adenosine deaminases that act on RNA. Annu Rev Biochem 2002; 71:817-846.
55. Schauer S, Jacobsen S, Meinke D et al. *DICER-LIKE1*: blind men and elephants in *Arabidopsis* development. Trends Plant Sci 2002; in press.
56. Jacobsen S, Running M, Meyerowitz E. Disruption of an RNA helicase/RNAse III gene in *Arabidopsis* causes unregulated cell division in floral meristems. Development 1999; 126:5231-5243.
57. Reinhart B, Weinstein E, Rhoades M et al. MicroRNAs in plants. Genes Dev 2002; 16:1616-1626.

58. Park W, Li J, Song R et al. CARPEL FACTORY, a Dicer homolog, and HEN1, a novel protein, act in microRNA metabolism. Curr Biol 2002; 12:1484-1495.
59. Finnegan EJ, Kovac K. Plant DNA methyltransferases. Plant Mol Biol 2000; 443:189-201.
60. Cao X, Jacobsen S. Role of the *Arabidopsis DRM* methyltransferases in de novo methylation and gene silencing. Curr Biol 2002; 12:1138-1144.
61. Kishimoto N, Sakai S, Jackson J et al. Site specificity of the *Arabidopsis* MET1 DNA methyltransferase demonstrated through hypermethylation of the superman locus. Plant Mol Biol 2001; 46:171-183.
62. Henikoff S, Comai L. A DNA methyltransferase homolog with a chromodomain exists in multiple polymorphic forms in *Arabidopsis*. Genetics 1998; 149:307-318.
63. Lindroth A, Cao X, Jackson J et al. Requirement of CHROMOMETHYLASE3 for maintenance of CpXpG methylation. Science 2001; 292:2077-2080.
64. Papa C, Springer N, Muszynski M et al. Maize chromomethylase *Zea methyltransferase2* is required for CpNpG methylation. Plant Cell 2001; 13:1919-1928.
65. Bartee L, Malagnac F, Bender J. *Arabidopsis cmt3* chromomethylase mutations block nonCG methylation and silencing of an endogenous gene. Genes Dev 2001; 15:1753-1758.
66. Lyko F. DNA methylation learns to fly. Trends Genet 2001; 17:169-172.
67. Pinarbasi R, Elliott J, Hornby D. Activation of a yeast pseudo DNA methyltransferase by deletion of a single amino acid. J Mol Biol 1996; 12:804-813.
68. Bestor T. The DNA methyltransferase of mammals. Hum Mol Genet 2000; 9:2395-2402.
69. Lorincz M, Groudine M. CmC(a/t)GG methylation: a new epigenetic mark in mammalian DNA?. Proc Natl Acad Sci USA 2001; 98:10034-10036.
70. Leonhardt H, Page A, Weier H.-U. et al. A targeting sequence directs DNA methyltransferase to sites of DNA replication in mammalian nuclei. Cell 1992; 71:865-873.
71. Akhtar A, Zink D, Becker P. Chromodomains are protein-RNA interaction modules. Nature 2000; 407:405-409.
72. Wheeler J, VanderZwan C, Xu X et al. Distinct in vivo requirements for establishment versus maintenance of transcriptional repression. Nat Genet Jul 29 (epub ahead of print) 2002.
73. Robyr D, Suka Y, Xenarios I et al. Microarray deacetylation maps determine genome-wide functions for yeast histone deacetylases. Cell 2002; 190:437-446.
74. Ramsahoye B, Biniszkiewicz D, Lyko F et al. NonCpG methylation is prevalent in embryonic stem cells and may be mediated by DNA methyltransferase 3a. Proc Natl Acad Sci USA 2000; 97:5237-5242.
75. Pal-Bhadra M, Bhadra U, Birchler J. RNAi related mechanisms affect both transcriptional and posttranscriptional transgene silencing in *Drosophilae*. Mol Cell 2002; 9:315-327.
76. Volpe T, Kidner C, Hall I et al. Regulation of heterochromatin silencing and histone H3 lysine-9 methylation by RNAi. Science 2002; in press.
77. Dudley N, Labbé J-C, Goldstein B. Using RNA interference to identify genes required for RNA interference. Proc Natl Acad Sci USA 2002; 99:4191-4196.
78. Dong A, Yoder J, Zhang X et al. Structure of human DNMT2, an enigmatic DNA methyltransferase homolog that displays denaturant-resistant binding to DNA. Nucl Acids Res 2001; 29:439-448.
79. Burgers W, Fuks F, Kouzarides T. DNA methyltransferases get connected to chromatin. Trends Genet 2002; 18:275-277.
80. Sleutels F, Barlow D, Lyle R. The uniqueness of the imprinting mechanism. Curr Opin Genet Devel 2000; 10:229-233.
81. Pfeifer K, Tilghman S. Allele-specific gene expression in mammals: the curious case of the imprinted RNAs. Genes Dev 1994; 8:1867-1874.
82. Matzke M, Mette MF, Aufsatz W et al. Host defenses to parasitic sequences and the evolution of epigenetic control mechanisms. Genetica 1999; 107:271-287.
83. Tabara H, Sarkissian, M. et al. The *rde-1* gene, RNA interference, and transposon silencing in *C. elegans* Cell 1999; 99:123-132.
84. Ketting R, Haverkamp T, van Leunen H et al. *mut-7* of *C. elegans*, required for transposon silencing and RNA interference, is a homolog of Werner syndrome helicase and RNaseD. Cell 1999; 99:133-141.

85. Okamoto H, Hirochika H. Silencing of transposable elements in plants. Trends Plant Sci 2001; 6:527-534.
86. Wu-Scharf D, Jeong B, Zhang C et al. Transgene and transposon silencing in *Chlamydomonas reinhardtii* by a DEAH-Box RNA helicase. Science 2000; 290:1159-1162.
87. Miura A, Yonebayashi S, Watanabe K et al. Mobilization of transposons by a mutation abolishing full DNA methylation in *Arabidopsis*. Nature 2001; 411:212-214.
88. Singer T, Yordan C, Martienssen R. Robertson's *Mutator* transposons in *A. thaliana* are regulated by the chromatin-remodeling gene *Decrease in DNA methylation* (*DDM1*). Genes Dev 2001; 15:591-602.
89. Tompa R, McCallum C, Delrow J et al. Genome-wide profiling of DNA methylation reveals transposon targets of CHROMOMETHYLASE3. Curr Biol 2001; 12:65-68.
90. Mourrain P, Béclin C, Elmayan T et al. *Arabidopsis SGS2* and *SGS3* genes are required for post-transcriptional gene silencing and natural virus resistance. Cell 2000; 101:533-542.
91. Matzke M, Mette MF, Matzke A. Transgene silencing by the host genome defense: implications for the evolution of epigenetic control mechanisms in plants and vertebrates. Plant Mol Biol 2000; 43:401-415.
92. Mazzone M, De Gregorio E, Lavitola A et al. Whole-genome organization and functional properties of miniature DNA insertion sequences conserved in pathogenic *Neisseriae*. Gene 2001; 278:211-222.
93. De Gregorio E, Abrescia C, Carlomagno et al. The abundant class of nemis provides RNA substrates for ribonuclease III in *Neisseria*. Biochim Biophys Acta 2002; 1576:39-44.
94. Ruvkun G. Glimpses of a tiny RNA world. Science 2001; 294:797-799.
95. Hamilton A, Baulcombe D. A species of small antisense RNA in posttranscriptional gene silencing in plants. Science 1999; 286: 950-952.

CHAPTER 11

Brain-Specific Nonmessenger RNAs

Jürgen Brosius, Alexander Hüttenhofer and Henri Tiedge

Abstract

RNAs that do not encode proteins, as do messenger RNAs, play much more prominent roles in the functioning of cells than we first anticipated–qualitatively and quantitatively. At least in Eukarya one has the impression that we have hardly left the RNA world. This RNA life is not only a remnant from the RNA/RNP worlds but is still ongoing, as even young RNAs have been shown to contribute to a variety of cellular tasks. The brain, in particular, exhibits a number of nonmessenger RNAs that are brain specific and in most cases even neuron-specific. This chapter lists prominent examples of neuron-specific RNAs that do not encode proteins, with emphasis on small nonmessenger RNAs (snmRNAs) and discusses their potential functions in those cells.

Introduction

Apart from ribosomal RNAs and transfer RNAs that are known to be involved in the process of translation, other nonmessenger RNAs (nmRNAs) were discovered around three to four decades ago (for a recent review, see ref. 1). When RNAs were first shown to exhibit enzymatic activity[2,3] and, thus, established the likelihood of an RNA world,[4] cellular nmRNAs were considered fossils or remnants from a bygone era that gradually were to be replaced by the "better" macromolecules, proteins. The discovery of nmRNAs including large and small nonmessenger RNAs (snmRNAs) that were restricted to certain orders, sub-orders, genera or even species, raised the possibility that novel nmRNA species arose even in relatively recent evolutionary times.[5-10] The alternative scenario, that the respective genes encoding such RNAs are ancient and were independently lost in all the other lineages is virtually impossible.

As we compare the total genomic sequences from more and more different species, we begin to realize that, at least in closely related species, we do not encounter large numbers of novel genes. Instead, we find that a major difference lies in the differential expression of orthologous genes that are shared between the species or perhaps of newly arisen paralogous genes (J. Brosius, in preparation). Many instances of differential gene expression will have been precipitated by the action of retronuons,[11] where a nuon is any definable stretch of a nucleic acid.[12-14] In the order of Primates, it seems to be the brain that has had the greatest "need" to recruit or exapt[12,15,16] genes that, in other organisms, were expressed at low levels or were altogether absent in that tissue. A comparison of proteins expressed in the brains of chimpanzee and man revealed that changes in gene expression of orthologous genes was most pronounced in the brain.[17]

The fact that many young nmRNAs are expressed exclusively or preferentially in brain, and are even restricted there to neurons, is testimony to the hypothesis that exaptation of

Noncoding RNAs: Molecular Biology and Molecular Medicine, edited by Jan Barciszewski and Volker A. Erdmann.

RNAs into the central nervous system was also instrumental in the molecular changes that underlie its expanded capabilities in certain lineages. Here we list and describe known brain-specific nmRNAs and speculate about their functional roles in brain.

Large nmRNAs in the Brain

Nonmessenger RNAs can arbitrarily be divided into two categories: small ones below and large ones above ~500 nucleotides. Large nmRNAs often resemble mRNAs in that they are polyadenylated and processed from a primary transcript by intron splicing but lack open reading frames (ORFs). Examples of such RNAs are summarized in recent reviews.[18-23] The similarity to mRNAs points to the possibility that most of these RNAs originated as mRNAs and lost their protein coding function because the protein products were no longer under positive selection. In this model, such RNAs were concomitantly coopted or exapted into novel functions or they maintained functions that they possessed in addition to protein coding (see below). An example of such a brain-specific nmRNA is KLHL1AS RNA. It partially overlaps (in the opposite orientation) an mRNA-yielding transcript that encodes a member of the family of actin-organizing proteins (KLHL1).[24] Interestingly, spinocerebellar ataxia type 8 and some major psychoses are caused by the KLHL1AS RNA, presumably via a CUG trinucleotide repeat extension that does not involve an associated polypeptide extension in a cognate protein, due to the absence of an ORF.[25,26] Another large nonmessenger RNA is Disrupted-in-schizophrenia 2 (DISC2) that is antisense to Disrupted-In-Schizophrenia 1 (DISC1) mRNA on a locus, disruption of which, might be involved in Schizophrenia.[27] In the rat, ntab has been reported as a brain–specific transcript of diffuse size (~1.2 – 5 kb). The hippocampus (CA3 region and dentate gyrus) features high levels of ntab expression and it is localized to dendritic processes. Although polyadenylated, the transcript does not appear to encode a protein.[28]

As will be discussed in more detail below, a number of the transcripts from the Prader-Willi/Angelman syndrome region of chromosome 15 (such as IPW and the RNA that is antisense to the 3' part of the UBE3A transcript) were host genes for snoRNAs. Likewise, the brain-specific Bsr RNA in rat[8] has been shown to be a host transcript for RBII-36 C/D box snoRNA.[29] Many large nonmessenger RNAs may turn out to be precursors for snoRNAs and perhaps other snmRNAs such as microRNAs (miRNAs).

snmRNAs in the Brain

Since the discovery in the eighties of the first snmRNA in the brain, BC1 RNA,[30,31] a number of additional such RNAs have been identified. We will discuss two groups that are distinguished by their subcellular localization, namely cytoplasmic and nuclear (including nucleolar) snmRNAs.

Cytoplasmic snmRNAs

BC1 RNA in rodents[5] and BC200 RNA in anthropoid primates[6,7,32] are similar but not homologous cytoplasmic RNAs that are almost exclusively expressed in the nervous system.[31,33] Depending on the species, BC1 RNA and BC200 RNA are about 150 and 200 nucleotides long, respectively; both are transcribed by RNA polymerase III.[6,31,33,34] They are both expressed from single loci ;[6,7,31,32] BC1 from distal chromosome 7 in mouse[35] and BC200 on chromosome 2p16 in man.[36] They arose by retroposition from $tRNA^{Ala}$ and a monomeric Alu element, respectively.[6,31,37,38] BC1 RNA is the first identified, direct master gene for short interspersed repetitive elements (SINEs), the subclass of rodent ID repetitive elements.[39,40] BC200 RNA was much less efficient as a master gene when compared to its founder, a monomeric Alu element, but still gave rise to more than 200 retronuons in the human genome.[32]

Both RNAs can be subdivided into three domains. A 5' domain is derived from the founder genes, $tRNA^{Ala}$ and a monomeric Alu element, respectively. The latter in turn represents the Alu domain of 7SL RNA from the signal recognition particle (SRP). The central A-rich domains (~50 nucleotides) have been acquired by both RNAs during the retropositional events. Both RNAs feature a short (~30 nucleotides) region at their 3' ends that have been coopted from the locus of integration. These regions are termed "unique", as they are not related to any of the ancestral repetitive elements or to the retronuons that are generated from BC1 and BC200 RNA master genes. As is the case for most RNA polymerase III transcripts, both coding regions terminate with a stretch of at least four uridine residues. The 5' region of BC1 RNA exhibits close to 80% sequence similarity to $tRNA^{Ala}$ but is not folded like a tRNA. Instead, secondary structure predictions as well as enzymatic and chemical probing revealed a long stem-loop with a few bulges.[41] The 5' end of BC200 RNA can be folded like the Alu domain of SRP RNA.[7,42]

As expected, BC1 and BC200 RNAs are complexed with proteins in the cell and sediment as particles of 8.7S and 11.4S, respectively.[43-45] Poly(A) binding protein is associated with both RNPs, an interaction that is mediated by their central A-rich regions.[46,47] Also not surprisingly, BC200 RNA binds SRP9/14 in vitro and in vivo.[42] This protein dimer is also contained in the native signal recognition particle, binding to the Alu domain that both 7SL RNA and BC200 RNA share. Identification of proteins that are bona fide components of a given RNP is not trivial. Numerous proteins bind more or less specifically to RNA. Fishing in cellular extracts with naked RNA for genuine binding partners can be problematic. The list of additional putative protein binding partners for BC1 RNA may serve as an example. Unless one proposes a number of temporal interactions, the list already exceeds the number of different protein species that could bind a particle of moderate size 8.7S or at most 10S.[43,44] For most of these proposed binding proteins there is only indirect or incomplete evidence. They include at least two species of Bp-1 proteins,[48] translin,[49] and pur alpha.[50] Binding of translin to BC1 RNA could not be shown in other laboratories;[46] (Muddashetty MS, Khanam T, Kremerskothen J, unpublished observations). Most recently, a 190 kD protein, p190RhoGEF, having previously been shown to interact with a domain on neurofilament light chain messenger RNA, has been reported to interact via its C-terminus with the 5' domain of BC1 RNA.[51] This would constitute a link between the BC1 RNP and microtubules. On the other hand, this binding may be artefactual or transient. Analogously, antibodies against the La protein immunoprecipitated both BC1 and BC200 RNPs that contained the respective RNA.[52] Nevertheless, this association might be artefactual or temporal, as we could not detect La in dendritic processes of neurons by immunocytochemistry (M. Bundman, unpublished).

This leads to another interesting parallel between both RNPs, their sub cellular location: BC1 and BC200 RNA are expressed in neurons but not in glial cells. They are not restricted to neuronal cell bodies but are also transported into the distal parts of dendritic processes[33,53-57] where components of the translational apparatus[58-60] as well as selected messenger RNAs[61] are also located. The only other significant locus of expression for both RNAs is the testes. BC1 RNA is found in rodent testes[40] and BC200 RNA in primate testes.[32] BC1 RNA is expressed in premeiotic spermatogonia, with particularly high amounts in syncytial ensembles of cells that are primed for synchronous spermatogenic differentiation.[62] Furthermore, BC1 RNA has been shown to be developmentally expressed.[56] Expression onset correlates with synapse formation in the respective neurons. In primary hippocampal cultures, BC1 RNA expression depends on transsynaptic activity as it is inhibited by tetrodotoxin.[63] Interestingly, the expression of both RNAs is deregulated in immortalized cell lines and certain tumors, including those of nonneuronal origin.[64,65]

BC1 RNA[66] and BC200 RNA (unpublished) are specifically transported to dendrites. BC1 RNA was found enriched in postsynaptic microdomains,[67] sites of local translation in dendrites. BC1 RNA was also identified in axons of a select subset of neurons,[68] and it has recently been shown to be targeted, in a two-step mechanism that requires the sequential participation of microtubules and actin filaments, to local axonal sites of protein synthetic capacity.[69] These and other data have prompted the hypothesis that BC1 RNA and BC200 RNA play a functional role in the translational control of gene expression in local neuronal domains.

This hypothesis has recently received experimental support from evidence that BC1 RNA is a specific repressor of translation in cell-free systems;[70] and (A. Kondrachov et al, in preparation). These data indicate that BC1 RNA represses translation by inhibiting assembly of 48S preinitiation complexes. Significantly, such repression is effective not only in the cap-dependent mode of translation initiation, but also in an internal ribosome entry mode of initiation provided the latter is mediated by the eIF4 group of translation initiation factors. The results suggest that BC1 RNA acts as a repressor of translation that targets a step in the initiation pathway at a point just before the system commits itself to elongation. We submit that such translational repression is in place in neuronal microdomains as a default mechanism, to prevent ectopic expression and to ensure that proteins are only produced when needed.[56,58]

Nucleolar snmRNAs in the Brain

Although some developmentally regulated isoforms of U1 small nuclear RNA were reported in the mid-eighties in various mouse tissues,[71] the first report of a neuron-specific nuclear RNA was that of several small nucleolar RNAs (snoRNAs) in mouse and human brain.[72] In vertebrates, most snoRNAs are encoded by introns and are processed from these by splicing and specific exo- and endonucleolytic cleavage, leading to the mature 5'- and 3'- ends of snoRNAs. The finding of brain-specific snoRNAs was surprising because the function of snoRNAs was assumed to be modification of nucleotides, mainly of ribosomal RNAs and to a lesser extent of some snRNAs and tRNAs.[73-78] Could there be cell-type-specific modification of rRNA?

Imprinting is an epigenetic phenomenon that restricts gene expression to only one chromosome, either the paternal or maternal allele.[79] Most snoRNAs that are preferentially expressed in brain were shown to be imprinted.[72] An apparent exception to this rule is HBI-36/MBI-36 snoRNA, which is also exclusively expressed in brain but is not subject to imprinting.[72,75] However, expression of this snoRNA gene occurs from one chromosome only since its host gene (serotonin receptor 5-HT2C) maps to the X chromosome. Thus it appears that all brain-specific snoRNAs detected so far are exclusively transcribed from one chromosome only. Southern blot analysis demonstrates that HBI-36/MBI-36 is present as a single copy gene[72] in the second intron of its host gene. Ubiquitous, intronic modification guide snoRNAs in vertebrates are generally hosted by genes coding for proteins that are directly involved in ribosome biogenesis or function. The data indicate that this peculiar gene organization could provide the basis for a coordinate production of functionally linked gene products.[80-82] In this context, it is tempting to speculate that the presence of brain-specific H/ACA box snoRNA HBI-36 within an intron of the serotonin 5-HT2C receptor might reflect some functional link between HBI-36 snoRNA and its host gene.

Brain-specific imprinted snoRNA genes have been recently detected at two human chromosomal loci, 15q11q13 and 14q32, where they share the same unusual genomic organization.[72,83] The novel snoRNAs are intron-encoded like all guide snoRNAs previously reported in vertebrates. However, the snoRNA-containing intron and flanking exons are tandemly repeated to scores of copies. The snoRNAs are processed from complex transcription units spanning the entire snoRNA repeat array which also give rise to spliced RNAs lacking open reading frames. The function, if any, of the spliced host-genes for snoRNAs remains unknown.

The first imprinted snoRNA genes, MBII-13, MBII-85 and MBII-52, were identified through a systematic search for small nonmessenger RNAs expressed in mouse brain.[72] Their genes map at mouse chromosome 7C and their human homologs, HBII-13, -85 and -52, at the syntenic 15q11q13 region. HBII-52 and HBII-85 are arranged into two tandem arrays of 27 and 47 snoRNA gene copies, respectively, each one embedded within an ~ 2 kb long repeat unit.[72,84] HBII-13 snoRNA is structurally related to HBII-52 but is encoded by a single copy gene. Within each of the two HBII-85 and HBII-52 clusters, copies of each snoRNA sequence are highly similar to each other (90 and 94% identity, respectively) whereas surrounding sequences in the repeat unit diverge substantially from each other. The snoRNA gene organization at the syntenic mouse locus is similar but its analysis remains incomplete.[72,85]

While the flanking sequences in human and rodents diverge extensively from one another, the three novel snoRNAs are strongly conserved, pointing to their biological importance.[72] The human 15q11q13 locus is associated with two very different human neurological disorders, Prader-Willi (PWS) or Angelman syndromes (AS), that result from the loss of paternal or maternal gene expression within this region, respectively.[86] The three C/D snoRNAs are not expressed in PWS patients with a paternal deletion of the imprinted locus or in a mouse model mimicking some aspects of the PWS phenotype, indicating they are only expressed from the paternal allele.[72]

The snoRNAs at this locus are processed from a very large transcript (>460 kb), where, in the antisense orientation, the 3' region reaches the maternally expressed UBE3A gene.[84] This, in turn, might regulate UBE3A paternal expression. While the three imprinted snoRNAs are processed from the same transcript, only HBII-52 is strictly brain-specific, an observation that suggests the involvement of tissue-specific RNA processing events. Using biocomputational methods, four additional snoRNAs (HBII-436, HBII-437, HBII-438A and HBII-438B) were identified within this 460 kb transcript.[84] Almost all of the snoRNAs are encoded within the introns of this large transcript. Northern blot analysis indicates that most if not all of these snoRNAs are indeed expressed by processing from these introns. As there is no evidence of other genes in this region, which appears to be critical for the neonatal Prader-Willi syndrome phenotype in mice, lack of all or some of these snoRNAs may be causally related to this disease.[84]

Remarkably, among the brain-specific snoRNAs from the PWS locus, C/D snoRNA HBII-52 displays an unusually long (18 nt), phylogenetically conserved antisense element to a mRNA - rather than to a ribosomal or to a spliceosomal RNA - that is specifically expressed in the brain.[72] This snoRNA could play a key role in the processing of its presumptive RNA target, the serotonin receptor 5-HT2c mRNA. Strikingly, the mRNA nucleotide potentially targeted for 2'-O-methylation by snoRNA MBII-52 is also the subject of a physiologically important adenosine-to-inosine editing.[87] In vitro, 2'-O-methylation of the adenosine to be edited dramatically inhibits its deamination to inosine,[88] suggesting a role for MBII-52 in the regulation of 5-HT2C mRNA editing. Alternatively, the same antisense element in snoRNA MBII-52 might also control alternative splicing of 5-HT2c premRNA at a nearby splice site, through steric occlusion instead of nucleotide modification.

A mere regulatory function, however, is difficult to reconcile with the high abundance of HBII-52 and HBII-85 snoRNAs. An alternative function of HBII-52, HBII-85 and similarly expressed snoRNAs could be regulating not only their own imprinted expression but also that of neighboring genes via interaction with their own genomic locus by RNA/DNA hybridization. Perhaps binding of the snoRNAs precludes binding of other macromolecules, such as proteins, to DNA or vice versa. Alternatively, in analogy to RNA, perhaps snoRNA-binding guides, the methylation of bases in DNA or guides demethylation or perhaps prevents one or the other.

Three additional brain-specific snoRNAs have been identified in mouse, designated as MBII-48, MBII-49 and MBII-78, by an experimental RNomics approach.[75] So far neither the genomic loci nor the potential functions of these brains specific snmRNAs have been elucidated.

Another brain-specific snoRNA, designated as RBII-36, was isolated by screening a rat C/D snoRNA library.[29] It is encoded within introns of a previously described, nonprotein coding gene, Bsr, that spans an array of ~ 100 snoRNA-containing tandem repeated units of 0.9 kb.[8] This situation is strongly reminiscent of the snoRNA gene organization observed at the PWS locus. RBII-36 snoRNA is exclusively expressed in neurons where it exhibits a nucleolar localization. While RBII-36 and its Bsr host-gene could only be detected in rat, the human and mouse loci syntenic to the rat Bsr-containing 6q32 locus, human 14q32 and mouse distal 12, respectively, also contain unrelated tandemly repeated arrays of novel, imprinted brain-specific C/D snoRNAs.[83] Interestingly, in contrast to PWS-encoded snoRNAs, these intronic snoRNAs are only expressed from the maternal allele.

MicroRNAs in the Brain

A plethora of short (21-23 nucleotides) RNAs have been discovered recently that partake in a variety of regulatory mechanisms in the cell. These microRNAs (miRNAs) - which also include short interfering RNAs - are transcribed from larger precursors and are processed by the enzyme, dicer. They may act as double and single-stranded RNAs mainly in modulating mRNA translation either by binding to 3' UTRs or by specifically degrading mRNAs. More detailed treatment of their biogenesis and mode of action can be found in several reviews.[77,89-91] Most recently, microRNAs have been identified in the yeast *Saccharomyces pombe* where they are proposed to be involved in heterochromatic silencing by methylating a lysine residue in histone H3.[92,93] The first members of the miRNA class, let-7 and lin-4, were described in *C. elegans* as small temporal RNAs (stRNAs) as they are developmentally regulated and themselves control developmental programs.[94-96] Subsequently, hundreds of apparently ubiquitous microRNAs were reported.[97-102] However, it was only a matter of time until microRNAs were reported that are preferentially expressed in certain tissues such as brain[103]. Presumably, such miRNAs target RNAs in specific cell types.

Outlook

Not only the numerical explosion of nonmessenger RNA discoveries but also their pervasive and varied functional roles in cells from all three domains of life should convince researchers that studying nontranslated RNAs is a worthwhile endeavor. The final push may come from recent findings that snmRNAs play a role in the etiology of disease. For example, mutations in the RNA component of the endoribonuclease ribonucleoprotein complex cause cartilage-hair hypoplasia.[104-106] Deletion analysis in yeast demonstrated a role of RNase MRP in cell cycle progression at the end of mitosis. Conservation of this function in humans may explain many of the pleiotropic phenotypes of cartilage hair hypoplasia.[107] A further link of a nonmessenger RNA to a genetic disease was established in Dyskeratosis congenita that is characterized by progressive bone-marrow failure, abnormal skin pigmentation and a predisposition to certain types of malignancy. Vuillamy et al demonstrated that X-linked and autosomal dominant forms of the disease result from mutations in two different genes encoding dyskerin, an H/ACA snoRNP protein, and telomerase RNA, respectively[108] The sheer number of nonmessenger RNAs in genomes will guarantee that these are not isolated cases.

Acknowledgements

A.H. and J.B. are supported by the German Human Genome Project through the BMBF (#01KW9966); H.T. by the National Institutes of Health (grant NS34158). We would like to thank Marsha Bundman for editing the manuscript.

References

1. Hüttenhofer A, Brosius J. Experimental RNomics. In: Galperin M, Koonin EV, eds. Functional Genomics. New York: Horizon Scientific Press, 2002.
2. Kruger K, Grabowski PJ, Zaug AJ et al. Self-splicing RNA: autoexcision and autocyclization of the ribosomal RNA intervening sequence of *Tetrahymena*. Cell 1982; 31:147-157.
3. Guerrier-Takada C, Gardiner K, Marsh T et al. The RNA moiety of ribonuclease P is the catalytic subunit of the enzyme. Cell 1983; 35:849-857.
4. Gesteland RF, Cech TR, Atkins JF. The RNA World. NY: Cold Spring Harbor Laboratory Press, 1999.
5. Martignetti JA, Brosius J. Neural BC1 RNA as an evolutionary marker: guinea pig remains a rodent. Proc Natl Acad Sci USA 1993; 90:9698-9702.
6. Martignetti JA, Brosius J. BC200 RNA: a neural RNA polymerase III product encoded by a monomeric Alu element. Proc Natl Acad Sci USA 1993; 90:11563-11567.
7. Skryabin BV, Kremerskothen J, Vassilacopoulou D et al. The BC200 RNA gene and its neural expression are conserved in *Anthropoidea* (Primates). J Mol Evol 1998; 47:677-685.
8. Komine Y, Tanaka NK, Yano R et al. A novel type of noncoding RNA expressed in the rat brain. Brain Res Mol Brain Res 1999; 66:1-13.
9. Gogolevskaya IK, Kramerov DA. Evolutionary history of 4.5SI RNA and indication that it is functional. J Mol Evol 2002; 54:354-364.
10. Wang W, Brunet FG, Nevo E et al. Origin of sphinx, a young chimeric RNA gene in *Drosophila melanogaster*. Proc Natl Acad Sci USA 2002; 99:4448-4453.
11. Brosius J. The contribution of RNAs and retroposition to evolutionary novelties. Genetica 2003; 188:in press.
12. Brosius J, Gould SJ. On "genomenclature": a comprehensive (and respectful) taxonomy for pseudogenes and other "junk DNA". Proc Natl Acad Sci USA 1992; 89:10706-10710.
13. Brosius J, Gould SJ. Molecular constructivity. Nature 1993; 365:102.
14. Brosius J. Gene duplication and other evolutionary strategies: from the RNA world to the future. J Struct Funct Genomics 2003; 3:1-17.
15. Gould SJ. The Structure of Evolutionary Theory. MA: Cambridge: The Belknap Press of Harvard University Press, 2002.
16. Gould SJ, Vrba ES. Exaptation - a missing term in the science of form. Paleobiology 1982; 8:4-15.
17. Enard W, Khaitovich P, Klose J et al. Intra- and interspecific variation in primate gene expression patterns. Science 2002; 296:340-343.
18. Erdmann VA, Barciszewska MZ, Szymanski M et al. The noncoding RNAs as riboregulators. Nucleic Acids Res 2001; 29:189-193.
19. Erdmann VA, Barciszewska MZ, Hochberg A et al. Regulatory RNAs. Cell Mol Life Sci 2001; 58:960-977.
20. Mattick JS. Noncoding RNAs: the architects of eukaryotic complexity. EMBO Rep 2001; 2:986-991.
21. Mattick JS, Gagen MJ. The evolution of controlled multitasked gene networks: the role of introns and other noncoding RNAs in the development of complex organisms. Mol Biol Evol 2001; 18:1611-1630.
22. Herbert A, Rich A. RNA processing and the evolution of eukaryotes. Nat Genet 1999; 21:265-269.
23. Eddy SR. Noncoding RNA genes and the modern RNA world. Nat Rev Genet 2001; 2:919-929.
24. Nemes JP, Benzow KA, Moseley ML et al. The SCA8 transcript is an antisense RNA to a brain-specific transcript encoding a novel actin-binding protein (KLHL1). Hum Mol Genet 2000; 9:1543-1551.
25. Koob MD, Moseley ML, Schut LJ et al. An untranslated CTG expansion causes a novel form of spinocerebellar ataxia (SCA8). Nat Genet 1999; 21:379-384.

26. Vincent JB, Yuan QP, Schalling M et al. Long repeat tracts at SCA8 in major psychosis. Am J Med Genet 2000; 96:873-876.
27. Millar JK, Wilson-Annan JC, Anderson S et al. Disruption of two novel genes by a translocation cosegregating with schizophrenia. Hum Mol Genet 2000; 9:1415-1423.
28. French PJ, Bliss TV, O'Connor V. Ntab, a novel noncoding RNA abundantly expressed in rat brain. Neuroscience 2001; 108:207-215.
29. Cavaillé J, Vitali P, Basyuk E et al. A novel brain-specific box C/D small nucleolar RNA processed from tandemly repeated introns of a noncoding RNA gene in rats. J Biol Chem 2001; 276:26374-26383.
30. Sutcliffe JG, Milner RJ, Bloom FE et al. Common 82-nucleotide sequence unique to brain RNA. Proc Natl Acad Sci USA 1982; 79:4942-4946.
31. DeChiara TM, Brosius J. Neural BC1 RNA: cDNA clones reveal nonrepetitive sequence content. Proc Natl Acad Sci USA 1987; 84:2624-2628.
32. Kuryshev VY, Skryabin BV, Kremerskothen J et al. Birth of a gene: locus of neuronal BC200 snmRNA in three prosimians and human BC200 pseudogenes as archives of change in the *Anthropoidea* lineage. J Mol Biol 2001; 309:1049-1066.
33. Tiedge H, Chen W, Brosius J. Primary structure, neural-specific expression, and dendritic location of human BC200 RNA. J Neurosci 1993; 13:2382-2390.
34. Martignetti JA, Brosius J. BC1 RNA: transcriptional analysis of a neural cell-specific RNA polymerase III transcript. Mol Cell Biol 1995; 15:1642-1650.
35. Taylor BA, Navin A, Skryabin BV et al. Localization of the mouse gene (Bc1) encoding neural BC1 RNA near the fibroblast growth factor 3 locus (Fgf3) on distal chromosome 7. Genomics 1997; 44:153-154.
36. Basile V, Vicente A, Martignetti JA et al. Assignment of the human BC200 RNA gene (BCYRN1) to chromosome 2p16 by radiation hybrid mapping. Cytogenet Cell Genet 1998; 82:271-272.
37. Brosius J, Tiedge H. Reverse transcriptase: mediator of genomic plasticity. Virus Genes 1995; 11:163-179.
38. Brosius J. Genomes were forged by massive bombardments with retroelements and retrosequences. Genetica 1999; 107:1209-238.
39. Kim J, Martignetti JA, Shen MR et al. Rodent BC1 RNA gene as a master gene for ID element amplification. Proc Natl Acad Sci USA 1994; 91:3607-3611.
40. Deininger PL, Tiedge H, Kim J et al. Evolution, expression, and possible function of a master gene for amplification of an interspersed repeated DNA family in rodents. In: Cohn WE, Moldave K, eds. Progr Nucleic Acids Res. 1996; 52:67-88.
41. Rozhdestvensky TS, Kopylov AM, Brosius J et al. Neuronal BC1 RNA structure: Evolutionary conversion of a tRNA-Ala domain into an extended stem-loop structure. RNA 2001; 7:722-730.
42. Kremerskothen J, Zopf D, Walter P et al. Heterodimer SRP9/14 is an integral part of the neural BC200 RNP in primate brain. Neurosci Lett 1998; 245:123-126.
43. Kobayashi S, Goto S, Anzai K. Brain-specific small RNA transcript of the identifier sequences is present as a 10 S ribonucleoprotein particle. J Biol Chem 1991; 266:4726-4730.
44. Cheng JG, Tiedge H, Brosius J. Identification and characterization of BC1 RNP particles. DNA Cell Biol 1996; 15:549-559.
45. Cheng JG, Tiedge H, Brosius J. Expression of dendritic BC200 RNA, component of a 11.4S ribonucleoprotein particle, is conserved in humans and simians. Neurosci Lett 1997; 224:206-210.
46. West N, Roy-Engel AM, Imataka H et al. Shared protein components of SINE RNPs. J Mol Biol 2002; 321:423-432.
47. Muddashetty R, Khanam T, Kondrashov A et al. Poly(A)-binding protein is associated with neuronal BC1 and BC200 ribonucleoprotein particles. J Mol Biol 2002; 321:433-445.
48. Kobayashi S, Tokuno T, Suzuki K et al. Developmental change in sub cellular location of Bp-1 protein with an ability to interact with both identifier sequence and its brain-specific transcript, BC-1 RNA. Biochem Biophys Res Commun 1992; 189:53-58.
49. Kobayashi S, Takashima A, Anzai K. The dendritic translocation of translin protein in the form of BC1 RNA protein particles in developing rat hippocampal neurons in primary culture. Biochem Biophys Res Commun 1998; 253:448-453.

50. Kobayashi S, Agui K, Kamo S et al. Neural BC1 RNA associates with pur alpha, a single-stranded DNA and RNA binding protein, which is involved in the transcription of the BC1 RNA gene. Biochem Biophys Res Commun 2000; 277:341-347.
51. Ge W, Wu J, Zhai J et al. Binding of p190RhoGEF to a destabilizing element on the light neurofilament mRNA is competed by BC1 RNA. J Biol Chem 2002; 277:42701-42705.
52. Kremerskothen J, Nettermann M, op de Bekke A et al. Identification of human autoantigen La/SS-B as BC1/BC200 RNA-binding protein. DNA Cell Biol 1998; 17:751-759.
53. Tiedge H, Fremeau RT, Weinstock PH et al. Dendritic location of neural BC1 RNA. Proc Natl Acad Sci USA 1991; 88:2093-2097.
54. Tiedge H, Dräger UC, Brosius J. Murine BC1 RNA in dendritic fields of the retinal inner plexiform layer. Neurosci Lett 1992; 141:136-138.
55. Lin Y, Brosius J, Tiedge H. Neuronal BC1 RNA: coexpression with growth-associated protein-43 messenger RNA. Neuroscience 2001; 103:465-479.
56. Brosius J, Tiedge H. In: Lipshitz HD, ed. Neural BC1 RNA: Dendritic localization and transport in Localized RNAs. TX, Austin: R.G. Landes, 1995:289-330.
57. Brosius J, Tiedge H. In: Richter D, ed. Neuronal BC1 RNA: Intracellular transport and activity-dependent modulation in Cell polarity and sub cellular RNA localization. Berlin: Springer Verlag, 2001:129-138.
58. Tiedge H, Brosius J. Translational machinery in dendrites of hippocampal neurons in culture. J Neurosci 1996; 16:7171-7781.
59. Gardiol A, Racca C, Triller A. Dendritic and postsynaptic protein synthetic machinery. J Neurosci 1999; 19:168-179.
60. Torre ER, Steward O. Protein synthesis within dendrites: glycosylation of newly synthesized proteins in dendrites of hippocampal neurons in culture. J Neurosci 1996; 16:5967-5978.
61. Job C, Eberwine J. Localization and translation of mRNA in dendrites and axons. Nat Rev Neurosci 2001; 2:889-898.
62. Muslimov IA, Lin Y, Heller M et al. A small RNA in testis and brain: implications for male germ cell development. J Cell Sci 2002; 115:1243-1250.
63. Muslimov IA, Banker G, Brosius J et al. Activity-dependent regulation of dendritic BC1 RNA in hippocampal neurons in culture. J Cell Biol 1998; 141:1601-1611.
64. Chen W, Bocker W, Brosius J et al. Expression of neural BC200 RNA in human tumours. J Pathol 1997; 183:345-351.
65. Chen W, Heierhorst J, Brosius J et al. Expression of neural BC1 RNA: induction in murine tumours. Eur J Cancer 1997; 33:288-292.
66. Muslimov IA, Santi E, Homel P et al. RNA transport in dendrites: a cis-acting targeting element is contained within neuronal BC1 RNA. J Neurosci 1997; 17:4722-4733.
67. Chicurel ME, Terrian DM, Potter H. mRNA at the synapse: analysis of a preparation enriched in hippocampal dendritic spines. J Neurosci 1993; 13:4054-4063.
68. Tiedge H, Zhou A, Thorn NA et al. Transport of BC1 RNA in hypothalamo-neurohypophyseal axons. J Neurosci 1993; 13:4114-4219.
69. Muslimov IA, Titmus M, Koenig E et al. Transport of Neuronal BC1 RNA in Mauthner Axons. J Neurosci 2002; 22:4293-4301.
70. Wang H, Iacoangeli A, Popp S et al. Dendritic BC1 RNA: Functional role in regulation of translation initiation. J Neurosci in press 2002.
71. Lund E, Kahan B, Dahlberg JE. Differential control of U1 small nuclear RNA expression during mouse development. Science 1985; 229:1271-1274.
72. Cavaillé J, Buiting K, Kiefmann M et al. Identification of brain-specific and imprinted small nucleolar RNA genes exhibiting an unusual genomic organization. Proc Natl Acad Sci USA 2000; 97:14311-14316.
73. Omer AD, Lowe TM, Russell AG et al. Homologs of small nucleolar RNAs in Archaea. Science 2000; 288:517-522.
74. d'Orval BC, Bortolin ML, Gaspin C et al. Box C/D RNA guides for the ribose methylation of archaeal tRNAs. The tRNATrp intron guides the formation of two ribose-methylated nucleosides in the mature tRNATrp. Nucleic Acids Res 2001; 29:4518-4529.

75. Hüttenhofer A, Kiefmann M, Meier-Ewert S et al. RNomics: an experimental approach that identifies 201 candidates for novel, small, nonmessenger RNAs in mouse. EMBO J 2001; 20:2943-2953.
76. Kiss T. Small nucleolar RNAs: an abundant group of noncoding RNAs with diverse cellular functions. Cell 2002; 109:145-148.
77. Hüttenhofer A, Brosius J, Bachellerie JP. RNomics: identification and function of small, nonmessenger RNAs. Curr Opin Chem Biol 2002 in press.
78. Marker C, Zemann A, Terhörst T et al. Experimental RNomics: identification of 140 candidates for small, nonmessenger RNAs in the plant *Arabidopsis thaliana*. Current Biology 2002 in press.
79. Reik W, Walter J. Genomic imprinting: parental influence on the genome. Nat Rev Genet 2001; 2:21-32.
80. Maxwell ES, Fournier MJ. The small nucleolar RNAs. Annu Rev Biochem 1995; 64:897-934
81. Smith CM, Steitz JA. Sno storm in the nucleolus: new roles for myriad small RNPs. Cell 1997; 89:669-672.
82. Pelczar P, Filipowicz W. The host gene for intronic U17 small nucleolar RNAs in mammals has no protein-coding potential and is a member of the 5'-terminal oligopyrimidine gene family. Mol Cell Biol 1998; 18:4509-4518.
83. Cavaillé J, Seitz H, Paulsen M et al. Identification of tandemly-repeated C/D snoRNA genes at the imprinted human 14q32 domain reminiscent of those at the Prader-Willi/Angelman syndrome region. Hum Mol Genet 2002; 11:1527-1538.
84. Runte M, Hüttenhofer A, Gross S et al. The IC-SNURF-SNRPN transcript serves as a host for multiple small nucleolar RNA species and as an antisense RNA for UBE3A. Hum Mol Genet 2001; 10:2687-2700.
85. de los Santos T, Schweizer J, Rees CA et al. Small evolutionarily conserved RNA, resembling C/D box small nucleolar RNA, is transcribed from PWCR1, a novel imprinted gene in the Prader-Willi deletion region, which Is highly expressed in brain. Am J Hum Genet 2000; 67:1067-1082.
86. Nicholls RD, Knepper JL. Genome organization, function, and imprinting in Prader-Willi and Angelman syndromes. Annu Rev Genomics Hum Genet 2001; 2:153-175.
87. Burns CM, Chu H, Rueter SM et al. Regulation of serotonin-2C receptor G-protein coupling by RNA editing. Nature 1997; 387:303-308.
88. Yi-Brunozzi HY, Easterwood LM, Kamilar GM et al. Synthetic substrate analogs for the RNA-editing adenosine deaminase ADAR-2. Nucleic Acids Res 1999; 27:2912-2917.
89. Ambros V. microRNAs: tiny regulators with great potential. Cell 2001; 107:823-826.
90. Grosshans H, Slack FJ. Micro-RNAs: small is plentiful. J Cell Biol 2002; 156:17-22.
91. Pasquinelli AE. MicroRNAs: deviants no longer. Trends Genet 2002; 18:171-173.
92. Reinhart BJ, Bartel DP. Small RNAs correspond to centromere heterochromatic repeats. Science 2002; 297:1831.
93. Volpe TA, Kidner C, Hall IM et al. Regulation of heterochromatic silencing and histone H3 lysine-9 methylation by RNAi. Science 2002; 297:1833-1837.
94. Lee RC, Feinbaum RL, Ambros V. The *C elegans* heterochronic gene lin-4 encodes small RNAs with antisense complementarity to lin-14. Cell 1993; 75:843-854.
95. Wightman B, Ha I, Ruvkun G. Posttranscriptional regulation of the heterochronic gene lin-14 by lin- 4 mediates temporal pattern formation in *C elegans*. Cell 1993; 75:855-862.
96. Reinhart BJ, Slack FJ, Basson M et al. The 21-nucleotide let-7 RNA regulates developmental timing in *Caenorhabditis elegans*. Nature 2000; 403:901-906.
97. Lagos-Quintana M, Rauhut R, Lendeckel W et al. Identification of novel genes coding for small expressed RNAs. Science 2001; 294:853-858.
98. Lau NC, Lim LP, Weinstein EG et al. An abundant class of tiny RNAs with probable regulatory roles in *Caenorhabditis elegans*. Science 2001; 294:858-862.
99. Lee RC, Ambros V. An extensive class of small RNAs in *Caenorhabditis elegans*. Science 2001; 294:862-864.
100. Llave C, Kasschau KD, Rector MA et al. Endogenous and silencing-associated small RNAs in plants. Plant Cell 2002; 14:1605-1619.
101. Mourelatos Z, Dostie J, Paushkin S et al. miRNPs: a novel class of ribonucleoproteins containing numerous microRNAs. Genes Dev 2002; 16:720-728.

102. Reinhart BJ, Weinstein EG, Rhoades MW et al. MicroRNAs in plants. Genes Dev 2002; 16:1616-1626.
103. Lagos-Quintana M, Rauhut R, Yalcin A et al. Identification of tissue-specific microRNAs from mouse. Curr Biol 2002; 12:735-739.
104. Ridanpaa M, van Eenennaam H, Pelin K et al. Mutations in the RNA component of RNase MRP cause a pleiotropic human disease, cartilage-hair hypoplasia. Cell 2001; 104:195-203.
105. Bonafe L, Schmitt K, Eich G et al. RMRP gene sequence analysis confirms a cartilage-hair hypoplasia variant with only skeletal manifestations and reveals a high density of single-nucleotide polymorphisms. Clin Genet 2002; 61:146-151.
106. Ridanpaa M, Sistonen P, Rockas S et al. Worldwide mutation spectrum in cartilage-hair hypoplasia: ancient founder origin of the major70A—>G mutation of the untranslated RMRP. Eur J Hum Genet 2002; 10:437-447.
107. Cai T, Aulds J, Gill T et al. The *Saccharomyces cerevisiae* RNase mitochondrial RNA processing is critical for cell cycle progression at the end of mitosis. Genetics 2002; 161:1029-1042.
108. Vulliamy T, Marrone A, Goldman F et al. The RNA component of telomerase is mutated in autosomal dominant dyskeratosis congenita. Nature 2001; 413:432-435.

Chapter 12

New Frontiers for the snoRNA World

Jean-Pierre Bachellerie and Jérôme Cavaillé

Abstract

Eukaryal rRNAs contain two prevalent types of modified nucleotides, 2'-O-methylated nucleotides and pseudouridines. The site of each of these modifications is accurately specified by two large families of snoRNAs (small nucleolar RNAs) termed box C/D and H/ACA snoRNAs, respectively, through formation of a hallmark, bimolecular RNA duplex structure spanning the rRNA nucleotide to be modified. However, both snoRNA guide families now appear larger than anticipated and their roles clearly are not restricted to ribosome synthesis. Recently, novel members of the C/D and H/ACA families have been identified that can direct nucleotide modifications onto cellular RNAs other than rRNA in Eukarya, including spliceosomal snRNAs and mRNAs, and an increasing number of "orphan" guide snoRNAs without known RNA targets - including intriguing brain-specific snoRNAs linked to the phenomenon of parental genomic imprinting - have been characterized in mammals. Taken together, these findings dramatically expand the range of biological functions that can be envisioned for both snoRNA families. Meanwhile, identification of archaeal homologs of both families of eukaryal snoRNAs provides new insights on the evolutionary origin of these two major families of noncoding RNAs. It also extends the diversity of cellular RNAs targeted by the modification guide system, which includes archaeal tRNAs.

Introduction

Assembly of eukaryotic ribosomes in the nucleolus involves endo- and exo-nucleolytic cleavages of the rRNA primary transcript to produce stoichiometric amounts of mature small and large subunit (SSU and LSU) rRNAs. Before its processing the common rRNA precursor undergoes an elaborate pattern of nucleoside modifications of two prevalent types, 2'-O-ribose methylation or pseudouridylation, restricted to its SSU and LSU rRNA sequences and located within the most conserved, functionally important rRNA domains.[1-5] Methylation of 2'-hydroxyl groups may protect RNA from hydrolytic degradation, enhance hydrophobic surfaces and stabilize helical stems. Pseudouridines, through their flexible C-C glycosyl bonds and increased capacity, relative to uridines, to form H-bonds, may significantly contribute to RNA tertiary structure. These modifications frequently conserved among distant eukaryotes are likely to affect rRNA folding, assembly of ribosomal proteins and ribosome activity. Most of them however are not individually required for cell viability and their precise roles remain elusive. Site-specific formation of 2'-O-methylated nucleotides and pseudouridines in rRNA is directed by two large families of snoRNAs, termed box C/D and H/ACA snoRNAs, respectively.[5-7] Each snoRNA in these families is specific of a rRNA modification site through the presence of an appropriate rRNA complementarity, termed antisense element, which allows

Noncoding RNAs: Molecular Biology and Molecular Medicine, edited by Jan Barciszewski and Volker A. Erdmann.

formation of an RNA duplex spanning the modification site. This RNA duplex structure, which substantially differs among the C/D and H/ACA families, accurately determines the rRNA nucleotide to be modified. The modification itself is catalyzed by a common protein enzyme, methylase or pseudouridine synthase, associated with each guide snoRNA which is assembled into a snoRNP particle including a small set of additional specific proteins. Since a eukaryotic ribosome contains about 50-100 modifications of each type the overall number of these snoRNAs was expected to approach 100-200 in most eukaryotic organisms.

However, the two snoRNA guide families now appear larger than anticipated with roles not restricted to the modification of ribosomal RNA. This is reflected in vertebrates by snoRNA guides that can target snRNAs[8-12] and probably even mRNAs,[13] and by an increasing number of "orphan" guides still devoid of identified RNA targets.[10,14] Intriguingly, several of the novel specimens in mammals are expressed in a tissue-specific fashion and submitted to genomic imprinting, with one of them probably involved in a control of pre-mRNA editing.[13,15,16] Meanwhile, identification of homologs of guide snoRNAs in organisms lacking a cell nucleus, Archaea, provides further insights into the evolutionary origin and function of these two large families of noncoding RNAs and reveals their range of RNA targets also includes archaeal tRNAs.[17-20] In this article, following an updated review of the basic properties of rRNA modification guides we will focus on recent breakthroughs revealing their unexpected structural and functional diversity in organisms ranging from from Archaea to Eukarya.

Two Large Families of Modification Guide snoRNAs

Except for the RNA component for RNase MRP, all snoRNAs fall into two major families, antisense box C/D and box H/ACA snoRNAs, based on conserved sequence motifs.[21,22] While a few members in each class are involved in pre-rRNA cleavage,[23] the vast majority of them guide the 2'-O-ribose methylations and pseudouridylations, respectively, of rRNA.[2,3,5] The two snoRNA guide families have been identified in a wide spectrum of eukaryotes, from metazoans to yeasts, plants, and kinetoplastid protozoans.

Antisense Box C/D snoRNAs

They contain two short sequence motifs, box C (5'PuUGAUGA3') and box D (5'CUGA3'), located only a few nucleotides away from the 5' and 3' ends, respectively, and generally brought together in a typical stem-box structure—including a 4-5 bp long 5'-3' terminal stem (Fig. 1)—critical for snoRNA biogenesis and nuclear localization.[2,24-28] In mature snoRNAs lacking a canonical 5'-3' terminal stem, an internal stem or a stem structure forming on the snoRNA precursor molecule ensures proper juxtapositioning of the two box motifs.[28,29] In their central portion and their 5' half, antisense box C/D snoRNAs also contain boxes C' and D', respectively, frequently imperfect copies of boxes C and D.[30,31] Finally, they exhibit immediately upstream from box D and/or box D' one (sometimes two) antisense elements(s). The antisense element corresponds to a 10-21-nt complementarity to a site of rRNA 2'-O-ribose methylation which allows formation of a hallmark guide RNA duplex at each modification site.[32-34] Each antisense element associated with the downstream box D (or D') is the sole determinant of the rRNA nucleotide targeted for modification, invariably opposite the fifth snoRNA nucleotide upstream from box D (or box D') in the guide RNA duplex[33,34] (Fig. 1). Owing to the presence of conserved box motifs and precisely positioned, relatively long rRNA complementarities antisense box C/D snoRNAs can be readily identified by computer search of eukaryal genomes of moderate size. A virtually full complement of rRNA methylation guides has been characterized in yeast *S. cerevisiae*[35] and plant *A. thaliana*.[36-38] Known box C/D snoRNAs collectively guide 51 of the 55 ribose methylations in *S. cerevisiae* rRNA.[35] While the precise number of 2'-O-methylated rRNA nucleotides has not been determined in *A. thaliana*, 66 C/D snoRNAs targeting a total of 86 rRNA ribose methylations have been identified in

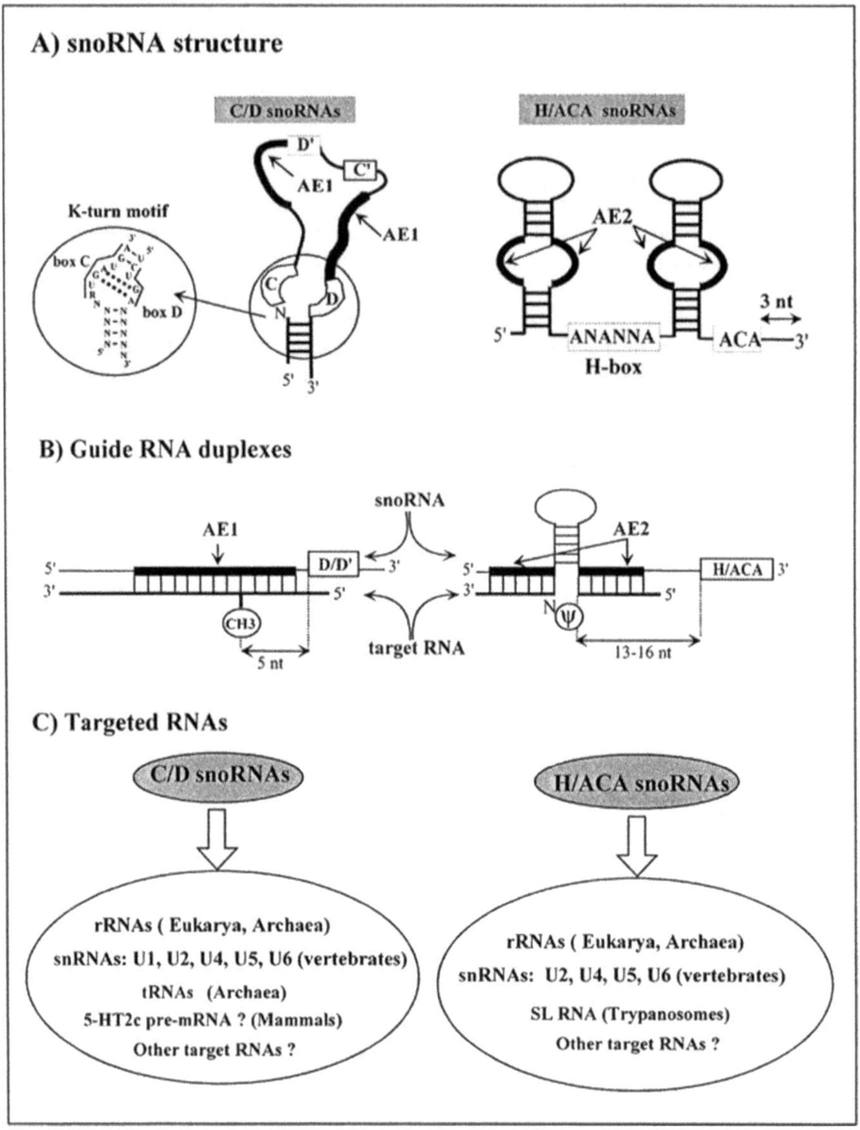

Figure 1. Structural features of the two families of modification guide snoRNAs. A) Schematic secondary structures of eukaryotic C/D and H/ACA snoRNAs. Conserved box motifs are indicated. Antisense elements, i.e., sequence tracts complementary to the cognate RNA target, are depicted by a thick line (AE1: uninterrupted antisense element; AE2: bipartite antisense element). For C/D snoRNAs (left), the 5'-3' terminal stem allowing formation of the box C/D structural motif, or K-turn, is represented. B) Canonical structure of each type of guide RNA duplex The conserved spacing between the nucleotide to be modified and the cognate box motif is indicated. C) Repertoire of cellular RNAs targeted by modification guide snoRNAs in different groups of organisms.

that organism (see Note Added in Proof 1).[36-38] While searches on complete mammalian genomes are hampered by large genomic sizes, analyses focused on introns have been fruitful.[32,39-41] Recent progress in mammals has essentially resulted from experimental screens, with 24 new C/D and 19 new H/ACA snoRNAs targeting mouse rRNA detected by an EST-like sequencing approach.[10] Altogether, snoRNA guides have been identified for 93 of the 105-107 2'-O-methylations in mammalian rRNAs. Methylation guides have also been identified in *Drosophila*[42] and Trypanosoma.[43-45] Interestingly, a universally conserved LSU rRNA ribose methylation for which a yeast snoRNA guide was predicted but not experimentally confirmed,[35] Um2918 in yeast 25S rRNA, is catalyzed in bacteria and mitochondria by a site-specific methyltransferase which has an homologue in the yeast nuclear genome.[46-48]

H/ACA snoRNAs

Box H/ACA snoRNAs exhibit a common secondary structure generally consisting of two large hairpin domains linked by a hinge and followed by a short single-strand tail.[21,49] Conserved motifs termed boxes H (ANANNA, where N stands for any nucleotide) and ACA (a trinucleotide always found 3 nucleotides away from the 3' end) are located in the hinge and tail, respectively (Fig. 1). Each H/ACA snoRNA contains one (sometimes two) bipartite antisense element(s) invariably located in the upper part of the internal loop of one (or both) of the two large hairpin domains.[3,50] In-between the two elements of the 9-13 bp bipartite guide duplex two rRNA nucleotides remain unpaired, the 5'-most of which corresponding to the uridine to be modified which is thus accessible for isomerization (Fig. 1). The distance between the target uridine and the downstream H or ACA box of the snoRNA, invariably 13-16 nt, is a critical determinant of the pseudouridylation site reminiscent of the target/box-D spacing rule observed for methylation guides.[50,51] Due to shorter box motifs and bipartite antisense elements, no computational search algorithm has been described for H/ACA snoRNAs, in contrast to antisense C/D snoRNAs. In yeast *S. cerevisiae* only 15 H/ACA snoRNAs are known, which collectively target 19 of the 44 rRNA pseudouridines.[50,52] H/ACA guides have been identified for 42 of the 91-93 mammalian rRNA pseudouridines.[10,50,53] Scores of H/ACA snoRNAs guiding rRNA pseudouridylations have been identified in *A. thaliana* and *D. melanogaster* by experimental screens for small nonmessenger RNAs (see Note Added in Proof 1). Even when the H/ACA snoRNA contains a single antisense element both hairpin domains are essential for rRNA pseudouridylation in yeast and mammals.[51] However, the first H/ACA snoRNA characterized in Trypanosoma, the shortest eukaryal rRNA pseudouridylation guide so far, consists of a single hairpin domain.[54] Although C/D snoRNAs also exhibit a bi-modular structure, with two structurally related halves, each containing a box C (or C') and a box D (or D') near the 5' and 3' ends, respectively, no C/D snoRNA restricted to a single module has been identified so far. However, consistent with the notion that each half may represent a basic structural and/or functional unit for C/D snoRNA too, the two-hairpin H/ACA snoRNA sequence in the H/ACA-C/D chimers recently detected in vertebrates (see below) is inserted between the two modules of a C/D snoRNA.[11,12]

snoRNP Biogenesis

The two families of guide snoRNAs share unique characteristics in terms of genic organization and biosynthetic pathways. In vertebrates guide snoRNAs are exclusively encoded within introns and are not independently transcribed but processed from the pre-mRNA introns, in most cases by exonucleolytic digestion of the debranched lariat (Fig. 2).[2,55] In other eukaryal groups the gene organization of guide snoRNAs shows more diversity. In yeast *S. cerevisiae*, intronic snoRNAs are only a minority: most snoRNAs are synthesized from independent mono-, di-, or polycistronic RNA transcripts processed by endo- and exonucleases.[56-58] In higher plants

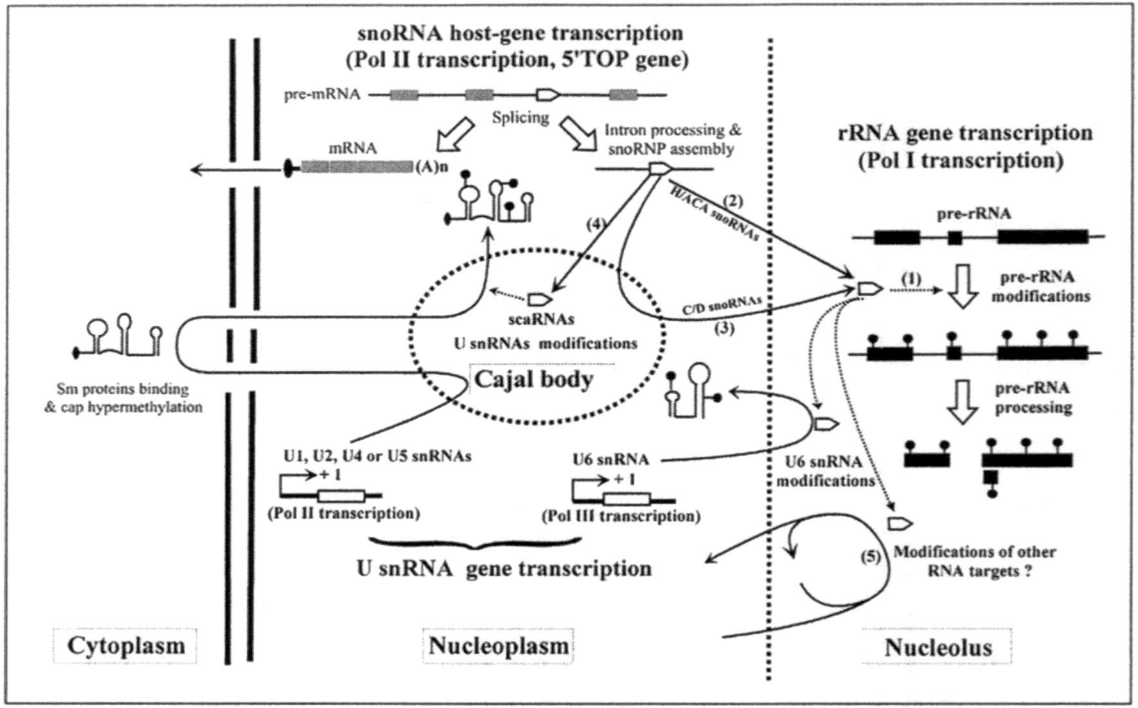

Figure 2: Biosynthesis and intranuclear traffic of the two snoRNA families. 1) snoRNA-guided modifications occur on the nascent prerRNA transcript. 2) and 3) C/D and H/ACA snoRNAs seem to follow different routes to the nucleoli following intron processing, based on the results of microinjections into Xenopus oocyte nuclei.[94] However, failure to detect H/ACA snoRNAs in Cajal Bodies might merely reflect more rapid transport or the need to study precursor RNAs.[91] 4) C/D and H/ACA guides directing the modification of RNA polymerase II-transcribed snRNAs which are exclusively localized in the Cajal Bodies are termed scaRNAs.[12] 5) other RNA targets synthesized in the nucleolus or nucleoplasm might be targeted by orphan snoRNA guides acting within or outside the nucleolus.

and Trypanosoma too, clustered snoRNA genes are widespread. They are transcribed as polycistronic precursors from which individual snoRNAs - sometimes of both C/D and H/ACA types—are processed.[36-38,43,54,59,60] Intriguingly, a novel yeast C/D snoRNA with a human intronic homolog, U86, is encoded within an open reading frame and its accumulation and that of the cotranscribed mRNA are mutually exclusive.[61]

In vertebrates all genes hosting an intronic snoRNA correspond to actively transcribed housekeeping genes belonging to the 5' TOP (terminal oligopyrimidine) gene family which includes ribosomal protein genes, which could provide the basis for a coordination of snoRNA biosynthesis at the transcriptional level.[62-64] In yeast, promoters of independent mono- or polycistronic snoRNA genes and of genes hosting intronic snoRNAs share common control elements, pointing to coordinated transcription.[58]

Whatever their gene organization both types of guide snoRNAs are assembled with a small set of specific proteins into small ribonucleoprotein particles (snoRNPs) (Fig. 3). Box C/D snoRNPs contain four evolutionarily conserved, essential core proteins, fibrillarin (Nop1p), Nop56p, Nop58p and Snu13p. Fibrillarin is the likely snoRNA-guided methylase as point mutations in its methylase-like domain disrupt all rRNA methylations.[65-67] Snu13p binds specifically to the hallmark C/D structural motif of this snoRNA family, also termed K-turn, which involves nucleotides in both boxes (Fig. 1). In this motif a 3-nt bulge is flanked by a regular stem on one side and stabilized by tandem sheared G.A pairs on the other, while a protuded pyrimidine nucleotide is important for protein recognition.[68-70] Remarkably, Snu13p is also an integral component of the U4/U6.U5 tri-snRNP complex spliceosomal complex in which it binds a K-turn motif in U4,[71] which could provide a link between mRNA splicing and snoRNA synthesis. Two other C/D snoRNP proteins, Nop56p and Nop58p, have very similar amino acid sequences but do not perform redundant functions since both genes are essential in yeast.[72-76] The four proteins are all required for normal localization of C/D snoRNAs and their absence leads to mislocalization of the snoRNA in the nucleoplasm.[77] Protein-protein interactions among fibrillarin, Nop56p and Nop58p have been observed in vivo and association of Nop56p with C/D snoRNAs depends on the presence of fibrillarin while Nop58p and fibrillarin associate independently with the snoRNAs.[74,78] In vitro reconstitution of archaeal homologs of C/D RNPs has recently provided new insights into their assembly process (see below).

Each H/ACA snoRNP particle contains two copies of four evolutionarily conserved proteins, Cbf5p (dyskerin), Gar1p, Nhp2p and Nop10p, all essential for pseudouridylation[79,80] (Fig. 3). Cbf5p is presumably the pseudouridine synthase catalyzing the snoRNA-guided reaction based on effects of mutations in signature motifs.[81,82] Mutations in the human homologue of Cbf5p, dyskerin, are a cause of dyskeratosis congenita, an inherited human disease linked to an altered telomerase activity,[83] reflecting the specific binding of dyskerin to human telomerase RNA which also contains an H/ACA domain (see below). Cbf5p is closely related to TruB which catalyzes pseudouridine formation in the T loops of virtually all tRNAs and both enzymes seem to recognize RNA similarly.[84] In an in vitro reconstitution system the four human core proteins directly contact the snoRNA, as shown by UV cross-linking.[85] The glycine-arginine-rich (GAR) domains of yeast Gar1p are not essential for in vitro binding to H/ACA snoRNAs.[86] Nhp2p, by far the most efficiently cross-linked protein, can directly bind yeast H/ACA snoRNAs in vitro but the interaction does not seem highly specific and neither the H nor ACA box motif are required for binding.[87]

Remarkably two snoRNP proteins belonging to different families, Snu13p and Nhp2p, are substantially similar to each other, which might provide the basis for coordinated production of both guide families.[88] The two proteins - together with eukaryal ribosomal proteins L7a and S12 - share a common archaeal homolog, ribosomal protein L7Ae, pointing to a common evolutionary origin of both snoRNP families (see below). Two related, highly conserved

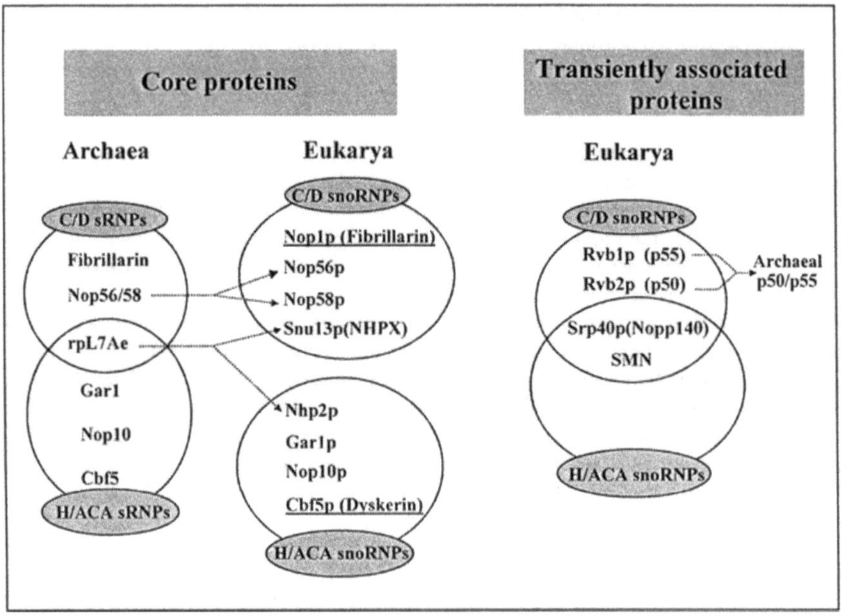

Figure 3. Protein components of box C/D and H/ACA snoRNPs. Each modification enzyme is underlined. Two pairs of core snoRNP proteins and a pair of transiently-associated snoRNP proteins in eukaryotes are represented by a single protein gene in archaea (dotted lines). Protein components shared by the two snoRNP families are denoted by an overlap.

nucleoplasmic (not nucleolar) proteins with DNA helicase activity and linked with chromatin remodeling and transcription, p50 (Rvb2) and p55 (Rvb1), bind a model box C/D snoRNA in vitro, suggesting they function at an early stage of snoRNP biogenesis in the nucleoplasm.[72,89] Depletion of p50 impairs assembly or trafficking of both C/D and H/ACA snoRNPs. Indirect evidence suggests that transient association with Nopp140 may assist assembly of both snoRNP families.[90] This highly conserved protein which colocalizes with snoRNP components in the nucleolus and Cajal body (see below) specifically interacts with both types of snoRNPs. Finally, the SMN protein, which interacts directly with Sm proteins and is involved in spliceosomal snRNP assembly, also interacts with a component of each snoRNP family, fibrillarin and Gar1p.[91]

Box motifs of both snoRNA types, unlike their antisense elements, represent essential cis-acting signals for snoRNA processing and nuclear localization.[24-27,29,87,92-94]

Novel Guides for snRNA Modifications

In mammals, snRNA 2'-O-methylations and pseudouridines amount to 4.2% and 3.3%, respectively, of snRNA nucleotides (vs. 1.5% and 1.4%, respectively, for rRNA).[1,95] Remarkably, these modifications mostly map within snRNA segments which have a crucial importance, i.e., those involved in intermolecular RNA-RNA interactions or conformational switches during spliceosome assembly and function.[95] Nucleotide modifications in the 5' terminal region of vertebrate U2 snRNA have been shown to be essential for proper assembly of snRNPs active in premRNA splicing.[96]

C/D guides for snRNA 2'-O-methylation have been first identified for vertebrate U6, a RNA polymerase III transcript.[8,9] Several C/D and H/ACA snoRNAs able to guide

2'-O-methylations and pseudouridylations onto snRNAs transcribed by RNA polymerase II, U1, U2, U4 and U5, have been subsequently identified.[7,10-12] Remarkably, modification of two adjacent nucleotides in the invariant loop of snRNA U5, 2'-O-methylation of position 45 and pseudouridylation of uridine 46, can be directed by human U85, an intriguing C/D-H/ACA chimeric snoRNA conserved in *Drosophila*.[11] Three additional human C/D-H/ACA chimers able to guide modifications onto snRNAs U4 and U5 have been detected recently.[12] All C/D-H/ACA chimers, composed of a box C/D and a H/ACA snoRNA domain, are associated with both box C/D and box H/ACA snoRNP proteins, as reflected by immunoprecipitation from a human HeLa cell extract using anti-fibrillarin and anti-GAR1 antibodies.[12]

No snoRNA guide for snRNA modification has been reported in yeast *S. cerevisiae* in contrast to fission yeast *S. pombe*.[97] *S. cerevisiae* snRNAs contain considerably less pseudouridines than vertebrate snRNAs and their 2'-O-methylation content remains unknown.[95,98] Pseudouridylation of *S. cerevisiae* snRNAs is not dependent on Cbf5p, which catalyzes the snoRNA-guided pseudouridylation of rRNA.[81,98] Remarkably, one of these pseudouridylations, ψ34 in *S. cerevisiae* U2, is conserved in vertebrates in which it is snoRNA-guided, showing modification of the same U2 nucleotide involves dramatically different mechanisms in *S. cerevisiae* and vertebrates. Formation of another conserved pseudouridine in yeast *S. cerevisiae* U2, equivalent of ψ43 in vertebrate U2, is catalyzed by Pus1p, a tRNA pseudouridine synthase.[98]

Remarkably, all guide RNAs—including the C/D-H/ACA chimers—involved in modification of vertebrate snRNAs transcribed by RNA polymerase II comprise a new class of cellular RNAs that localize exclusively to the Cajal body, not the nucleolus.[7,12] This was demonstrated by in situ hybridization microscopy with fluorescent antisense RNA probes which colocalize perfectly with p80 coilin, the only unambiguous marker for the Cajal body (or coiled body, CB). Frequently located close to the nucleolus, CB, the most extensively studied nuclear organelle besides the nucleolus, is involved in the biogenesis of snRNAs and snoRNAs as well.[27,94,99] Newly synthesized snRNAs are proposed to transit through CB before and after their cytoplasmic stage of snRNP assembly, before reaching the speckles, the nuclear structures where splicing is detected.[100] CB must therefore provide the cellular locale for modification of pol II-specific snRNAs by guide RNAs, either before or after the cytoplasmic stage of snRNP assembly (Fig. 2). The mechanisms responsible for CB rather than nuclear localization of this subset of guide RNAs is unknown. Since typical snoRNAs, at least for the C/D type, transit through the CB before accumulating in the nucleolus, this subset of guides could be selectively retained in CB through the binding of retention factor(s) to specific RNA motifs which remain to be identified.

Orphan Guides snoRNAs in Search of RNA Targets

Further illustrating the unexpected complexity of guide snoRNA functions, telomerase RNA in vertebrates, not in yeast, contains a typical H/ACA domain at its 3' end[101] and the four evolutionarily conserved core proteins of H/ACA snoRNPs are also parts of human telomerase.[102,103] In contrast yeast telomerase is associated with Sm proteins, like spliceosomal snRNPs. No presumptive RNA target for pseudouridylation has been identified for the H/ACA domain of vertebrate telomerase RNA which might have some distinctive features relative to typical H/ACA snoRNAs.[85] Moreover, several ubiquitously expressed "orphan" snoRNAs, i.e., snoRNAs belonging to the box H/ACA or C/D families in mammals but devoid of antisense element to rRNA or snRNAs, have been reported recently together with a single C/D yeast specimen in this category.[10,14,61] Some of them could function in pre-rRNA cleavage instead of modification, similar to a few C/D and H/ACA snoRNAs characterized in vertebrates and yeast.[23,104-107] Some might even have a dual function in rRNA cleavage and modification of another RNA. SnoRNAs of the C/D and H/ACA types with a dual function in pre-rRNA modification and cleavage have been reported previously.[108,109] Alternatively, orphan snoRNAs might target a wide range of RNAs transiting through the nucleolus and not directly related to ribosome biogenesis,[110,111] including mRNAs. None of the novel specimens seems able to target any of three stable noncoding RNAs trafficking through the nucleolus, telomerase RNA, RNase P or SRP RNA. A novel brain-specific C/D snoRNA, MBII-52, displays an outstanding antisense element—18-nt long—to a mRNA.[13] Indirect evidences point to the biological relevance of the presumptive snoRNA.mRNA interaction, suggesting for this snoRNA a novel role in premRNA processing (Fig. 4). Like the snoRNA, the potential mRNA target—encoding serotonin receptor 5-HT2C—is also exclusively expressed in the brain, and the 18-bp snoRNA.mRNA complementarity is perfectly conserved in mammals. Moreover, the nucleotide predicted to be targeted for 2'-O-methylation in this mRNA is also the site of a physiologically important adenosine-to-inosine editing.[112] Remarkably, 2'-O-methylation of the adenosine to be edited blocks its deamination to inosine in vitro,[113] suggesting for this brain-specific snoRNA a role in the control of 5-HT2C mRNA editing (see also below for additional information on this snoRNA). Alternatively or additionally, the snoRNA base-paired interaction might also modulate alternative splicing of the 5-HT2c pre-mRNA at a nearby splice site, through steric occlusion instead of nucleotide modification (Fig. 4). Additional orphan guide snoRNAs exclusively expressed in the brain (see below) might provide further insights onto the range of snoRNA functions. Meanwhile, observations in an organism with an unusual mRNA processing pathway, Trypanosoma, further support the notion that mRNAs may be targeted by snoRNA guides. Trans-splicing is a process initially discovered in kinetoplastid protozoans and subsequently identified in a wide range of metazoans where a common, relatively short terminal 5' exon, the spliced leader (SL RNA), is linked to all or a variable fraction of mRNAs in an organism.[114] Trypanosoma SL RNA can be cross-linked to another small RNA, SLA1, that folds as an H/ACA snoRNA with a single hairpin structure.[115] Based on the canonical rules for H/ACA guides SLA1 is predicted to direct an experimentally verified pseudouridylation in SL RNA, 12 nucleotides upstream from the 5' splice site.[54] As expected, mutations in the SL RNA disrupting its interaction domain with SLA1 abolished pseudouridylation. The conserved pseudouridine might fine-tune trans-splicing, nuclear export of spliced mRNAs or even its activity during translation. Remarkably, SLA1 RNA which seems to shuttle between the nucleolus and the nucleoplasm where it is likely to bind its SL RNA target also carries a Sm-like site and is associated with core Sm proteins. It thus represents also a bona fide snRNA which might have an additional function for trans- splicing.[115,116]

Brain-Specific Imprinted snoRNAs

Whereas most orphan snoRNAs are constitutively expressed like snoRNAs guiding rRNA or snRNA modifications, a few specimens of the C/D family are exclusively (or mainly) detected within the brain.[13,15,16,117-120] Remarkably, their genes are subject to genomic imprinting, an epigenetic phenomenon of elusive biological significance that restricts gene expression to only one chromosome, either the paternal or maternal allele.[121] Most of the about 50 imprinted genes reported in mammals seem involved in the regulation of fetal/placental growth, cell cycle and brain development. Genetic alterations such as chromosomal deletions or uniparental disomy can lead to loss of expression of the active allele or abnormal expression of the silent allele, giving rise to human genetic syndromes (Prader-Willi, Angelmans, Beckwith-Wiedemann), behavioural disorders (autism, Turner and Tourette syndromes) or cancers (Wilm's tumors).[121] Several imprinted snoRNA genes at two different human chromosomal loci, 15q11q13 and 14q32, exhibit a similar outstanding genic organization (Fig. 5). The coding region of each snoRNA species, generally in an intron, is part of a basic unit encompassing the whole snoRNA-containing intron and its flanking exon(s)—sometimes together with an additional intron - and this basic unit is tandemly repeated to dozens of

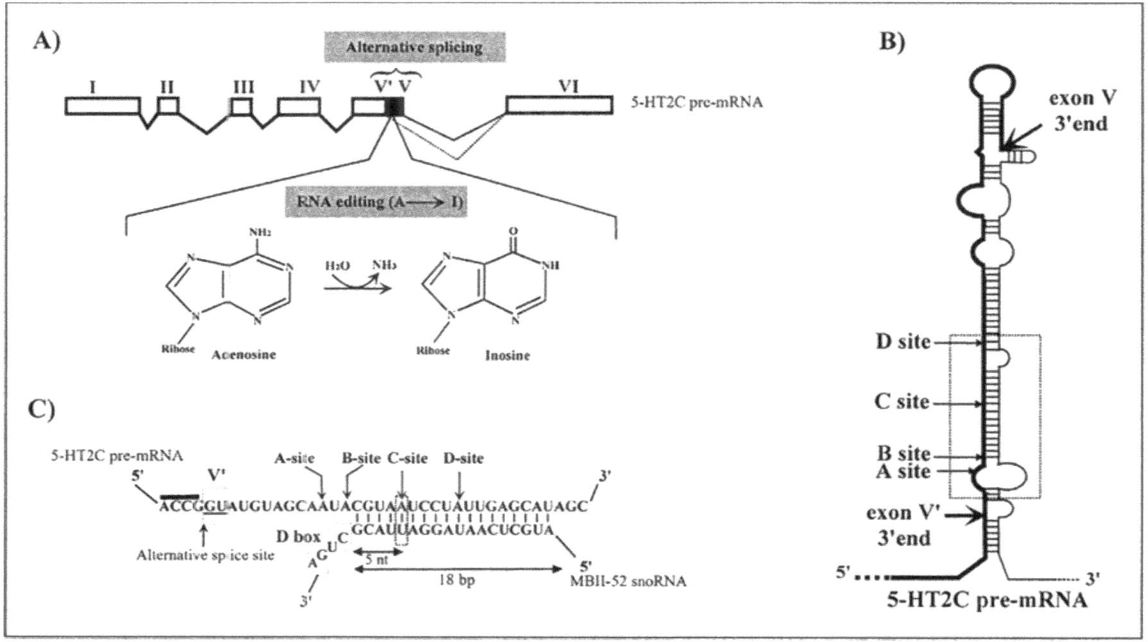

Figure 4. A potential mRNA target for brain-specific C/D snoRNA MBII-52. A) Schematic structure of the human serotonin receptor 5-HT2C premRNA, with delineation of the exon V segment undertaking both alternative splicing and an adenosine-to-inosine editing reactions at four nucleotide positions.[13,112] B) Partial secondary structure of 5-HT2C pre-mRNA presumed to be important for the editing reaction at four sites A-D.[142] It involves the alternative exon and the 5' end of the downstream intron which is required for in vivo editing in rat (exon: thick line; intron: thin line). C) Potential base pairing between the antisense element of MBII-52 and the segment of human 5-HT2C exon V undergoing adenosine-to-inosine editing at sites A-D. This interaction and part of the secondary structure in B (dotted box) are mutually exclusive. Based on the spacing rule applying to C/D snoRNAs, RNA editing C-site (dotted) located 5-nt upstream from box D in the canonical RNA duplex is predicted to be targeted for 2'-O-methylation by snoRNA MBII-52. The nearby site of alternative splicing is also indicated.

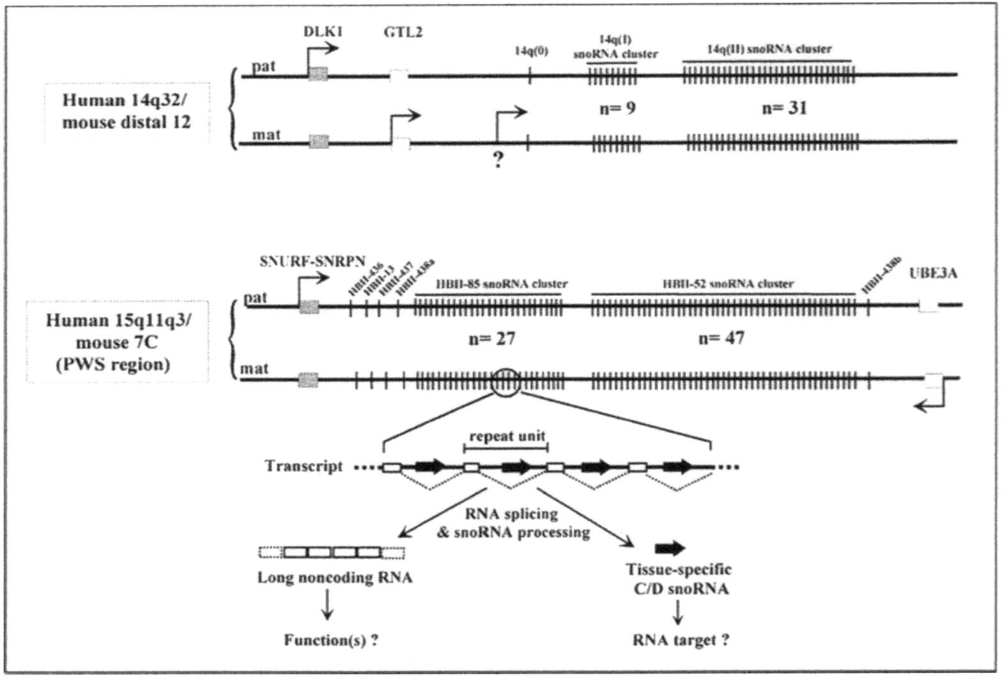

Figure 5. Schematic representation of clusters of tandemly repeated snoRNA genes at two human imprinted loci.[13,16] SnoRNA coding regions are denoted by vertical bars. Each cluster of repeats encodes a different type of C/D snoRNA with a specific antisense element (the name and copy number of each snoRNA in a cluster is indicated above and below the cluster, respectively). In imprinted loci 14q32 and 15q11q13 (only a portion of which is represented) snoRNA genes are expressed from the maternally- or paternally-inherited alleles, respectively. Start sites and direction of transcription of imprinted genes, including the large transcripts spanning the snoRNA gene clusters, are indicated by arrows. The typical structure of a snoRNA-containing repeat unit in the large transcript is enlarged below (the snoRNA is denoted by a black arrow, noncoding exons by open boxes, and intronic sequences by a line) and its RNA processing pathway is schematized.

copies.[13,16,117,119] These snoRNAs are processed from huge transcripts spanning the entire repeat array, which also gives rise to spliced RNAs lacking open reading frames of elusive function. Imprinted snoRNA genes MBII-13, MBII-85 and MBII-52, map at mouse Chr. 7C and their human homologs, HBII-13, -85 and -52, at the syntenic 15q11q13 region. HBII-85 and HBII-52 are among the most abundant C/D snoRNAs in mouse brain. While HBII-13 - structurally related to HBII-52 - is encoded by a single gene copy, HBII-52 and HBII-85 genes are arranged into two tandem arrays of 27 and 47 copies, respectively, each one embedded within a ~ 2 kb long repeat unit.[13,117,119,120] Within each array of tandem repeats in humans, the snoRNA copies are very similar to each other unlike the rest of the repeat unit. Likewise, the snoRNA sequences are strongly conserved between human and mouse, in contrast to the flanking intronic sequence. Two very different human neurological disorders are associated with human 15q11q13 locus: Prader-Willi (PWS) or Angelmans syndromes (AS) that result from the loss of paternal or maternal gene expression within this region, respectively.[122] The three C/D snoRNAs are not expressed in a PWS patient (with a large paternal deletion of the whole imprinted locus) or in a mouse model indicating they are only expressed from the paternal allele.[13] While they are processed from a common very large transcript (>460 kb), an antisense RNA to the maternally expressed UBE3A gene[120] which might regulate UBE3A paternal expression (Fig. 5), only HBII-52 is strictly brain-specific, suggesting snoRNA processing or accumulation might be regulated in a tissue-specific fashion. Genes of another brain-specific C/D snoRNA exclusively detected in rat, RBII-36, are part of 0.9 kb unit tandemly repeated ~ 100 times within a nonprotein coding gene, Bsr.[15,123] This snoRNA exclusively detected in neurons is generated by superimposition of two mutually exclusive processing pathways of its host gene transcript.[15,123] The human and mouse loci syntenic to the rat Bsr-containing locus also contain tandemly repeated arrays of novel, imprinted tissue-specific intronic C/D snoRNAs only expressed from the maternal allele[16] (Fig. 5). Illustrating the possibility that the tandemly-repeated gene organization can provide the basis for a snoRNA functional diversification, antisense elements of this specimen are substantially divergent among repeats in each array. These snoRNAs and/or their intriguing host-genes could play a direct role in the imprinting mechanism. C/D snoRNP proteins p55 and p50 are linked with chromatin remodeling and transcription complexes and Nop56p/ Nop58p interact with matrix-attached regions (MARs) in plants.[89,124] The snoRNP proteins could be driven to an imprinted chromosomal locus by high local concentration of C/D snoRNAs, independent of their antisense sequences. Alternatively, the repeated snoRNAs might affect brain function and/or development by guiding modification of a specific RNA, in line with the proposal that HBII-85 snoRNA is a candidate gene for PWS neurological defects.[125] However, except for HBII-52 (see above), potential targets remain elusive. Intriguingly, the only brain-specific H/ACA snoRNA, HBI-36, is not subject to imprinting. However, it is also expressed from one chromosome only: its host gene maps to the X chromosome.[13]

Archaeal Homologs of Modification Guide snoRNAs

Among prokaryotic organisms, Archaea are more particularly related to Eukarya by their replication, transcription and translation machineries.[126] Ribosomal RNA of typical bacterium *E. coli* contains only four 2'-O-methylations and 10 pseudouridines and each of these modifications is catalyzed by a site-specific, RNA-independent protein enzyme.[4,47,127] Conversely, the high number of rRNA 2'-O-methylations in archaeon *Sulfolobus solfataricus* is very similar to that of eukaryal rRNAs[128] and archaeal genomes contain ORFs encoding unambiguous homologs of all eukaryotic C/D and H/ACA snoRNP core proteins.[68,129-131] Archaeal homologs of antisense box C/D snoRNAs have been identified. They correspond to minimalists C/D specimens, probably reflecting strong size constraints exerted on archaeal genomes.[17,18] More homogeneous in size than their eukaryotic counterparts, they exhibit slightly extended consensus motifs for box C and C', and generally two (instead of one) rRNA antisense elements. Remarkably, the two antisense elements generally match two target nucleotides very close to each other in rRNA structure, strongly suggesting the two guide duplexes might form simultaneously and mediate an additional chaperone function in prerRNA folding.[17-19] Archaeal C/D sRNA genes are not clustered but widely distributed in the genome and encoded in both DNA strands, usually (but not always) between ORFs. Some sRNAs might be transcribed from independent sRNA promoters embedded within an ORF or through differential processing of chimeric sRNA/mRNA transcripts. Two *Pyrococcus* C/D sRNAs seem to be processed from a common dicistronic transcript.[17] Although only a few predicted archaeal rRNA 2'-O-methylations have been verified so far their number may widely vary among Archaea, even among hyperthermophiles, as illustrated by results of experimental screens and genomic searches in *A. fulgidus* which detected only a few C/D guide RNAs in this organism.[18,132] Archaeal C/D sRNAs appear to guide methylation through the same spacing rule as their eukaryotic homologs. They can assemble into functional RNPs in the eukaryotic nucleus and direct site-specific 2'-O-methylation of eukaryotic rRNA.[133] Remarkably, an archaeal C/D sRNP active as methylation guide has been reconstituted in vitro from its cloned components in *Sulfolobus solfataricus*[134] In vitro reconstitution of an eukaryal C/D snoRNP has not been reported so far, which could reflect an intrinsic inability of the eukaryal particle for self-assembly, possibly reminiscent of what has been observed for the eukaryal ribosome. The 2'-O-methylation is directed to the nucleotide in the fifth base pair upstream from box D in the duplex formed with archaeal target RNA and the reaction depends on the integrity of the intermolecular RNA duplex, as in eukaryotes (Fig. 1). Gel electrophoresis retardation shows that among the three archaeal particle proteins only L7Ae can bind directly to the sRNA, in line with the conserved presence of a K-turn motif in archaeal C/D sRNAs.[134] While aNOP56 binds a L7Ae/sRNA complex fibrillarin assembly depends on prior formation of a ternary complex, L7Ae/sRNA/aNOP56.[134] Archaeal fibrillarin lacks an N-terminal glycine-arginine-rich domain that is present in Eukarya. Its shorter N-terminal domain contains a novel fold that in the crystal appears to mediate the formation of a homodimer.[67] Even though this domain is likely to be present in eukaryal fibrillarins based upon sequence homology, whether or not it mediates dimerization has not been determined.

The first H/ACA small RNAs in Archaea have been recently identified through an experimental screen for small nonmessenger RNAs in hyperthermophile, sulphate-reducing archaeon *A. fulgidus* and the predicted rRNA pseudouridylations have been verified.[132] Three of the *A. fulgidus* H/ACA sRNAs are strongly reminiscent of the single-hairpin pseudouridylation guides detected in Trypanosoma.[54] Recognition of each targeted uridine in *A. fulgidus* rRNA involves a bipartite RNA duplex typical of eukaryal H/ACA snoRNAs and obeying the same target-H/ACA spacing rule (Fig. 1). While archaeal Gar1p is predicted to lack the glycine-arginine-repeats found at the N- and C-terminal regions of its eukaryal homologs, the putative Nop10p and Cbf5p sequences from Archaea and Eukarya are virtually the same length.[131] Remarkably, H/ACA snoRNP Nhp2p and C/D snoRNP Snu13p share a common archaeal homolog, L7Ae, which is also a ribosomal protein, suggesting both snoRNA families could well have their origin in primordial ribosomes (see Note Added in Proof 2).[88]

Archaeal tRNAs Are Targeted Too

Several C/D sRNAs with an antisense element to a tRNA instead of rRNA have been detected by genomic search and experimentally verified in different archaeal organisms.[18-20] While in most cases 2'-O-methylation at the expected tRNA position has not been tested experimentally the tRNA guide function of some of these C/D sRNAs is supported by phylogenetic evidence. *Pyrococcus* C/D RNAs, sR47, sR48 and sR49—predicted to target the wobble position in the anticodon of three different tRNAs, Leu(CAA), Leu(UAA) and elongator-Met, respectively—exhibit 11-nt long antisense elements conserved among three *Pyrococcus* species[20]. In addition, for sR47 and sR49 the presumptive guide RNA duplexes are preserved through compensatory changes among more distant archaea.[20] Unlike sR47 and sR48, sR49 targets the precursor, not mature tRNA, through a guide duplex spanning its 5' exon/intron junction, suggesting methylation occurs before pretRNA splicing, like some eukaryal tRNA modifications.[135] Strikingly, an archaeal C/D sRNA targeting a tRNA, sR50, corresponds to the intron of its own pretRNA target[20] (Fig. 6). The unusually large pretRNA-Trp intron - so far the sole archaeal intron with box C/D sRNA hallmarks - can direct two distinct, experimentally verified 2'-O-methylations, on the first position of the anticodon and on position 39, in the 5' and 3' exons, respectively. Both guide RNA duplexes, supported by comparative evidence in Archaea, span an exon-intron junction in the precursor, strongly suggesting the guide function is performed in cis-. While mechanisms of the intramolecularly guided reaction and its potential relationship with tRNA splicing in vivo[136] remain unclear the predicted function of the C/D intron, has been experimentally verified in vitro using a *H. volcanii* extract.[20] Remarkably, the pretRNA-Trp molecule must undergo major structural rearrangements after the two intron-guided methylations before splicing can occur, since formation of the BHB (bulge-helix-bulge) motif required for archaeal splicing is dependent on dissociation of both intramolecular guide duplexes (Fig. 6A). This unique biological system could illuminate an elusive aspect of modification guide function in Archaea and Eukarya as well, i.e., their intrinsic role of RNA chaperones directing the folding of their RNA targets.[2,5]

Archaeal rRNA processing involves cleavage at BHB motifs within the long stems flanking pre16S and pre23S processing intermediates. Unexpectedly, cleavage is followed by religation of pre-rRNA spacers flanking pre-rRNA intermediates,[137] similar to religation of spliced archaeal exons. Intriguingly, the religated pre-rRNA spacers harbor box C/D hallmarks in representatives of the two major archaeal lineages, *A. fulgidus* and *S. solfataricus* (Fig. 6B,C). The spliced C/D RNAs, which remain without presumptive RNA target, significantly differ from typical archaeal C/D sRNAs guiding rRNA or tRNA methylation. However, their C and D boxes form a typical K-turn efficiently recognized by the L7Ae protein.[137] They might belong to a subset of C/D small RNAs controlling prerRNA folding and processing in cis instead of trans such as essential eukaryotic snoRNAs U3, U8 or U22.[23,104,105]

Conclusions

The complex system of RNA guides used by eukaryal and archaeal organisms to ensure site-specific formation of 2'-O-methylations and pseudouridines must reflect the biological importance of the two prevalent types of RNA modifications. Consistent with this notion, a substantial fraction of them are phylogenetically conserved and the guide system preserved despite severe constraints on genome size in the archaeal domain of life. The two types of RNA modifications can alter RNA local conformation, thereby fine-tuning a wide range of RNA-RNA and RNA-protein interactions. In rRNA they could modulate the assembly and activity of the ribosome, even if the rRNA 2'-O-methylations and pseudouridylations analyzed so far are dispensable for cell viability or growth. In snRNAs they are likely to affect the activity of the splicing apparatus. The identification in higher organisms of many snoRNA guides devoid of rRNA or snRNA antisense elements suggests that the range of cellular RNAs targeted for modifications is probably larger than anticipated. However, experimental verification and functional dissection of the predicted modifications, which could mediate a variety of cellular processes, remains a challenge for the future, particularly for modifications involving a short-lived RNA precursor (as for the serotonin receptor pre-mRNA mentioned above).

Both families of snoRNA guides appear of very ancient origin. In Archaea they share a common RNP protein which is also a ribosomal protein, hinting at a common evolutionary origin in primordial ribosomes.[88] Based on simple modular structures both snoRNA families

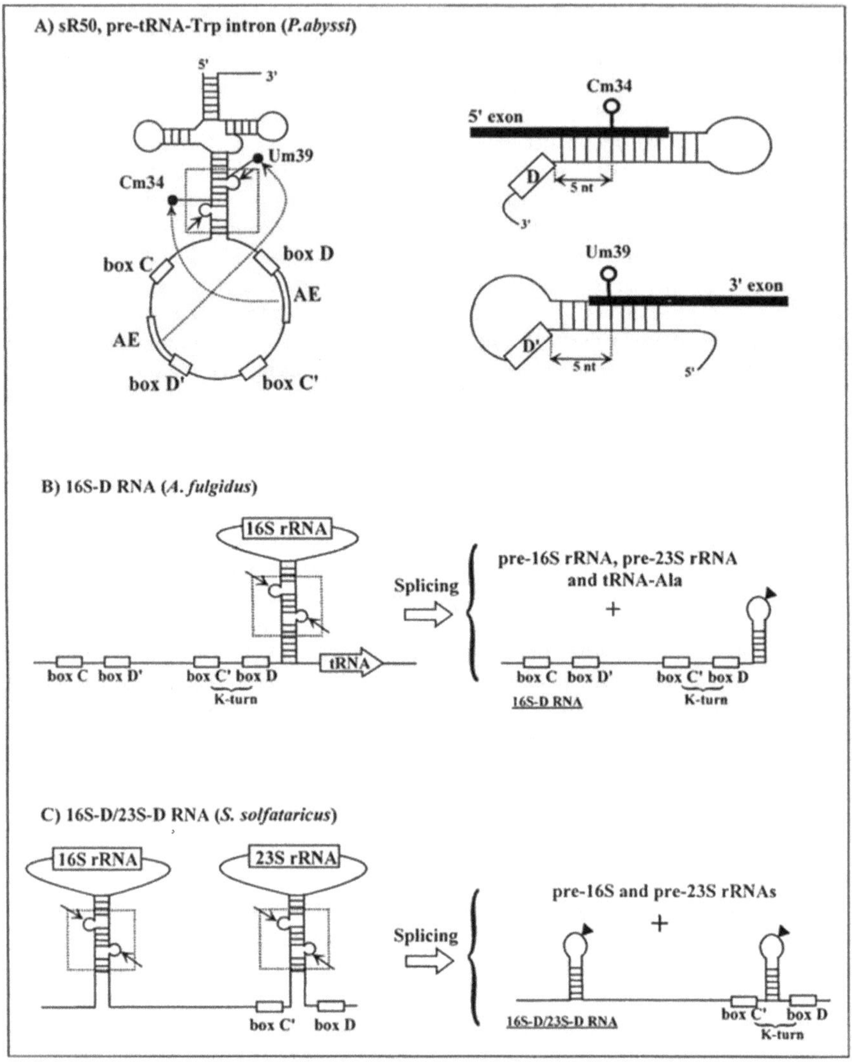

Figure 6. Novel types of C/D small RNAs linked with archaeal RNA splicing. A) sR50, a box C/D RNA in the intron of archaeal pretRNA-Trp.[20] Left: Splicing-competent form of pre-tRNA folding. Hallmark box motifs and antisense elements (AE) in the intron are indicated.The canonical BHB (bulge-helix-bulge) splicing motif (dotted box), splice sites (arrows) and the two sites of intron-guided 2'-O-ribose methylation (filled circles) in the anticodon stem-loop are also denoted. Right: Intramolecular RNA duplexes guiding two 2'-O-methylations in the tRNA antidon stem-loop. Each duplex spans an exon/intron junction (intronic and exonic sequences are depicted by thin and thick lines, respectively). B) A C/D RNA generated by religation of flanking prerRNA spacers following cleavage at the processing stem delineating pre-16S rRNA in euryarchaeon *A. fulgidus*.[137] C' and D motifs able to form a canonical K-turn motif are denoted. In the religated RNA the splicing site is shown by an arrowhead. Other symbols as in (A). C) A related, longer C/D RNA generated in crenarchaeon S. solfataricus by religation of pairs of prerRNA spacers surrounding both pre-16S and pre-23S rRNAs following cleavage at both processing stems. Accumulation of the longer RNA product probably reflects the lack of a tRNA sequence in the 16S-23S interval, in contrast to *A. fulgidus* prerRNA (B). See ref. 132 for further information.

seem to have exploited a remarkable potential for diversification in the evolution of organisms. This is illustrated in vertebrates by the detection of a H/ACA domain in telomerase RNA, C/D-H/ACA chimeric snoRNAs and scores of ubiquitous orphan snoRNAs. This is also reflected by the presence in mammals of intriguing tissue-specific, imprinted snoRNAs encoded by clusters of tandemly repeated gene copies which are likely to provide a reservoir for functional novelty. In the modern RNA world, a diversity of posttranscriptional controls involving steric occlusion of functionally important sites are achieved by sequence-specific base complementarity between a target RNA and a cognate regulatory RNA, as illustrated by microRNAs or *E. coli* riboregulators.[138-141] Clearly, the two ancient snoRNA families appear as a paradigm of noncoding RNAs highly adapted to the modern RNA world, which have probably revealed only a fraction of their diversity. We may expect that many C/D and H/ACA domains remain to be discovered in novel RNA contexts, particularly in multicellular organisms. To identify a full repertoire of guide snoRNAs the computational comparative analysis of closely related, completely sequenced genomes undoubtedly represent a powerful approach, which should also play a crucial role for identifying RNA targets of "orphan" snoRNAs. Through their antisense element(s) modification guide snoRNAs have intrinsic properties of RNA chaperones. Distinguishing between the tightly linked functions of modification guide and chaperone also represents a challenge for the future. Progress in this direction could illuminate our understanding of the biological significance of these nucleotide modifications. The unanticipated diversity of gene controls which could be mediated by the two major snoRNA families opens an exciting era for noncoding RNA research.

Acknowledgments

We thank Alex Hüttenhoffer and members of his laboratory for in-depth collaboration over several aspects of the work reviewed in this chapter. We also thank Christine Gaspin, Patrice Vitali, Hervé Seitz, Béatrice Clouet d'Orval and Marie-Line Bortolin who have contributed recent original reports summarized in this review. Work in the authors' laboratory was supported by funds from the Centre National de la Recherche Scientifique and Université Paul Sabatier, Toulouse, by grants from Association pour la Recherche sur le Cancer, the Toulouse Genopole and the Ministère de l'Education Nationale, de la Recherche et de la Technologie (Programme de Recherche Fondamentale en Microbiologie et Maladies Infectieuses et Parasitaires, 2001–2002) to J.P.B., and a grant from the Programme Interdisciplinaire du C.N.R.S. "Dynamique et réactivité des Assemblages Biologiques" to J.C.

Note Added in Proof

1. Several additional C/D snoRNAs have been recently reported in *A. thaliana*.[143]
2. The L7Ae protein binds both C/D and H/ACA RNAs in Archaea.[144]

References

1. Maden BE. The numerous modified nucleotides in eukaryotic ribosomal RNA. Prog Nucleic Acid Res Mol Biol 1990; 39:241-303.
2. Bachellerie JP, Cavaillé J. Small nuclear RNAs guide the ribose methylations of eukaryotic rRNAs. In: Grosjean H, Benne R, eds. Modification and Editing of RNA: The Alteration of RNA Structure and Function. Washington, DC: ASM Press, 1998:255-272.
3. Ofengand J, Fournier MJ. The pseudouridine residues of rRNA: number, location, biosynthesis and function. In: Grosjean H, Benne RE, eds. Modification and Editing of RNA: the Alteration of RNA Structure and Function. Washington, DC: ASM Press, 1998:229-253.
4. Ofengand J, Rudd K. Bacterial, archaea, and organellar RNA pseudouridines and methylated nucleosides and their enzymes. In: Garrett R, Douthwaite S, Liljas A, Matheson A, Moore PB, Noller HE, eds. Ribosome: Structure, Function, Antibiotics, and Cellular Interactions. Washington, DC: ASM Press, 2000:175-190.

5. Bachellerie JP, Cavaillé J, Qu LH. Nucleotide modifications of eukaryotic rRNAs : The world of small nuclear RNAs revisited. In: Garrett R, Douthwaite S, Liljas A, Matheson A, Moore PB, Noller HE, eds. Ribosome: Structure, Function, Antibiotics, and Cellular Interactions. Washington, DC: ASM Press, 2000:191-203.
6. Kiss T. Small nuclear RNA-guided posttranscriptional modification of cellular RNAs. EMBO J 2001; 20:3617-3622.
7. Kiss T. Small nuclear RNAs: an abundant group of noncoding RNAs with diverse cellular functions. Cell 2002; 109:145-148.
8. Tycowski KT, You ZH, Graham PJ et al. Modification of U6 spliceosomal RNA is guided by other small RNAs. Mol Cell 1998; 2:629-638.
9. Ganot P, Jady BE, Bortolin ML et al. Nuclear factors direct the 2'-O-ribose methylation and pseudouridylation of U6 spliceosomal RNA. Mol Cell Biol 1999; 19:6906-6917.
10. Huttenhofer A, Kiefmann M, Meier-Ewert S et al. RNomics: an experimental approach that identifies 201 candidates for novel, small, nonmessenger RNAs in mouse. EMBO J 2001; 20:2943-2953.
11. Jady BE, Kiss T. A small nuclear guide RNA functions both in 2'-O-ribose methylation and pseudouridylation of the U5 spliceosomal RNA. EMBO J 2001; 20:541-551.
12. Darzacq X, Jady B, Verheggen C et al. Cajal body-specific small nuclear RNAs: a novel class of 2'-O-methylation and pseudouridylation guide RNAs. EMBO J 2002; 21:2746-2756.
13. Cavaille J, Buiting K, Kiefmann M et al. Identification of brain-specific and imprinted small nuclear RNA genes exhibiting an unusual genomic organization. Proc Natl Acad Sci USA 2000; 97:14311-14316.
14. Jady BE, Kiss T. Characterization of the U83 and U84 small nuclear RNAs: two novel 2'- O-ribose methylation guide RNAs that lack complementarities to ribosomal RNAs. Nucleic Acids Res 2000; 28:1348-1354.
15. Cavaille J, Vitali P, Basyuk E et al. A novel brain-specific box C/D small nuclear RNA processed from tandemly repeated introns of a noncoding RNA gene in rats. J Biol Chem 2001; 276:26374-26383.
16. Cavaille J, Seitz H, Paulsen M et al. Identification of tandemly-repeated C/D snoRNA genes at the imprinted human 14q32 domain reminiscent of those at the Prader-Willi/Angelman syndrome region. Hum Mol Genet 2002; 11:1527-1538.
17. Gaspin C, Cavaille J, Erauso G et al. Archaeal homologs of eukaryotic methylation guide small nuclear RNAs: lessons from the *Pyrococcus* genomes. J Mol Biol 2000; 297:895-906.
18. Omer AD, Lowe TM, Russell AG et al. Homologs of small nuclear RNAs in Archaea. Science 2000; 288:517-522.
19. Dennis PP, Omer A, Lowe T. A guided tour: small RNA function in Archaea. Mol Microbiol 2001; 40:509-519.
20. Clouet d'Orval B, Bortolin ML, Gaspin C et al. Box C/D RNA guides for the ribose methylation of archaeal tRNAs. The tRNATrp intron guides the formation of two ribose-methylated nucleosides in the mature tRNATrp Nucleic Acids Res 2001; 29:4518-4529.
21. Balakin AG, Smith L, Fournier MJ. The RNA world of the nucleolus: two major families of small RNAs defined by different box elements with related functions. Cell 1996; 86:823-834.
22. Bachellerie JP, Michot B, Nicoloso M et al. Antisense snoRNAs: a family of nuclear RNAs with long complementarities to rRNA. Trends Biochem Sci 1995; 20:261-264.
23. Tollervey D. Trans-acting factors in ribosome synthesis. Exp Cell Res 1996; 229:226-232.
24. Cavaille J, Bachellerie JP. Processing of fibrillarin-associated snoRNAs from premRNA introns: an exonucleolytic process exclusively directed by the common stem-box terminal structure. Biochimie 1996; 78:443-456.
25. Caffarelli E, Fatica A, Prislei S et al. Processing of the intron-encoded U16 and U18 snoRNAs: the conserved C and D boxes control both the processing reaction and the stability of the mature snoRNA. EMBO J 1996; 15:1121-1131.
26. Lange TS, Borovjagin A, Maxwell ES et al. Conserved boxes C and D are essential nuclear localization elements of U14 and U8 snoRNAs. EMBO J 1998; 17:3176-3187.
27. Samarsky DA, Fournier MJ, Singer RH et al. The snoRNA box C/D motif directs nuclear targeting and also couples snoRNA synthesis and localization. EMBO J 1998; 17:3747-3757.

28. Villa T, Ceradini F, Bozzoni I. Identification of a novel element required for processing of intron-encoded box C/D small nuclear RNAs in *Saccharomyces cerevisiae*. Mol Cell Biol 2000; 20:1311-1320.
29. Darzacq X, Kiss T. Processing of intron-encoded box C/D small nuclear RNAs lacking a 5',3'-terminal stem structure. Mol Cell Biol 2000; 20:4522-4531.
30. Kiss-Laszlo Z, Henry Y, Kiss T. Sequence and structural elements of methylation guide snoRNAs essential for site-specific ribose methylation of prerRNA. EMBO J 1998; 17:797-807.
31. Tycowski KT, Shu MD, Steitz JA. A mammalian gene with introns instead of exons generating stable RNA products. Nature 1996; 379:464-466.
32. Nicoloso M, Qu LH, Michot B et al. Intron-encoded, antisense small nuclear RNAs: the characterization of nine novel species points to their direct role as guides for the 2'-O- ribose methylation of rRNAs. J Mol Biol 1996; 260:178-195.
33. Kiss-Laszlo Z, Henry Y, Bachellerie JP et al. Site-specific ribose methylation of preribosomal RNA: a novel function for small nuclear RNAs. Cell 1996; 85:1077-1088.
34. Cavaille J, Nicoloso M, Bachellerie JP. Targeted ribose methylation of RNA in vivo directed by tailored antisense RNA guides. Nature 1996; 383:732-735.
35. Lowe TM, Eddy SR. A computational screen for methylation guide snoRNAs in yeast. Science 1999; 283:1168-1171.
36. Barneche F, Gaspin C, Guyot R et al. Identification of 66 box C/D snoRNAs in Arabidopsis thaliana: extensive gene duplications generated multiple isoforms predicting new ribosomal RNA 2'-O-methylation sites. J Mol Biol 2001; 311:57-73.
37. Qu LH, Qing M, Zhou H et al. Identification of 10 novel snoRNA gene clusters from *Arabidopsis thaliana*. Nucleic Acids Res 2001; 29:1623-1630.
38. Brown JW, Clark GP, Leader DJ et al. Multiple snoRNA gene clusters from Arabidopsis. RNA 2001; 7:1817-1832.
39. Nicoloso M, Caizergues-Ferrer M, Michot B et al. U20, a novel small nuclear RNA, is encoded in an intron of the nucleolin gene in mammals. Mol Cell Biol 1994; 14:5766-5776.
40. Qu LH, Nicoloso M, Michot B et al. U21, a novel small nuclear RNA with a 13 nt. complementarity to 28S rRNA, is encoded in an intron of ribosomal protein L5 gene in chicken and mammals. Nucleic Acids Res 1994; 22:4073-4081.
41. Qu LH, Henry Y, Nicoloso M et al. U24, a novel intron-encoded small nuclear RNA with two 12 nt long, phylogenetically conserved complementarities to 28S rRNA. Nucleic Acids Res 1995; 23:2669-2676.
42. Tycowski KT, Steitz JA. Noncoding snoRNA host genes in *Drosophilae*: Expression strategies for modification guide snoRNAs. Eur J Cell Biol 2001; 80:119-125.
43. Xu Y, Liu L, Lopez-Estrano C et al. Expression studies on clustered trypanosomatid box C/D small nuclear RNAs. J Biol Chem 2001; 276:14289-14298.
44. Dunbar DA, Chen AA, Wormsley S et al. The genes for small nuclear RNAs in Trypanosoma brucei are organized in clusters and are transcribed as a polycistronic RNA. Nucleic Acids Res 2000; 28:2855-2861.
45. Dunbar DA, Wormsley S, Lowe TM et al. Fibrillarin-associated box C/D small nuclear RNAs in Trypanosoma brucei. Sequence conservation and implications for 2'-O-ribose methylation of rRNA. J Biol Chem 2000; 275:14767-14776.
46. Bugl H, Fauman EB, Staker BL et al. RNA methylation under heat shock control. Mol Cell 2000; 6:349-360.
47. Caldas T, Binet E, Bouloc P et al. Translational defects of Escherichia coli mutants deficient in the Um(2552) 23S ribosomal RNA methyltransferase RrmJ/FTSJ. Biochem Biophys Res Commun 2000; 271:714-718.
48. Pintard L, Bujnicki JM, Lapeyre B et al. MRM2 encodes a novel yeast mitochondrial 21S rRNA methyltransferase. EMBO J 2002; 21:1139-1147.
49. Ganot P, Caizergues-Ferrer M, Kiss T. The family of box ACA small nuclear RNAs is defined by an evolutionarily conserved secondary structure and ubiquitous sequence elements essential for RNA accumulation. Genes Dev 1997; 11:941-956.
50. Ganot P, Bortolin ML, Kiss T. Site-specific pseudouridine formation in preribosomal RNA is guided by small nuclear RNAs. Cell 1997; 89:799-809.

51. Bortolin ML, Ganot P, Kiss T. Elements essential for accumulation and function of small nuclear RNAs directing site-specific pseudouridylation of ribosomal RNAs. EMBO J 1999; 18:457-469.
52. Ni J, Tien AL, Fournier MJ. Small nuclear RNAs direct site-specific synthesis of pseudouridine in ribosomal RNA. Cell 1997; 89:565-573.
53. Ofengand J, Bakin A. Mapping to nucleotide resolution of pseudouridine residues in large subunit ribosomal RNAs from representative eukaryotes, prokaryotes, archaebacteria, mitochondria and chloroplasts. J Mol Biol 1997; 266:246-268.
54. Liang XH, Liu L, Michaeli S. Identification of the first trypanosome H/ACA RNA that guides pseudouridine formation on rRNA. J Biol Chem 2001; 276:40313-40318.
55. Maxwell ES, Fournier MJ. The small nuclear RNAs. Annu Rev Biochem 1995; 64:897-934.
56. Chanfreau G, Legrain P, Jacquier A. Yeast RNase III as a key processing enzyme in small nuclear RNAs metabolism. J Mol Biol 1998; 284:975-988.
57. Chanfreau G, Rotondo G, Legrain P et al. Processing of a dicistronic small nuclear RNA precursor by the RNA endonuclease Rnt1. EMBO J 1998; 17:3726-3737.
58. Qu LH, Henras A, Lu YJ et al. Seven novel methylation guide small nuclear RNAs are processed from a common polycistronic transcript by Rat1p and RNase III in yeast. Mol Cell Biol 1999; 19:1144-1158.
59. Leader DJ, Clark GP, Watters J et al. Clusters of multiple different small nuclear RNA genes in plants are expressed as and processed from polycistronic presnoRNAs. EMBO J 1997; 16:5742-5751.
60. Leader DJ, Clark GP, Boag J et al. Processing of vertebrate box C/D small nuclear RNAs in plant cells. Eur J Biochem 1998; 253:154-160.
61. Filippini D, Renzi F, Bozzoni I et al. U86, a novel snoRNA with an unprecedented gene organization in yeast. Biochem Biophys Res Commun 2001; 288:16-21.
62. Pelczar P, Filipowicz W. The host gene for intronic U17 small nuclear RNAs in mammals has no protein-coding potential and is a member of the 5'-terminal oligopyrimidine gene family. Mol Cell Biol 1998; 18:4509-4518.
63. Smith CM, Steitz JA. Classification of gas5 as a multi-small-nuclear-RNA (snoRNA) host gene and a member of the 5'-terminal oligopyrimidine gene family reveals common features of snoRNA host genes. Mol Cell Biol 1998; 18:6897-6909.
64. Yoshihama M, Uechi T, Asakawa S et al. The human ribosomal protein genes: sequencing and comparative analysis of 73 genes. Genome Res 2002; 12:379-390.
65. Niewmierzycka A, Clarke S. S-Adenosylmethionine-dependent methylation in *Saccharomyces cerevisiae*. Identification of a novel protein arginine methyltransferase .J Biol Chem 1999; 274:814-824.
66. Tollervey D, Lehtonen H, Jansen R et al. Temperature sensitive mutations demonstrate roles for yeast fibrillarin in prerRNA processing, prerRNA methylation, and ribosome assembly. Cell 1993; 72:443-457.
67. Wang H, Boisvert D, Kim KK et al. Crystal structure of a fibrillarin homologue from Methanococcus jannaschii, a hyperthermophile, at 1.6 A resolution. EMBO J 2000; 19:317-323.
68. Watkins NJ, Segault V, Charpentier B et al. A common core RNP structure shared between the small nucleoar box C/D RNPs and the spliceosomal U4 snRNP. Cell 2000; 103:457-466.
69. Vidovic I, Nottrott S, Hartmuth K et al. Crystal structure of the spliceosomal 15.5kD protein bound to a U4 snRNA fragment. Mol Cell 2000; 6:1331-1342.
70. Klein DJ, Schmeing TM, Moore PB et al. The kink-turn: a new RNA secondary structure motif. EMBO J 2001; 20:4214-4221.
71. Nottrott S, Hartmuth K, Fabrizio P et al. Functional interaction of a novel 15.5kD [U4/U6.U5] tri-snRNP protein with the 5' stem-loop of U4 snRNA. EMBO J 1999; 18:6119-6133.
72. Newman DR, Kuhn JF, Shanab GM et al. Box C/D snoRNA-associated proteins: two pairs of evolutionarily ancient proteins and possible links to replication and transcription. RNA 2000; 6:861-879.
73. Filippini D, Bozzoni I, Caffarelli E. p62, a novel Xenopus laevis component of box C/D snoRNPs. RNA 2000; 6:391-401.
74. Lafontaine DL, Tollervey D. Synthesis and assembly of the box C+D small nuclear RNPs. Mol Cell Biol 2000; 20:2650-2659.
75. Lafontaine DL, Tollervey D. Nop58p is a common component of the box C+D snoRNPs that is required for snoRNA stability. RNA 1999; 5:455-467.

76. Wu P, Brockenbrough JS, Metcalfe AC et al. Nop5p is a small nuclear ribonucleoprotein component required for pre18 S rRNA processing in yeast. J Biol Chem 1998; 273:16453-16463.
77. Verheggen C, Mouaikel J, Thiry M et al. Box C/D small nuclear RNA trafficking involves small nuclear RNP proteins, nuclear factors and a novel nuclear domain. EMBO J 2001; 20:5480-5490.
78. Gautier T, Berges T, Tollervey D et al. Nuclear KKE/D repeat proteins Nop56p and Nop58p interact with Nop1p and are required for ribosome biogenesis. Mol Cell Biol 1997; 17:7088-7098.
79. Henras A, Henry Y, Bousquet-Antonelli C et al. Nhp2p and Nop10p are essential for the function of H/ACA snoRNPs. EMBO J 1998; 17:7078-7090.
80. Watkins NJ, Gottschalk A, Neubauer G et al. Cbf5p, a potential pseudouridine synthase, and Nhp2p, a putative RNA- binding protein, are present together with Gar1p in all H BOX/ACA-motif snoRNPs and constitute a common bipartite structure. RNA 1998; 4:1549-1568.
81. Lafontaine DL, Bousquet-Antonelli C, Henry Y et al. The box H + ACA snoRNAs carry Cbf5p, the putative rRNA pseudouridine synthase. Genes Dev 1998; 12:527-537.
82. Zebarjadian Y, King T, Fournier MJ et al. Point mutations in yeast CBF5 can abolish in vivo pseudouridylation of rRNA. Mol Cell Biol 1999; 19:7461-7472.
83. Vulliamy T, Marrone A, Goldman F et al. The RNA component of telomeres is mutated in autosomal dominant dyskeratosis congenita. Nature 2001; 413:432-435.
84. Hoang C, FerreD'Amare AR. Cocrystal structure of a tRNA Psi55 pseudouridine synthase: nucleotide flipping by an RNA-modifying enzyme. Cell 2001; 107:929-939.
85. Dragon F, Pogacic V, Filipowicz W. In vitro assembly of human H/ACA small nuclear RNPs reveals unique features of U17 and telomeres RNAs. Mol Cell Biol 2000; 20:3037-3048.
86. Bagni C, Lapeyre B. Gar1p binds to the small nuclear RNAs snR10 and snR30 in vitro through a nontypical RNA binding element. J Biol Chem 1998; 273:10868-10873.
87. Henras A, Dez C, Noaillac-Depeyre J et al. Accumulation of H/ACA snoRNPs depends on the integrity of the conserved central domain of the RNA-binding protein Nhp2p. Nucleic Acids Res 2001; 29:2733-2746.
88. Kuhn JF, Tran EJ, Maxwell ES. Archaeal ribosomal protein L7 is a functional homolog of the eukaryotic 15.5kD/Snu13p snoRNP core protein. Nucleic Acids Res 2002; 30:931-941.
89. King TH, Decatur WA, Bertrand E et al. A well-connected and conserved nucleoplasmic helicase is required for production of box C/D and H/ACA snoRNAs and localization of snoRNP proteins. Mol Cell Biol 2001; 21:7731-7746.
90. Yang Y, Isaac C, Wang C et al. Conserved composition of mammalian box H/ACA and box C/D small nuclear ribonucleoprotein particles and their interaction with the common factor Nopp140. Mol Biol Cell 2000; 11:567-577.
91. Terns MP, Terns RM. Small nuclear RNAs: versatile trans-acting molecules of ancient evolutionary origin. Gene Expr 2002; 10:17-39.
92. Lange TS, Ezrokhi M, Amaldi F et al. Box H and box ACA are nuclear localization elements of U17 small nuclear RNA. Mol Biol Cell 1999; 10:3877-3890.
93. Narayanan A, Lukowiak A, Jady BE et al. Nuclear localization signals of box H/ACA small nuclear RNAs. EMBO J 1999; 18:5120-5130.
94. Narayanan A, Speckmann W, Terns R et al. Role of the box C/D motif in localization of small nuclear RNAs to coiled bodies and nucleoli. Mol Biol Cell 1999; 10:2131-2147.
95. Massenet S, Mougin A, Branlant C. Posttranscriptional modifications in the U snRNAs. In: Grosjean H, Benne RE, eds. Modification and Editing of RNA: The Alteration of RNA Structure and Function. Washington DC: ASM Press, 1998:201-228.
96. Yu YT, Shu MD, Steitz JA. Modifications of U2 snRNA are required for snRNP assembly and premRNA splicing. EMBO J 1998; 17:5783-5795
97. Zhou H, Chen YQ, Du YP et al. The Schizosaccharomyces pombe mgU6-47 gene is required for 2'-O-methylation of U6 snRNA at A41. Nucleic Acids Res 2002; 30:894-902.
98. Massenet S, Motorin Y, Lafontaine DL et al. Pseudouridine mapping in the *Saccharomyces cerevisiae* spliceosomal U small nuclear RNAs (snRNAs) reveals that pseudouridine synthase pus1p exhibits a dual substrate specificity for U2 snRNA and tRNA. Mol Cell Biol 1999; 19:2142-2154.
99. Matera AG. Nuclear bodies: multifaceted subdomains of the interchromatin space. Trends Cell Biol 1999; 9:302-309.

100. Sleeman JE, Lamond AI. Newly assembled snRNPs associate with coiled bodies before speckles, suggesting a nuclear snRNP maturation pathway. Curr Biol 1999; 9:1065-1074.
101. Mitchell JR, Cheng J, Collins K. A box H/ACA small nuclear RNA-like domain at the human telomeres RNA 3' end. Mol Cell Biol 1999; 19:567-576.
102. Pogacic V, Dragon F, Filipowicz W. Human H/ACA small nuclear RNPs and telomeres share evolutionarily conserved proteins NHP2 and NOP10. Mol Cell Biol 2000; 20:9028-9040.
103. Dez C, Henras A, Faucon B et al. Stable expression in yeast of the mature form of human telomeres RNA depends on its association with the box H/ACA small nuclear RNP proteins Cbf5p, Nhp2p and Nop10p. Nucleic Acids Res 2001; 29:598-603.
104. Hughes JM. Functional base-pairing interaction between highly conserved elements of U3 small nuclear RNA and the small ribosomal subunit RNA. J Mol Biol 1996; 259:645-654.
105. Sharma K, Tollervey D. Base pairing between U3 small nuclear RNA and the 5' end of 18S rRNA is required for prerRNA processing. Mol Cell Biol 1999; 19:6012-6019.
106. Morrissey JP, Tollervey D. Yeast snR30 is a small nuclear RNA required for 18S rRNA synthesis. Mol Cell Biol 1993; 13:2469-2477.
107. Tollervey D. A yeast small nuclear RNA is required for normal processing of preribosomal RNA. EMBO J 1987; 6:4169-4175.
108. Enright CA, Maxwell ES, Eliceiri GL et al. 5'ETS rRNA processing facilitated by four small RNAs: U14, E3, U17, and U3. RNA 1996; 2:1094-1099.
109. Liang WQ, Clark JA, Fournier MJ. The rRNA-processing function of the yeast U14 small nuclear RNA can be rescued by a conserved RNA helicase-like protein. Mol Cell Biol 1997; 17:4124-4132.
110. Pederson T. The plurifunctional nucleolus. Nucleic Acids Res 1998; 26:3871-3876.
111. Bertrand E, Houser-Scott F, Kendall A et al. Nuclear localization of early tRNA processing. Genes Dev 1998; 12:2463-2468.
112. Burns CM, Chu H, Rueter SM et al. Regulation of serotonin-2C receptor G-protein coupling by RNA editing. Nature 1997; 387:303-308.
113. Yi-Brunozzi HY, Easterwood LM, Kamilar GM et al. Synthetic substrate analogs for the RNA-editing adenosine deaminase ADAR-2. Nucleic Acids Res 1999; 27:2912-2917.
114. Nielsen H, Orum H, Engberg J. A novel class of nuclear RNAs from Tetrahymena. FEBS Lett 1992; 307:337-342.
115. Liang XH, Xu YX, Michaeli S. The spliced leader-associated RNA is a trypanosome-specific sn(o)RNA that has the potential to guide pseudouridine formation on the SL RNA. RNA 2002; 8:237-246.
116. Palfi Z, Xu GL, Bindereif A. Spliced leader-associated RNA of trypanosomes. Sequence conservation and association with protein components common to trans-spliceosomal ribonucleoproteins. J Biol Chem 1994; 269:30620-30625.
117. de los Santos T, Schweizer J, Rees CA et al. Small evolutionarily conserved RNA, resembling C/D box small nuclear RNA, is transcribed from PWCR1, a novel imprinted gene in the Prader- Willi deletion region, which Is highly expressed in brain. Am J Hum Genet 2000; 67:1067-82.
118. Filipowicz W. Imprinted expression of small nuclear RNAs in brain: time for RNomics. Proc Natl Acad Sci USA 2000; 97:14035-14037.
119. Meguro M, Mitsuya K, Nomura N et al. Large-scale evaluation of imprinting status in the Prader-Willi syndrome region: an imprinted direct repeat cluster resembling small nuclear RNA genes. Hum Mol Genet 2001; 10:383-394.
120. Runte M, Huttenhofer A, Gross S et al. The IC-SNURF-SNRPN transcript serves as a host for multiple small nuclear RNA species and as an antisense RNA for UBE3A. Hum Mol Genet 2001; 10:2687-2700.
121. Reik W, Walter J. Genomic imprinting: parental influence on the genome. Nat Rev Genet 2001; 2:21-32.
122. Nicholls RD, Knepper JL. Genome organization, function, and imprinting in Prader-Willi and Angelman syndromes. Annu Rev Genomics Hum Genet 2001; 2:153-175.
123. Komine Y, Tanaka NK, Yano R et al. A novel type of noncoding RNA expressed in the rat brain. Brain Res Mol Brain Res 1999; 66:1-13.

124. Hatton D, Gray JC. Two MAR DNA-binding proteins of the pea nuclear matrix identify a new class of DNA-binding proteins. Plant J 1999; 18:417-429.
125. Wirth J, Back E, Huttenhofer A et al. A translocation breakpoint cluster disrupts the newly defined 3' end of the SNURF-SNRPN transcription unit on chromosome 15. Hum Mol Genet 2001; 10:201-210.
126. Olsen GJ, Woese CR. Archaeal genomics: an overview. Cell 1997; 89:991-994.
127. Gustafsson C, Reid R, Greene PJ et al. Identification of new RNA modifying enzymes by iterative genome search using known modifying enzymes as probes. Nucleic Acids Res 1996; 24:3756-3762.
128. Noon KR, Bruenger E, McCloskey JA. Posttranscriptional modifications in 16S and 23S rRNAs of the archaeal hyperthermophile *Sulfolobus solfataricus.* J Bacteriol 1998; 180:2883-2888.
129. Amiri KA. Fibrillarin-like proteins occur in the domain Archaea. J Bacteriol 1994; 176:2124-2127.
130. Lafontaine DL, Tollervey D. Birth of the snoRNPs: the evolution of the modification-guide snoRNAs. Trends Biochem Sci 1998; 23:383-388.
131. Watanabe Y, Gray MW. Evolutionary appearance of genes encoding proteins associated with box H/ACA snoRNAs: cbf5p in Euglena gracilis, an early diverging eukaryote, and candidate Gar1p and Nop10p homologs in archaebacteria. Nucleic Acids Res 2000; 28:2342-2352.
132. Tang T, Bachellerie JP, Rozhdestvensky T et al. Identification of 86 candidates for small nonmessenger RNAs from the archaeon Archaeoglobus fulgidus. Proc Natl Acad Sci USA 2002; 99:7536-7541.
133. Speckmann WA, Li ZH, Lowe TM et al. Archaeal Guide RNAs Function in rRNA Modification in the Eukaryotic Nucleus. Curr Biol 2002; 12:199-203.
134. Omer AD, Ziesche S, Ebhardt H et al. In vitro reconstitution and activity of a C/D box methylation guide ribonucleoprotein complex. Proc Natl Acad Sci USA 2002; 99:5289-5294.
135. Grosjean H, Szweykowska-Kulinska Z, Motorin Y et al. Intron-dependent enzymatic formation of modified nucleosides in eukaryotic tRNAs: a review. Biochimie 1997; 79:293-302.
136. Nieuwlandt DT, Carr MB, Daniels CJ. In vivo processing of an intron-containing archael tRNA. Mol Microbiol 1993; 8:93-99.
137. Tang TH, Rozhdestvensky TS, Clouet d'Orval B et al. RNomics in Archaea reveals a further link between splicing of archaeal introns and rRNA processing. Nucleic Acids Res 2002; 30:921-930.
138. Lee RC, Ambros V. An extensive class of small RNAs in *Caenorhabditis elegans.* Science 2001; 294:862-864.
139. Lau NC, Lim LP, Weinstein EG et al. An abundant class of tiny RNAs with probable regulatory roles in *Caenorhabditis elegans.* Science 2001; 294:858-862.
140. Lagos-Quintana M, Rauhut R, Lendeckel W et al. Identification of novel genes coding for small expressed RNAs. Science 2001; 294:853-858.
141. Eddy SR. Noncoding RNA genes and the modern RNA world. Nat Rev Genet 2001; 2:919-929.
142. Rueter SM, Emeson RB. Adenosine-to-inosine conversion in mRNA. In: Grosjean H, Benne R, eds. Modification and Editing of RNA: The Alteration of RNA Structure and Function. Washington, DC: ASM Press, 1998:343-361.
143. Marker C, Zemann A, Terhörst T et al. Experimental RNomics: Identification of 140 candidates for small, non messenger RNAs in the plant *Arabidopsis thaliana.* Current Biol 2002; 12:2002 2013.
144. Rozhdestvensky TS, Tang TH, Tchirkova I et al. Binding of L7Ae protein to the K-turn of archaeal snoRNAs: A shared RNA binding motif for C/D and H/ACA box snoRNAs in Archaea. Nucl Acids Res 2003; 31:869-877.

CHAPTER 13

New Perspectives on Noncoding or Short ORF-Encoding RNAs in Plants

Martin Crespi, Anna Campalans, Claude Thermes and Adam Kondorosi

Abstract

mRNAs that do not contain a long open reading frame (longer than 100 amino acids; sORF-RNAs) have received considerable attention in recent years. These genes are generally not detected by usual sequence analysis and a large number of them likely remain to be discovered. Their functions may involve only the RNA molecule itself (noncodingRNAs), and/or the short ORF-encoded oligopeptides. In eukaryotes, expression studies have revealed a striking diversity of these mRNAs in many cell types from various organisms, which are induced at specific stages of development. This review describes several examples of plant sORF-RNA genes that have been shown as induced during development as well as in stress conditions and discusses the possible outcome of the analysis of their mechanisms of action. During nodule development, the early nodulin gene *enod40* is rapidly induced in the dividing cortical cells of the symbiotic root nodule primordium. Over expression of *enod40* resulted in cortical cell division in *Medicago* roots and accelerated nodulation whereas cosuppression of this gene blocked nodule development. Using a transient expression assay, *enod40* activity was found to be dependent on the translation of two sORFs (spanning two conserved nucleotide boxes) as well as an inter-ORF region (spanning a potential RNA structure). Another example is the cucumber *CR20* gene that is repressed by the phytohormone cytokinin in various plant organs. Other plant genes of the same class are induced during stress conditions such as dehydration processes (*cdt-1*), phosphate starvation (*tps11*) and cycloheximide treatment (*gut15*). Over expression of the *cdt-1* gene conferred dehydration tolerance to callus tissues in *Craterostigma* plants. Recently, several T-DNA insertions in genes without large ORFs displaying very strict tissue-specificity have been identified in the model plant *Arabidopsis thaliana* using systematic insertional mutagenesis. In general, these mRNAs are fairly abundant polyadenylated transcripts but little is known about the molecular mechanisms involved in their action.

These results suggest that sORF-encoded peptides and RNA structures might be active elements of the regulatory mechanisms involving those RNAs. Polypeptide signalling is an emerging field in plant biology and several biologically active peptides have been purified from plant tissues such as phytosulphokines, systemins and cysteine-rich pollen peptides involved in self-recognition. Even though peptides are generally produced by cleavage of larger precursors, certain sORF-RNAs may indeed code for small peptides to match the wide variety of putative peptide ligand receptors present in plant genomes. In addition, it was shown that small noncoding RNAs are involved in cell-cell signalling and gene silencing, suggesting that diverse RNAs may participate in various regulatory processes. Indeed, it is likely that numerous sORF-RNAs

Noncoding RNAs: Molecular Biology and Molecular Medicine, edited by Jan Barciszewski and Volker A. Erdmann.

genes are present in the genome of higher plants. Studies on their encoded oligopeptide or RNA signals may uncover novel functions involved in translational control, cell-to-cell communication and growth regulation in plants.

mRNAs without Long ORFs Exist in Plants and Animals

In eukaryotic cells, and specially in mammals, complex responses to external factors are governed by posttranscriptional processes allowing rapid cellular adaptations to the new conditions. Small regulatory molecules such as oligopeptides or small RNAs may play critical roles in the control of these processes both at intra- and intercellular levels, such as the well-characterized role for peptide hormones in long-distance signalling in mammals[1,2] and the tiny *let-7* RNA in worm development.[3]

Definition of sORF-RNAs

Using classical bioinformatic approaches, genes are predicted in general based on an ORF-encoding capacity of at least 100 amino acids. Additional criteria such as homology to known RNA genes (e.g., transfer RNA) may overcome this limitation. However, several mRNAs that do not contain a long ORF (larger than 100 amino acids) have been identified (sORF-RNAs). These sORF-RNAs are classified as noncoding or oligopeptide-encoding RNAs. In fact, our ability to predict genes lacking significant coding capacity is limited to relatively large and evolutionarily conserved genes. Other criteria also applied are splice site predictions, presence of polyadenylation sites and TATA boxes, analysis of GC content and/or codon usage.[4] Nevertheless, the identification of a significant ORF is still required for gene annotation even with these additional tools. Several genes have been ignored with these automated approaches as demonstrated recently by the discovery of a large variety of small microRNAs (miRNAs) in *Caenorhabditis elegans*, *Drosophila* and human cells using direct biochemical characterization.[5-8] Indeed, by using the genomic sequences available from a variety of organisms to infer precursors and determine genetic locations, it was shown that these miRNAs correspond to new genes and are not breakdown products of previously annotated mRNAs.[5]

Identification of sORF-RNAs

In eukaryotes, several sORF-RNAs are induced at specific stages of development suggesting their participation in various differentiation processes.[9-14] It is very difficult to estimate the number of genes belonging to this class that might exist in a genome.[4] In yeast, by searching for transcripts encoded in regions classified as intergenic larger than 2 kb, 18 new genes of this class were found.[15] However, using Serial Analysis of Gene Expression (SAGE) approaches, around 170 tags could not be linked to a predicted ORF in the yeast genome.[16] Even though SAGE tags may correspond to long untranslated regions of coding transcripts, it seems a reasonable estimate that about 0.4-2% of the genome correspond to sORF-RNAs.

In plants, identification of 50 cDNAs from a 300 kb contig of chromosome 1 from *Arabidopsis*, revealed seven noncoding genes among which four are antisense transcripts.[17] Bioinformatical searches ignore antisense transcripts, but they may also contribute significantly to the population of sORF-RNAs. In plants, naturally occurring antisense RNAs have been identified and are believed to control gene expression by annealing to the complementary sense transcript yielding a double stranded RNA.[18] These dsRNA molecules may affect RNA stability, transcription, translation or induce "silencing" mechanisms.[6] Taken into account this data in the model plant *Arabidopsis* with an estimate of 22000 genes, we expect at least 200 to 400 genes (1-2%) coding for sORF-RNAs but the list could be even much larger. For example, a search for genes homologous to transcripts encoding small peptides such as Clavata 3 or SCR (involved in meristem development and self-recognition, respectively) in the *Arabidopsis* genome yield more than 200 genes.[19-21] The majority of them do not contain a significantly long ORF recognizable exclusively by a bioinformatical approach without previous gene identification. Finally, it should be considered that isolation of cDNAs corresponding to RNAs without significant ORFs is necessary but not sufficient to propose a new gene since several cellular processes, such as RNA processing or transcription of retrotransposons, may yield noncoding RNA species. Experimental approaches (e.g., mutational analysis) as well as thorough analysis of nucleotide/amino acid sequence conservation are required to explore possible gene products and their functions. Indeed, there are several examples where RNAs originally described as noncoding have been shown to code for small peptides and vice versa.[4]

Towards a Genome-Wide Analysis of sORF-RNAs

The functional exploration of this novel class of genes requires parallel bioinformatic and biochemical approaches coupled to reverse genetics in a model plant system such as *Arabidopsis*. The genome sequence of this plant has been determined, large scale genetic and genomic resources have been developed, more than 200,000 insertional mutants are available and transgenic plants are easily produced allowing to modify gene expression in specific ways. Hence, it becomes possible to explore in detail the role of sORF-RNA genes in differentiation and signalling in this plant. Moreover, several cellular processes, such as alternative splicing, nuclear architecture, differentiation of stem cells, post-transcriptional regulation and gene complexity, can be studied in this plant which offers major advantages in comparison to yeast for the dissection of related mechanisms in human cells.

The availability of large EST collections and the *Arabidopsis* genomic sequence prompted MacIntosh et al to search for those genes using bioinformatic approaches.[22] Fifteen putative noncoding RNAs have been annotated in *Arabidopsis* based on the literature. Except for the signal recognition particle 7SRNA, all appear to be plant-specific. By using computational tools to filter protein-coding genes from genes corresponding to 20000 expressed sequence tags, MacIntosh et al identified 19 clones without any ORF but containing highly conserved nucleotide sequence regions, 9 putative peptide-encoding RNAs (conserved sORFs) and 11 RNAs containing putative sORFs without any significant homology to sequences found in databases. This latter class cannot be classified to either of these categories. Again, none of these new cDNAs had homologs outside of the plant kingdom.[22] We have carried out a similar search using other EST libraries and found 200 candidate genes only containing sORFs, encoded in intergenic regions of the *Arabidopsis* genome. Analysis of their coding capacity with various statistical tools suggests that among them there are several sORF-encoding RNAs, nonORF RNAs and small antisense RNAs (our unpublished results). These results open wide perspectives for the identification of sORF-RNA genes in plants. Based on the conservation of sequence elements present in those genes through the different plants and kingdoms, the possible role of nucleotide boxes and/or encoded sORFs will be assessed across large evolutionary ranges. In any case, sORF-RNA genes may reveal a novel unexplored region of the genome in plants and animals.[23]

mRNAs Only Containing sORFs Are Involved in Plant Growth and Development

In plants, expression studies have identified several sORF-RNAs induced at specific stages of development. Various examples of these genes have been also shown as induced under stress conditions. However, little is known about the roles of the different gene products, peptides and RNA, and the molecular mechanisms involved in their function.

Developmental Processes Involving sORF-RNAs

The *enod40* genes code for peculiar RNAs (around 700 bp) containing only short ORFs in their sequences (10-37 amino acids). Computer predictions suggested that they code for structured RNAs,[24] moreover the *enod40* RNA did not copurify with polysomes.[25] In the *enod40* genes two highly conserved regions were distinguished: box I in the 5' end, spanning a short conserved ORF (sORF I), and box II in the central part of the gene.[26] It has then been proposed that sORF I might encode an active oligopeptide. The recognition of the sORF I ATG was demonstrated in vivo using reporter gene fusions (in roots and tobacco protoplasts) and by in vitro translation.[26-28] Translational analysis of *enod40* indicated that the presence of sORFs did not prevent reinitiation at downstream AUG codons suggesting that different *enod40* sORFs are likely to be translated with variable efficiency in different species.[27,28]

Leguminous plants have the ability to enter in symbiosis with N_2-fixing bacteria (collectively called rhizobia) to form the root nodule. Development of this symbiotic organ depends on the expression of plant and bacterial genes.[29] At the initial stage of nodule organogenesis, the early nodulin gene *enod40* is rapidly induced in the dividing cortical cells of the nodule primordium and the vascular bundles of roots and nodules.[24,25,30,31] Expression was also detected in stem vascular tissues at specific stages of development and in lateral roots and flowers from legumes. Moreover, homologous genes containing the two conserved *enod40* regions have been found in the nonlegumes rice and tobacco.[26,32] Though *enod40* function is not exclusively associated to nodulation, it is expressed at much higher levels than in any other processes.

Overexpression of *enod40* in *M. truncatula* yield plants showing extensive proliferation of cortical cells accompanied by accelerated nodulation, further suggesting that *enod40* induction is a limiting step in nodule initiation. Among these transgenic plants, two lines showing cosuppression of *enod40*, formed few nodules with arrested meristems, indicating that its function is required for nodule morphogenesis.[33,34] Overexpression of this gene may then allow a large group of cells to become sensitive to the nodule-initiating signal regulating primordium formation. These results strongly suggest that *enod40* is a key regulator of the spatial control of nodule initiation. A biological assay related to this function was developed based on the introduction of *enod40* by microtargeting in the roots of alfalfa seedlings.[27] Microtargeting of constructs transiently expressing *enod40* induced cell division in the inner cortex, similarly to the action of rhizobia on these cells at the onset of nodulation. This cell-specific response depended on the translation of two sORFs spanning the conserved regions as shown by using selected point mutants. Noteworthy, translation, size and correct amino acid sequence of the 13-amino-acid-long sORF I were necessary for the biological activity of *enod40*. In addition, the RNA region between the two conserved nucleotide boxes was also required for the elicitation of cortical cell division, most likely without perturbing translation of the sORFs. Thus, both the RNA itself and the translation of the sORF-encoded peptides were required for the function of *enod40* in the root cortex. One hypothesis to explain these results is that the RNA region may be required for determining the subcellular site where *enod40* is translated. Interestingly, RNA signals present in the 3'UTR regions of several genes are involved in the localization of mRNA translation,[35] a feature that may be essential when the encoded gene products are unstable (such as the *enod40* oligopeptides). An alternative hypothesis is that sORF-mediated translational control of the transcript may also be exerted to allow stability or export of the RNA into the cytoplasm (as shown in animals[36]).

Another peculiar gene, the cucumber *CR20* gene was identified as repressed by the phytohormone cytokinin in various plant organs.[13] It was also repressed during development and stress conditions. This gene contains three exons and seems to generate at least three transcripts by alternative splicing of the second intron. However, none of the mRNAs contain a long ORF. Identification of several homologues in *Arabidopsis* and tobacco showed that no sORF-encoded oligopeptide was conserved among these sequences.[22] In contrast, a region of 180 nucleotides was highly conserved among different genes and seemed to form a stable secondary RNA structure. The *CR20* genes belong to the same family as NtGUT15, encoding one of the most unstable transcripts in tobacco cell cultures.[37] The fact that transcripts of this family are hormonally regulated and are unstable suggest that they may play regulatory roles in plant signalling but their real function is unknown.

sORF-RNAs Are Induced during Stress Conditions

The identification of genes induced during phosphate starvation allowed the characterization of the TPSI1/Mt4 family in *M. truncatula* and tomato.[9,22,38] These genes encode transcripts only containing sORFs and identification of an *Arabidopsis* homologue allowed to characterize a four-amino-acid-long sORF conserved among all these genes, coding for an oligopeptide whose amino acid sequence is MAIP.[39] Longer ORFs present in these transcripts are not conserved. At nucleotide level these genes share larger regions of similarity, including a core stretch of 22 nucleotides which is identical in all species (except for a single mismatch).

These genes were also systemically repressed in the whole root system after localized application of phosphate, suggesting a function in phosphate starvation signalling in plants. In *Arabidopsis*, phosphate deprivation induces AtTPSI1 expression in all cells (and not only in roots, as shown in *M. truncatula*). In mutants affected in phosphate loading from the xylem (*pho1* mutant), AtTPSI1 expression in roots was delimited to the endodermis. Hence, this gene is regulated both by biotic (cytokinins) and abiotic (phosphate starvation) signals and participates in long-distance systemic phosphate starvation responses.[39]

The sORF-RNA gene *cdt-1* was isolated during screening for regulatory genes involved in desiccation tolerance in the resurrection plant *Craterostigma plantagineum*, using a T-DNA activation tagging approach.[10] The *cdt-1* gene codes for a 0.9 kb intronless gene and the transcribed polyA+RNA contains only sORFs. Only a 22 amino-acid-long sORF was detected starting at an AUG. The absence of homologues in other species prevented further analysis, however, in the *C. plantagineum* genome, the *cdt* genes form a very large family. The presence of direct repeats in the genomic sequences flanking these genes suggested that they may be a family of potential retrotransposons.

Constitutive overexpression of *cdt-1* conferred desiccation tolerance to *C. plantagineum* callus tissues as well as constitutive expression of transcripts characteristic of the ABA signalling pathway. Thus, *cdt-1* is an intermediate in the ABA transduction pathway in this plant. This gene was also induced throughout the desiccation process and its expression was suppressed as soon as water was supplied to the dried tissues.[10] Since it is known that stress increases the transcription of retrotransposons, it is conceivable that *cdt-1* transcription during water stress could also lead to retrotransposition of *cdt*-like sequences.

T-DNA Tagging Revealed Peculiar Genes of Arabidopsis

In recent years, several GUS-promoter trap T-DNA insertions displaying strict tissue-specificity have been identified in *A. thaliana* using systematic insertional mutagenesis.[40] Several of these insertions activating GUS expression, are not flanked by easily identifiable genes (P. Gallois and K. Lindsey, personal communication). This may be due to the presence of cryptic promoters that were activated by the T-DNA insertion or genes with large number of exons, whose structure is difficult to predict. One of these insertion lines was identified as showing expression in the embryonic and seedling root and the T-DNA was inserted in a gene (designated *polaris, pls*) coding for a low abundance 500 bp transcript.[41] A sORF of 36 amino acids present in the *polaris* transcript was inactivated by the T-DNA insertion. Thus, the *polaris* gene seems to encode a small peptide that is involved in both cytokinin and auxin responses.

The gene is auxin-inducible and several markers of cytokinin and auxin action showed perturbed regulation in the *pls* mutants as well as changes in the extent of root growth. Complementation of the mutant with the sORF-encoding region and overexpression experiments further reinforces the idea that the peptide is involved in gene function. However, complementation of the root growth phenotype was not complete, suggesting that other regions of the transcript may be required for full activity or regulation. These results suggest that *pls* plays a role in fine-tuning regulation of cytokinin action in roots and further supports that sORF-encoded peptides play functions in plants.[19,41]

Using activation tagging, Weigel et al[42] identified 23 dominant mutants among 25000 *Arabidopsis* plants. These gain of function mutations may deregulate a pathway that is not expressed during normal growth conditions. Two of these lines, *jaw1* and *jaw2* were allelic and separated by a 1.6 kb of genomic DNA without any detectable gene in the intervening sequence. An RNA coding for a 0.55 kb transcript, was found to be overexpressed in these lines when probed with a 2-kb fragment derived from the T-DNA flanking regions. Even though no cDNA corresponding to this gene could be identified, overexpression of this RNA reproduced the observed phenotype further reinforcing that it is not a particular exon of a highly fragmented gene. The *jaw* mutants had deeply serrated leaves, slightly late flowering and their petals were light green.[42] No gene was predicted in this region of the *Arabidopsis* genome, suggesting that the JAW sORF-RNA encodes a very short ORF or does not encode a protein.

Potential Mechanisms of Action

sORF-RNAs may be active as such and play roles as catalytic or structural parts of ribonucleoprotein complexes through complex 3-D structures or through their sORF-encoded oligopeptides.

RNA

Several RNAs exhibit a function without being translated into proteins, for example tRNAs, rRNAs, RNAs in ribozymes and small nuclear RNAs from spliceosomes.[43] As mentioned above, relatively long antisense transcripts, likely involved in controlling the expression of the "sense" mRNA, have been reported.[18] In addition, certain sORF-RNAs have been shown to participate in the organization of the cytoplasm, in mRNA localization mechanisms to spatially regulate translation or stability of other transcripts, in protein secretion and in silencing large DNA regions in *cis*.[23,35,44,45] In addition, small regulatory RNAs produced during post-transcriptional gene silencing, PTGS, (25 bp siRNA molecules) have been associated to the systemic establishment of PTGS in plants suggesting that they may travel through the plant vascular tissues.[46] Although the nature of the systemic silencing signal is not yet known, this molecule must carry sequence specific information. Small RNA molecules presumably forming part of larger ribo nucleoprotein complexes are good candidates for this activity.[46] Small RNAs may also have novel intracellular roles as shown for the intron-encoded snoRNAs in nucleotide modification of cellular RNAs[45] or the miRNAs in translational controls of target transcripts.[3,6,47]

These results suggest that mRNAs and/or their derived products may play diverse roles in the cell.

sORF-Encoded Oligopeptides

sORF-encoded oligopeptides may be involved in the function of sORF-RNAs and we can classed them in two groups: signalling peptides and intracellular sORF-encoded peptides.

Signalling peptides are generally encoded in large precursors although the precursor itself may be encoded in a sORFs smaller than 100 amino acids. This is the case for the majority of *clavata* 3 homologues identified in *Arabidopsis*.[19,21] The presence of signal sequences or related

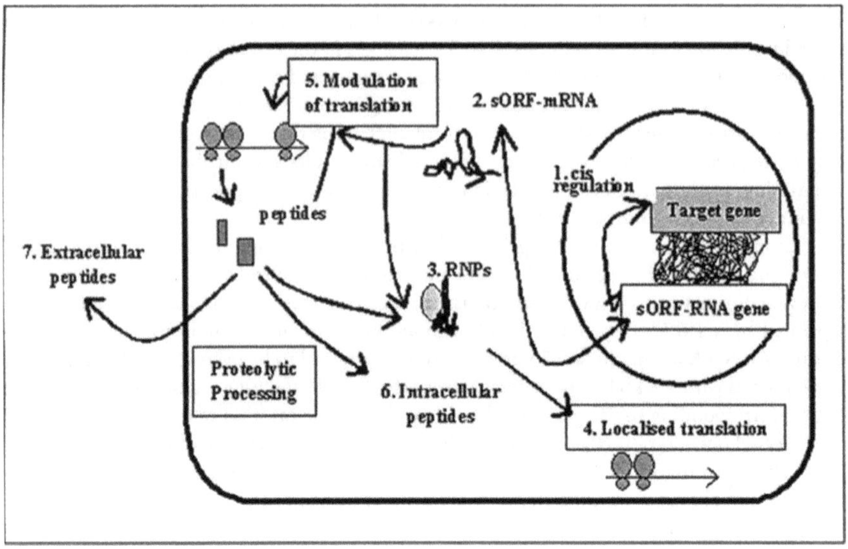

Figure 1. Potential mechanisms of action of sORF-RNAs. sORF-RNA genes may act in *cis* regulation (1) of target genes[44] and accumulate (2) in the cytoplasm (e.g., *enod40*,[25,31,32,34] *CR20*,[13] *GUT15*,[37] *TPS11*[9,39]). Certain transcripts may interact with ribonucleoprotein particles (3. RNPs) to regulate localised translation (4) and/or localisation of the RNP.[6,35] sORF-encoded oligopeptides may modulate translation (5. cis effects[56,57]) or eventually bind to the ribosome itself.[52] Translated peptides (e.g., *enod40*,[26,28] PLS[19,41]) may have intracellular targets (6) or generally yield signalling peptides (7) through proteolytic processing (e.g., PSK,[50] CLV3[21]).

features suggest that these sORF-RNAs encode secreted biologically active oligopeptides. Several reports suggest that plants do make use of peptides as signals in development, since both peptidic signals and their putative receptors have been found.[2] Moreover, mutations in several of these receptors influenced plant differentiation[48,49] and genome analysis in *Arabidopsis* indicate that this receptor-like kinase family is the largest family of receptors in plants.[21] In addition, the peptide product may require certain posttranslational modifications as shown for the phytosulphokine peptide growth factor.[50] Its primary translation product requires both processing from a larger precursor and tyrosyl sulfatation to function. In contrast, for tobacco systemins a large precursor yields several active peptides and are not encoded in a sORF-RNA.[51]

Alternatively, sORF-encoded small peptides may act as primary translation products. These small peptides may be able to diffuse out of the cell or, alternatively, they may have intracellular targets that could be reached directly after translation. It is likely that the PLS peptide has an intracellular target since no signal sequences were found in the encoding sORF.[41] Moreover, translation of sORFs present in sORF-RNAs may occur even though the main function of the gene lies in the RNA product. This is the case for a five amino acids sORF encoded in the 23S ribosomal RNA in *E. coli* , where the pentapeptide is likely produced then immediately binds its intracellular target, the ribosomal RNA.[52] This sORF-encoded pentapeptide was recently shown to render *E. coli* ribosomes resistant to a specific antibiotic. Hence, even the classical rRNA could be translated, further supporting that oligopeptides and RNAs can be gene products involved in the function of a particular gene. Another example is the *H19* sORF-RNA whose main function seems to lie in the mRNA molecule rather than in an encoded protein though a putative protein product of the *H19* RNA was detected using immunological approaches.[44,53]

Finally, several examples have been reported on *cis* effects of upstream sORFs on 3' translation mediated through binding to ribosomes.[54,55] Indeed, translation of upstream sORFs regulates expression of the 3' ORF corresponding to the gene product.[56,57] At the same time, very little is known about the fate of the encoded oligopeptides in the cell.

The possible mechanisms of action discussed above are summarized in (Fig. 1)

Perspectives: Small Regulatory RNAs and sORF-Encoded Oligopeptides as Novel Regulators of Gene Expression

The diverse biological processes involving sORF-RNAs suggest that they play different functions in plants, although little is known about the molecular mechanisms in which these genes are involved. sORF-encoded peptides and RNA structures might be active elements of the regulatory mechanisms involving sORF-RNAs. It can be foreseen that utilization of synthetic mimicking molecules corresponding to regulatory small noncoding RNAs or sORF-encoded peptides may serve for controlling those mechanisms.

On one hand, polypeptide signalling is an emerging field in plant biology. Several biologically active peptides display potent biological activities (antimicrobial, activating defence or affecting root growth; in the micromolar to nanomolar range[2,19]) that have been related to defence responses, growth and development in various plants. The dissection of their signal transduction pathways are paradigms for the analysis of cellular responses to external signals.

On the other hand, accumulating data indicate that, in plants, small noncoding RNAs are involved in cell-cell signalling, gene silencing, and mRNA translation. Small RNAs may target other nucleic acids with high specificity and, in this way, regulate gene expression and/or function through simple mechanisms.[46,47] These results suggest that small regulatory RNAs, eventually derived from certain sORF-RNA genes, may participate in various cellular processes.

sORF-encoded oligopeptides and structured RNA signals seem to be important elements in the molecular mechanisms involving sORF-RNAs. Further studies may uncover novel functions involved in translational control, cell-to-cell communication and growth regulation in plants. In addition, they may serve to develop novel small regulatory molecules which may be applied for agricultural and environmental purposes.

References

1. Alberts B, Bray D, Lewis J et al. Molecular Biology of the Cell. New York: Garland Science Publishing. 1994.
2. Franssen HJ, Bisseling T. Peptide signaling in plants. Proc Natl Acad Sci USA 2001; 98(23):12855-12856.
3. Reinhart BJ, Slack FJ, Basson M et al. The 21-nucleotide *let-7* RNA regulates developmental timing in *Caenorhabditis elegans*. Nature 2000; 403(6772):901-906.
4. Eddy SR. Computational genomics of noncoding RNA genes. Cell 2002; 109(2):137-140.
5. Lee RC, Ambros V. An extensive class of small RNAs in *Caenorhabditis elegans*. Science 2001; 294(5543):862-864.
6. Ruvkun G. Glimpses of a tiny RNA world. Science 2001; 294(5543):797-799.
7. Lagos-Quintana M, Rauhut R, Lendeckel W et al. Identification of novel genes coding for small expressed RNAs. Science 2001; 294(5543):853-858.
8. Lau NC, Lim LP, Weinstein EG et al. An abundant class of tiny RNAs with probable regulatory roles in *Caenorhabditis elegans*. Science 2001; 294(5543):858-862.
9. Burleigh SH, Harrison MJ. A novel gene whose expression in *Medicago truncatula* roots is suppressed in response to colonization by vesicular-arbuscular mycorrhizal (VAM) fungi and to phosphate nutrition. Plant Mol Biol 1997; 34(2):199-208.
10. Furini A, Koncz C, Salamini F et al. High level transcription of a member of a repeated gene family confers dehydration tolerance to callus tissue of *Craterostigma plantagineum*. EMBO J 1997; 16(12):3599-3608.

11. Hao Y, Crenshaw T, Moulton T et al. Tumour-suppressor activity of *H19* RNA. Nature 1993; 365(6448):764-767.
12. Tam W, Ben-Yehuda D, Hayward WS. *bic*, a novel gene activated by proviral insertions in avian leukosis virus-induced lymphomas, is likely to function through its noncoding RNA. Mol Cell Biol 1997; 17(3):1490-1502.
13. Teramoto H, Toyama T, Takeba G et al. Noncoding RNA for *CR20*, a cytokinin-repressed gene of cucumber. Plant Mol Biol 1996; 32(5):797-808.
14. Yoshida H, Kumimoto H, Okamoto K. *dutA* RNA functions as an untranslatable RNA in the development of *Dictyostelium discoideum*. Nucleic Acids Res 1994; 22(1):41-46.
15. Olivas WM, Muhlrad D, Parker R. Analysis of the yeast genome: identification of new noncoding and small ORF-containing RNAs. Nucleic Acids Res 1997; 25(22):4619-4625.
16. Velculescu VE, Zhang L, Zhou W et al. Characterization of the yeast transcriptome. Cell 1997; 88(2):243-251.
17. Kato A, Suzuki M, Kuwahara A et al. Isolation and analysis of cDNA within a 300 kb *Arabidopsis thaliana* genomic region located around the 100 map unit of chromosome 1. Gene 1999; 239(2):309-316.
18. Terryn N, Rouze P. The sense of naturally transcribed antisense RNAs in plants. Trends Plant Sci 2000; 5(9):394-396.
19. Lindsey K, Casson S, Chilley P. Peptides: new signalling molecules in plants. Trends Plant Sci 2002; 7(2):78-83.
20. Vanoosthuyse V, Miege C, Dumas C et al. Two large *Arabidopsis thaliana* gene families are homologous to the *Brassica* gene superfamily that encodes pollen coat proteins and the male component of the self-incompatibility response. Plant Mol Biol 2001; 46(1):17-34.
21. Cock JM, McCormick S. A large family of genes that share homology with CLAVATA3. Plant Physiol 2001; 126(3):939-942.
22. MacIntosh GC, Wilkerson C, Green PJ. Identification and analysis of *Arabidopsis* expressed sequence tags characteristic of noncoding RNAs. Plant Physiol 2001; 127(3):765-776.
23. Joyce GF. The antiquity of RNA-based evolution. Nature 2002; 418(6894):214-221.
24. Crespi MD, Jurkevitch E, Poiret M et al. *enod40*, a gene expressed during nodule organogenesis, codes for a nontranslatable RNA involved in plant growth. EMBO J 1994; 13(21):5099-5112.
25. Asad SY, Fang KL, Wycoff AMH. Isolation and characterization of cDNA and genomic clones of Ms*ENOD40*; transcripts are detected in meristematic cells of alfalfa Protoplasma 1994; 183:10-23.
26. van de Sande K, Pawlowski K, Czaja I et al. Modification of phytohormone response by a peptide encoded by *ENOD40* of legumes and a nonlegume. Science 1996; 273(5273):370-373.
27. Sousa C, Johansson C, Charon C et al. Translational and structural requirements of the early nodulin gene *enod40*, a short-open reading frame-containing RNA, for elicitation of a cell-specific growth response in the alfalfa root cortex. Mol Cell Biol 2001; 21(1):354-366.
28. Rohrig H, Schmidt J, Miklashevichs E et al. Soybean *ENOD40* encodes two peptides that bind to sucrose synthase. Proc Natl Acad Sci USA 2002; 99(4):1915-1920.
29. Schultze M, Kondorosi A. Regulation of symbiotic root nodule development. Annu Rev Genet 1998; 32:33-57.
30. Kouchi H, Hata S. Isolation and characterization of novel nodulin cDNAs representing genes expressed at early stages of soybean nodule development. Mol Gen Genet 1993; 238(1-2):106-119.
31. Yang WC, Katinakis P, Hendriks P et al. Characterization of Gm*ENOD40*, a gene showing novel patterns of cell-specific expression during soybean nodule development. Plant J 1993; 3(4):573-585.
32. Kouchi H, Takane K, So RB et al. Rice *ENOD40*: isolation and expression analysis in rice and transgenic soybean root nodules. Plant J 1999; 18(2):121-129.
33. Charon C, Johansson C, Kondorosi E et al. *enod40* induces dedifferentiation and division of root cortical cells in legumes. Proc Natl Acad Sci USA 1997; 94(16):8901-8906.
34. Charon C, Sousa C, Crespi M et al. Alteration of *enod40* expression modifies *Medicago truncatula* root nodule development induced by *Sinorhizobium meliloti*. Plant Cell 1999; 11(10):1953-1966.
35. Oleynikov Y, Singer RH. RNA localization: different zipcodes, same postman? Trends Cell Biol 1998; 8(10):381-383.
36. Hentze MW, Kulozik AE. A perfect message: RNA surveillance and nonsense-mediated decay. Cell 1999; 96(3):307-310.

37. Taylor CB, Green PJ. Identification and characterization of genes with unstable transcripts (GUTs) in tobacco. Plant Mol Biol 1995; 28(1):27-38.
38. Liu C, Muchhal US, Raghothama KG. Differential expression of TPS11, a phosphate starvation-induced gene in tomato. Plant Mol Biol 1997; 33(5):867-874.
39. Martin AC, del Pozo JC, Iglesias J et al. Influence of cytokinins on the expression of phosphate starvation responsive genes in *Arabidopsis*. Plant J 2000; 24(5):559-567.
40. Topping JF, Lindsey K. Promoter trap markers differentiate structural and positional components of polar development in *Arabidopsis*. Plant Cell 1997; 9(10):1713-1725.
41. Casson SA, Chillen PM, Topping JF et al. The *POLARIS* gene of *Arabidopsis* encodes a predicted peptide required for correct root growth and leaf vascular patterning. The Plant Cell 2002; 8:1705-1721.
42. Weigel D, Ahn JH, Blazquez MA et al. Activation tagging in *Arabidopsis*. Plant Physiol 2000; 122(4):1003-1013.
43. Cech TR, Bass BL. Biological catalysis by RNA. Annu Rev Biochem 1986; 55:599-629.
44. Leighton PA, Ingram RS, Eggenschwiler J et al. Disruption of imprinting caused by deletion of the *H19* gene region in mice. Nature 1995; 375(6526):34-39.
45. Kiss T. Small nucleolar RNAs: an abundant group of noncoding RNAs with diverse cellular functions. Cell 2002; 109(2):145-148.
46. Waterhouse PM, Wang MB, Lough T. Gene silencing as an adaptive defence against viruses. Nature 2001; 411(6839):834-842.
47. Hannon GJ. RNA interference. Nature 2002; 418(6894):244-251.
48. Becraft PW, Stinard PS, McCarty DR. CRINKLY4: A TNFR-like receptor kinase involved in maize epidermal differentiation. Science 1996; 273(5280):1406-1409.
49. Torii KU, Mitsukawa N, Oosumi T et al. The *Arabidopsis* ERECTA gene encodes a putative receptor protein kinase with extracellular leucine-rich repeats. Plant Cell 1996; 8(4):735-746.
50. Yang H, Matsubayashi Y, Nakamura K et al. Diversity of *Arabidopsis* genes encoding precursors for phytosulfokine, a peptide growth factor. Plant Physiol 2001; 127(3):842-851.
51. Pearce G, Moura DS, Stratmann J et al. Production of multiple plant hormones from a single polyprotein precursor. Nature 2001; 411(6839):817-820.
52. Tenson T, DeBlasio A, Mankin A. A functional peptide encoded in the *Escherichia coli* 23S rRNA. Proc Natl Acad Sci USA 1996; 93(11):5641-5646.
53. Leibovitch MP, Nguyen VC, Gross MS et al. The human ASM (adult skeletal muscle) gene: expression and chromosomal assignment to 11p15. Biochem Biophys Res Commun 1991; 180(3):1241-1250.
54. Lovett PS, Rogers EJ. Ribosome regulation by the nascent peptide. Microbiol Rev 1996; 60(2):366-385.
55. Mize GJ, Ruan H, Low JJ et al. The inhibitory upstream open reading frame from mammalian S-adenosylmethionine decarboxylase mRNA has a strict sequence specificity in critical positions. J Biol Chem 1998; 273(49):32500-32505.
56. Futterer J, Hohn T. Translation in plants—rules and exceptions. Plant Mol Biol 1996; 32(1-2):159-189.
57. Wang L, Wessler SR. Inefficient reinitiation is responsible for upstream open reading frame-mediated translational repression of the maize R gene. Plant Cell 1998; 10(10):1733-1746.

Chapter 14

The Noncoding Developmentally Active and Stress Inducible *hsrω* Gene of *Drosophila melanogaster* Integrates Post-Transcriptional Processing of Other Nuclear Transcripts

Subhash C. Lakhotia

Summary

The *93D* or the *hsr-omega* (*hsrω*) gene of *Drosophila melanogaster* became an interesting gene more than 3 decades ago in view of its unique inducibility with a brief benzamide treatment. Subsequent studies revealed many unusual features of this gene, a homologue of which is present in all *Drosophila* species examined. This gene is developmentally active in nearly all cells types of *Drosophila*, is induced by heat shock along with the other heat shock genes but is singularly induced by a variety of amides, all of which also inhibit general chromosomal transcription. The *hsrω* gene in all species of *Drosophila* has a characteristic architecture with two exons and an intron and a long stretch (>5 to ~15 kb) of tandem repeats on the 3' end of the gene. Like many other noncoding genes, the base sequence of the unique as well as the tandem repeat region of the *hsrω* gene is not conserved in different species. However, in all species examined, two primary nucleus-limited transcripts, ~2 kb and >10 kb, respectively, are produced but none of them carry any significant open-reading frame. The ~2 kb transcript is spliced to generate a 1.2 kb cytoplasmic transcript, which has a translatable ORF of 23-27 aa. Translation of this short ORF perhaps helps monitor the "health" of cellular translational machinery.

The large nucleus-limited >10kb *hsrω*-n transcript is so far the only known eukaryotic large RNA, which shows a speckled distribution in the nucleoplasm. These transcripts are present, besides at the site of transcription, as many nucleoplasmic speckles close to the chromatin domains. The various nuclear hnRNPs and some other proteins such as Sxl remain bound with the various transcriptionally active chromatin sites, and with the nucleoplasmic speckles formed by the *hsrω*-n transcripts. These speckles, designated as "omega speckles", are distinct from the well-known inter-chromatin granule clusters. The *hsrω*-n transcripts have an essential role in organizing the omega speckles, which serve to dynamically regulate the availability of hnRNPs and related proteins for RNA processing activities at any given time. Mutants that mis-express the *hsrω* gene and thus affect the omega speckles have diverse phenotypic consequences, presumably because of aberrant processing of various nuclear premRNAs due to altered availability of hnRNPs etc.

Noncoding RNAs: Molecular Biology and Molecular Medicine, edited by Jan Barciszewski and Volker A. Erdmann.

The involvement of the noncoding transcripts of *hsrω* gene in metabolism of nuclear hnRNPs and in monitoring the "health" of ribosomal machinery provides new paradigms for roles of noncoding transcripts in integration of cellular regulation. Such activities are vital to all eukaryotic cells and therefore, genomes of all of them should include noncoding genes performing comparable functions.

Introduction

Much of the very exciting progress in Biology during the past four decades has been propelled by the belief that a "gene", to be functional, must produce RNA, which must be translated into a protein.[1] Studies using this paradigm have enabled us to move from genetic engineering to genomics and now to the post-genomic era. Parallel and equally exciting developments in cell and developmental biology and other areas enabled remarkable correlations between gene expression, protein synthesis and the mechanistic details of cell function. Notwithstanding this most remarkable progress, one of the persisting riddles is the role/s of the bulk of the genomic DNA in eukaryotes. Almost all eukaryotes have much more DNA than accounted for by the protein coding "genes" and a significant proportion of the "noncoding" genomic DNA is nevertheless transcribed. Such "noncoding" sequences have often been brushed aside as "selfish" or "junk" because of the continuing emphasis on the "coding sequences". However, in recent years, the noncoding "genes" are also beginning to gain appreciation as meaningful components of genomes.[2-21]

The *93D* or *hsrω* gene of *Drosophila* was among the few "noncoding genes" known in the early 1980s. The unique organization of polytene chromosomes in dipteran insects and the phenomenon of puffing in polytene chromosomes provide a uniquely convenient approach to identify specific genes, their expression patterns and the proteins they produce. This indeed was responsible for the seminal identification of heat shock genes in *Drosophila*.[22] Subsequent identification of the heat shock proteins[23] provided the much needed demonstration of the causal relationship between gene activity (puffing), transcription and protein synthesis. One of the major heat shock induced puffs, the 93D puff, in polytene cells of *Drosophila melanogaster*[24] was found by Lakhotia and Mukherjee[25] to be uniquely inducible by brief treatment with benzamide and this made it an interesting system for further analysis. Studies during the past few decades revealed this locus to be a rather unusual noncoding gene in *Drosophila*. Unlike some other later discovered noncoding genes like *Xist*[26,27] or *rox*,[28-30] the unusual properties of this gene's expression and regulation and its functions remained enigmatic for a long time (for reviews see refs. 31-34) and it is only in the past few years that the functions of this gene are beginning to be understood.[35,36] These suggest novel roles for RNA molecules in regulating cellular activities.

The *93D* or the *hsrω* Gene in *Drosophila* Displays Unique and Conserved Inducibility but Apparently Does Not Produce a Protein

Initially this gene was named as *93D* because of its location in the 93D cytogenetic region of polytene chromosomes of *Drosophila melanogaster* but it was renamed later as the *hsrω* after its transcription products, heat shock RNA omega.[37] As a member of the heat shock gene family, the *93D* or the *hsrω* gene is one of the most active genes[24,38] following heat shock as well as after treatment with carbon dioxide, 2-4 dinitrophenol, arsenic compounds etc or after recovery from anoxia.[39] However, the most interesting feature of this gene, which attracted our attention was its singular inducibility with benzamide (Fig. 1).[25,40] It is now known that many other amides like, colchicine, colcemide, formamide, nicotinamide etc, also singularly induce the 93D puff in salivary glands of *D. melanogaster* larvae.[41,42] All species of *Drosophila* examined to date carry a homologue of the *93D* gene of *D. melanogaster*, since in all of them one of

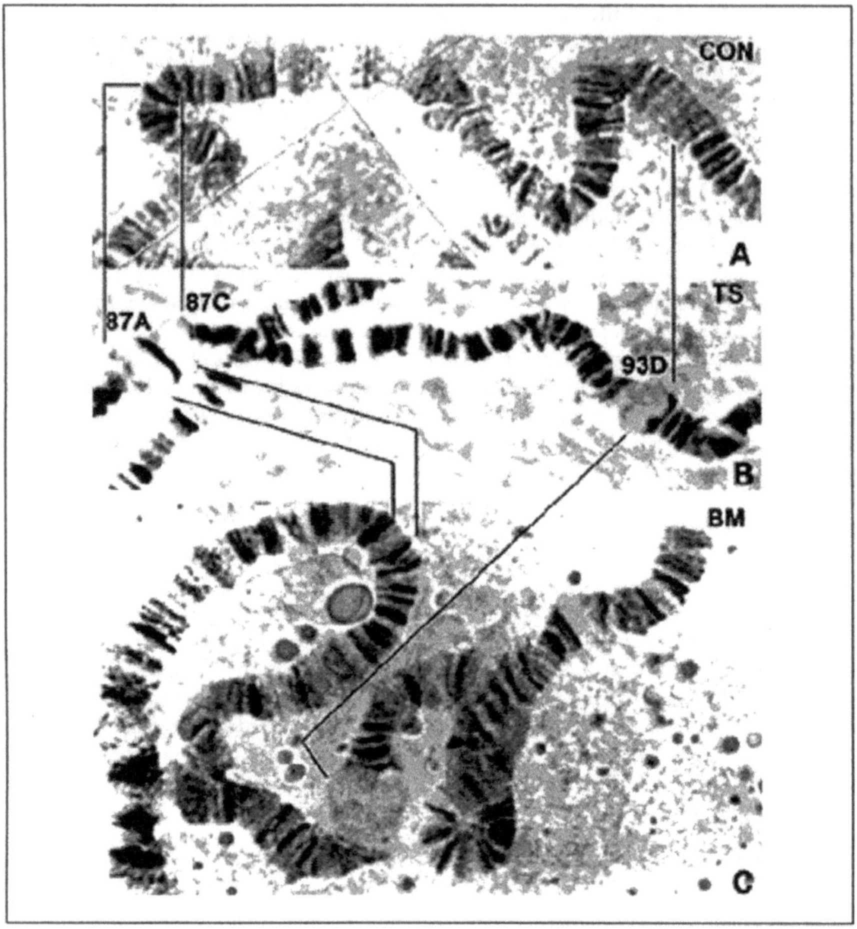

Figure 1. The 93D puff in salivary gland polytene chromosomes of *Drosophila melanogaster* is induced along with other members of the heat shock family following temperature shock (TS, B) or alone by benzamide (BM, C). The chromosome in A is from untreated control (CON) salivary gland. The 87A and 87C puffs, carrying multiple copies of the *hsp70* genes and the 9D puff are marked. Reprinted with permission from ref. 82.

the major heat shock genes, located at evolutionarily comparable region of the genome, is uniquely inducible by amides (Fig. 2).[43,44] Although an amide inducible heat shock puff has not been found in *Chironomus* and *Anopheles*,[45] heat shock inducible telomeric Balbiani rings in some species of *Chironomus* show properties that are in some ways comparable to the *hsrω* locus of *Drosophila*.[33,46-48]

Studies on activation of this locus in *Drosophila* documented a number of interesting, but as yet little understood, aspects of this gene's activation in larval salivary glands (for details, see reviews in refs. 31,32). Observations on RNA metabolism at the 93D locus by Spradling et al[49] and Lengyel et al[50] suggested that this gene harbors repetitive sequences and that bulk of its transcripts do not leave the nucleus. A more definitive suggestion that this gene may be a

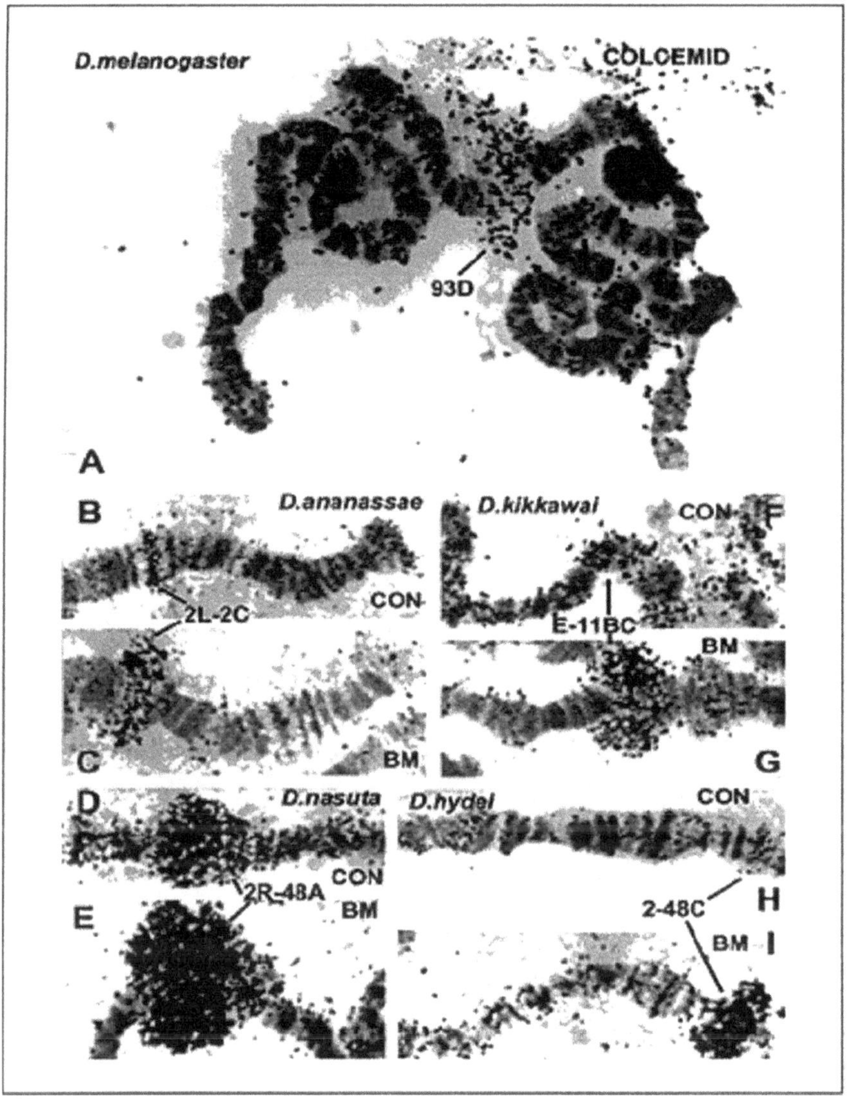

Figure 2. Amides selectively induce one member of the heat shock puffs in different species as revealed by ^{3}H-uridine incorporation following colcemid (A) or benzamide (BM, C, E, G. I) treatment of salivary glands of different species, as mentioned in each case. The amide-inducible puff is also active in untreated (CON) glands (B, D, F, H). Reprinted with permission from ref. 82.

noncoding one came from the studies by Lakhotia and Mukherjee,[51] who examined protein synthesis following the selective activation of the 93D puff in benzamide treated larval salivary gland cells and concluded that in spite of this gene's evolutionary conservation, it does not code for a protein.

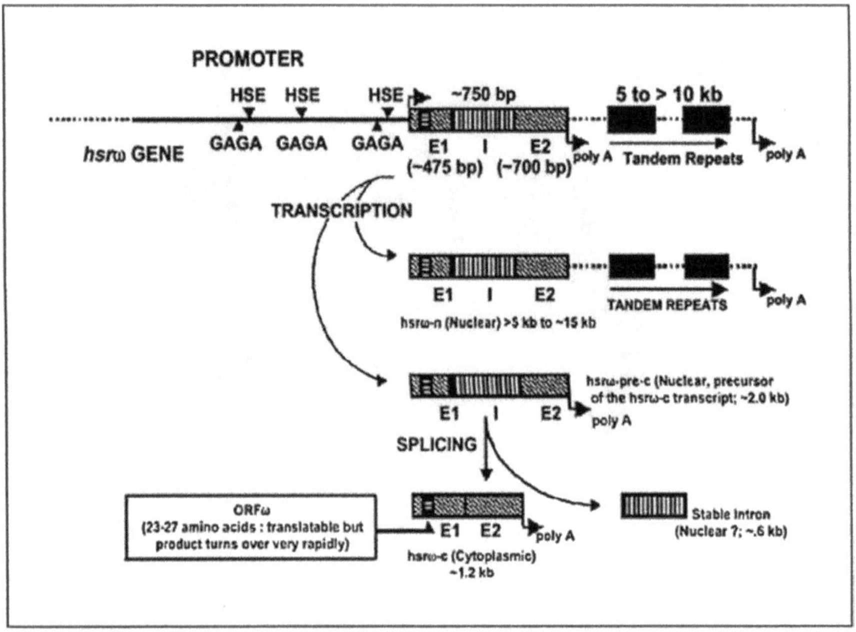

Figure 3. Architecture of the *hsrw* gene and its transcripts in different species of *Drosophila* (for details see text and the references mentioned therein)

The *hsrω* Gene Shows a Unique Conserved Architecture but with Little Conservation of Base Sequence in *Drosophila* Species

An unusually strong conservation of the organization of the gene (see Fig. 3) but little conservation of the base sequence is evident from a comparison of the *hsrω* homologues from *D.hydei*,[52] *D. melanogaster*[53,54] and *D. pseudoobscura*.[55] In all the three species, the *hsrω* transcription unit comprises a characteristic unique region (~2.5 kb) at the 5' end, which includes 2 exons (~500 bp and ~700 bp long, respectively) and one intron (~700 bp), with a poly-A site at ~1.9 kb from the transcription start site. This poly-A site is followed by another unique region (~600 bp) and then by a long stretch (varying from 5 kb to ~15 kb) of short tandem repeats unique to this locus.[55-58] In all species, the *hsrω* gene has two transcription termination sites, the first after the 2nd exon (~2 kb from the transcription start point) and the second after the last tandem repeat unit[56] and accordingly, two primary transcripts are produced (see later). Neither of these transcripts or their processed product has any significant open reading frame. The rapid divergence of the base sequence in its transcribed region in different species of *Drosophila*, in spite of its conserved architecture, further supports the noncoding nature of this gene.[52,54,56,59,60] A strong conservation of the "architecture" but not so much of the primary base sequence at the *hsrω* locus is comparable with the noncoding *Xist* gene in mammals.[27,61]

Short stretches of high conservation interspersed between regions of high degree of sequence divergence are notable in the nonrepeated part (till the first transcription termination site) in *D. melanogaster*, *D. hydei* and *D. pseudoobscura*.[55] The most striking regions of sequence conservation include the acceptor and donor splice sites (the conservation extends beyond the splice junction),[55] the polyadenylation site and transcription start site. Significance of these conserved regions is not known. Finally, an open reading frame (ORFω) coding for 23 to 27

Figure 4. Amino acids coded by the ORFω in three species of *Drosophila*. Note that only the first four amino acid residues and two others (diagonal shading) are conserved in all the three species. A few others (horizontal shading) show identity only in any two species. Data from ref. 55.

amino acids is present in all the three species at about +120 position from the transcription start site. Intriguingly however, except for the conservation of the first 4 amino acids and two others interspersed in between, the rest of the amino acids are not conserved, although a few of them are identical between two species (see Fig. 4). This ORF is translatable but the translated product is not detectable.[62]

Base sequence of the repeat units is also remarkably diverged between species, although within a species a very high conservation exists.[52,56] Thus the repeats in the *hsrω* genes of *D. hydei* and *D. melanogaster* are highly diverged. The repeat unit in *D. hydei* is 115 bp while in *D. melanogaster* it is 280 bp. Interestingly a nonamer (ATAGGTAGG) is the only sequence motif that seems to be conserved between the *hsrω* repeats of the two species, occurring once per ~115 nucleotides in both cases.[34] The length of the stretch of tandem repeats at the *hsrω* gene of *D. melanogaster* varies between 5 and 15 kb in different populations but for a given population the length seems to remain fairly constant.[58]

The *hsrω* Gene Produces Multiple Noncoding Transcripts

The *hsrω* gene does not code for any protein[51,54,57] although the small ORFω is translatable.[62] *hsrω* gene produces two primary transcripts (Fig. 3), viz., the *hsrω*-n transcript of ~10 to 15 kb length, which spans the entire transcription unit and remains localized in the nucleus, and the *hsrω*-prec transcript of ~2 kb length, spanning only the 5' region comprising of the two exons and the intron. The ~2 kb *hsrω*-prec transcript is spliced to give rise to the cytoplasmic 1.2 kb *hsrω*-c transcript.[37,56,57,62] Splicing occurs at the site of transcription[63] and the spliced out intron seems to be relatively stable.[37] The rapid sequence divergence at this locus in *Drosophila* species compares with most other noncoding genes. The production of transcripts of comparable properties suggests that the structure of the transcripts is more important than the base sequence itself.[33,35]

The different *hsrω* transcripts show inducer specific profiles. Heat shock causes marked increase in levels of all the three transcripts while the amide treatment leads to a significant increase of only the >10 kb long nuclear *hsrω*-n.[37,62] This difference seems to be related to the different effects of heat shock and amides on transcriptional and translational activities. Heat shock affects transcriptional as well as translational activities while the amides affect only chromosomal transcription without any effect on translational activity.[40,51] The *hsrω*-n as well as the *hsrω*-c transcripts display rapid turnover in normal cells but inhibition of transcription and translation has different effects on their stability.[37,64] The *hsrω*-n is stabilized when new transcription is inhibited with Actinomycin D,[64] while the *hsrω*-c is stabilized by inhibitors of protein synthesis.[37,65] The differential sensitivity of the two *hsrω*-n and *hsrω*-c transcripts to inhibitors of transcription and translation correlates with their nuclear and cytoplasmic roles, respectively (see later).

The *hsrω* Gene Shows Widespread Developmental Expression

Several in situ RNA localization studies[65-67] established that this gene is expressed under normal developmental conditions in almost all cells except the early embryonic pole cells and the spermatocytes in testis. The level of its transcripts in different cell types varies and heat shock induced increase in the amount of these transcripts in different cell types is generally proportional to its normal level of expression. It is interesting that the pole cells and the spermatocytes do not express this gene even after heat shock.[67]

The *hsrω* gene responds to ecdysone.[65,66] The level of this gene's developmental expression is tightly regulated from very low to very high levels in a cell type specific manner.[67] In embryos, *hsrω* transcripts are particularly abundant in the developing central nervous system; likewise the larval and adult brain ganglia also show high levels of *hsrω* gene expression.[67,68] In this context it is interesting that an allele of *hsrω* has been reported to enhance ataxin-1-induced neurodegeneration in *Drosophila* model.[69]

Promoter Region of the *hsrω* Gene Is Complex

The *hsrω* gene produces independently regulated multiple transcripts, and, besides being developmentally expressed, is inducible, either as a member of the heat shock gene family or individually, by a variety of experimental conditions. Therefore as expected, experimental studies revealed *hsrω* gene's promoter to be complex. Flies carrying transgenes with defined lengths of the *hsrω* promoter and the *lacZ* reporter gene suggested that while the heat shock inducibility and developmental expression are regulated by elements located within the proximal 845 bp of the promoter region, the amide inducibility was not included in this interval.[66,70] These and other[67] studies also indicated existence of regulators/enhancers that regulate this gene's expression in individual cell types/tissues. In a unique approach using small chromosomal deficiencies, Lakhotia and Tapadia[71] mapped the amide-response element/s to be more than 21 kb upstream of the transcription start point. Another aspect of the complexity of the *hsrω* gene's promoter region has been revealed through studies on an enhancer-trap line (*hsrω*^{05241}) in which a P-element with the *lacZ* reporter gene is inserted at –130bp position of the *hsrω* gene.[67,72] In spite of the insertion of ~13 kb of the P-transposon construct in the promoter region at –130 bp position, expression of the *hsrω* gene remains essentially unaffected in embryonic, larval and adult tissues of the *hsrω*^{05241} homozygotes.[67] Only the cyst cells of adult testis show a distinctly perturbed expression of the *hsrω* transcripts in the *hsrω*^{05241} homozygotes.[72] Interestingly, the *lacZ* reporter gene expression pattern in this enhancer-trap line is also comparable to the expression of the *hsrω* gene itself in most of the nonpolytene cells in embryonic, larval and adult stages. However, several larval polytene cell types do not show any expression of the lacZ, neither under normal developmental conditions nor after heat shock.[67] Yet, the *hsrω* transcripts are strongly induced by heat shock in nonpolytene cells as well as in all the polytene cell types, in spite of the fact that only one heat shock element (HSE), located at –57 to -43,[70] is close to the *hsrω* transcription unit while the other two (at -466 to –452 and at -250 to –235 nucleotide positions, respectively[70]) are far removed due to the insertion of P-transposon in this chromosome. It remains to be seen if only the proximal HSE is enough for the *hsrω* transcription unit or the more distal ones are still able to exert their role. Notwithstanding the long P-transposon insertion, the *hsrω*^{05241} allele is also strongly induced by amides.[67] Apparently, the amide response element/s, already located more than 21 kb upstream,[71] can still act even after further displacement due to the insertion.

The *hsrω* Gene Is Functional in Spite of Its Noncoding Transcripts

Conservation of the *93D* or *hsrω* gene, as a heat and amide-inducible gene in different species of *Drosophila* suggested its importance for flies although its noncoding nature and rapid sequence divergence caused persistent doubts about its functional capability. The following evidence, however, clearly establish that the noncoding transcripts of the *hsrω* gene have important functions.

Mutant Phenotype

An intensive screen for mutations at the 93D locus by Mohler and Pardue[73,74] did not recover any point mutations and for nonbelievers in the functions of a noncoding gene, this was as expected of a "junk" or "selfish" DNA. In hindsight, however, this appears to be related to the noncoding nature of this gene so that base changes are well tolerated, as have actually taken place at this locus during evolution of the different species. Mohler and Pardue identified two small deletions, viz. *Df(3R)e*^{Gp4} and *Df(3R)GC14*, whose overlap specifically defines the HS and amide-inducible *hsrω* locus.[73-75] It is significant that most of the *Df(3R)e*^{Gp4}/*Df(3R)GC14* trans-heterozygotes, which are nullisomic only for the *hsrω* gene, die as embryo; a few (~20%) embryos are delayed in hatching but develop as weak flies unable to properly walk or fly, and which die within a few days.[35,73,74,76,77] Imaginal discs from the *hsrω*-nullisomic larvae fail to differentiate in vitro in response to ecdysone.[31] The *hsrω* nullisomic flies are essentially sterile: the males produce motile sperms but are physically too weak to mate. On the other hand, oogenesis in the *hsrω* nullisomic females is affected so that only a few (~3%) of their eggs actually develop into viable progeny.[35] This correlates with high expression of the *hsrω* gene in the ovarian nurse cells.[66] Compared to wild type flies, *hsrω*-nullisomics are poorer in acquiring thermo tolerance and do not survive when grown at 31°C.[31,78] A role of the *hsrω* gene in thermo tolerance is further supported by changes in *hsrω* allele frequencies in unselected lines and those selected for resistance to knockdown by 39°C heat stress with or without prior hardening.[79] McKechnie et al[80] also observed significant differences in the constitutive levels of the nuclear and cytoplasmic *hsrω* transcripts between these lines.

The *hsrω*^{05241} enhancer-trap line, with a *P-lacZ* transposon insertion at –130 bp position (see above) displays an interesting phenotype. All stages of development of *hsrω*^{05241} heterozygotes as well as homozygotes appear normal, except that the *hsrω*^{05241} homozygous males are sterile.[72] As noted earlier, the expression of *hsrω* transcripts in different tissues and developmental stages of *hsrω*^{05241} homozygotes is more or less comparable to that in wild type.[67] An exception is the cyst cells in adult testis. A pair of cyst cells remains associated with a bundle of 64 developing male germ cells and these cells have significant functions in proper maturation and individualization of the sperms.[81] The promoter dysfunction due to the P-transposon insertion at –130 bp position causes the *hsrω*-n transcripts to be much more abundant in cyst cells of *hsrω*^{05241} homozygotes. This seems to prevent individualization so that bundles of nonmotile sperms are produced, resulting in male sterility.[72]

Interaction of hsrω *with Other Genes*

Interaction of the *hsrω* gene with a number of other genes has been studied either through studies on puffing in polytene chromosomes or through modification of the mutant phenotype of the target gene.

Interaction with hsp70 *Gene Loci*

D. melanogaster has two clusters of nearly identical two and three *hsp70* genes at the 87A and 87C cytogenetic regions, respectively.[82,83] Under the usual heat shock condition (37°C for 30 min), the 87A and 87C sites form nearly equal sized puffs in salivary gland polytene nuclei with comparable ^{3}H-uridine incorporation.[38] However, whenever the 93D puff fails to actively transcribe during heat shock, transcription at the twin puffs at 87A and 87C sites is affected (see reviews in refs. 31-33). As discussed later, the *hsrω* gene's transcripts regulate movements of heterogeneous nuclear RNA-binding proteins (hnRNPs) between active and inactive compartments in the nucleus. Our unpublished studies (Prasanth KV and Lakhotia SC) on the distribution of some hnRNPs show that their binding on the 87A and 87C puff sites also gets affected when the 93D puff is not induced by heat shock. Apparently, in the absence of the *hsrω* gene's continuing transcriptional activity during heat shock the hnRNP movement gets affected and this in turn affects RNA metabolism at the *hsp70* gene loci. A recent study in our laboratory[83] revealed that in a variety of embryonic, larval and adult cells, the *hsp70* genes at 87A and 87C loci respond differently even after a typical heat shock. It is not known if this is in some way related to the *hsrω* transcripts also being induced to different levels by heat shock in these cell types.[67]

Other Gene Mutations as Dominant Enhancers of Embryonic Lethality of hsrω*-Nullisomics*

As noted above, about 20% of *Df(3R)e*^{Gp4}*/Df(3R)GC14* (*hsrω*-nullisomic) embryos survive to adulthood. However, heterozygosity for recessive mutations at the *hsp83* gene[77] or at one of the Ras loci[84] causes 100% death of *hsrω*-nullisomic embryos. The *hsp83* or the Ras mutant alleles used in these studies by themselves do not cause any lethality in heterozygous condition. The nature of interaction between the *hsrω*, *hsp83* and Ras genes is not clear but this seems to be related to the fact that Hsp83 specifically binds with the *hsrω* locus in polytene nuclei after heat shock[48] and that *hsp83* mutations enhance phenotypes of genes involved in the Ras signaling pathway.[85]

Enhancement of Poly-Q Induced Neurodegeneration

Several poly-Q repeat expansion induced neurodegenerative disorders are known in man[86] and for some of them elegant *Drosophila* models have been established.[69,87] Two of the P-insertion mutant alleles of *hsrω* have been found to enhance the neurodegeneration caused by either over-expression of normal or poly-Q expansion carrying mutant allele of human SCA1 (spinocerebellar ataxia type 1) in transgenic *Drosophila* model.[69] The mechanism of enhancing effect of the *hsrω* mutant alleles noted in this study is not clear. Another aspect that needs further analysis is the nature of the two *hsrω* mutant alleles used by Fernandez-Funez et al[69] While little is known about the functional status of the P292 allele of *hsrω*, the other allele, *hsrω*^{05241} appears to be the same as studied in our laboratory.[67,72] Although Fernandez-Funez et al[69] identified this allele as a loss of function allele, our studies showed that this is rather a gain of function allele, at least in cyst cells of testis.[72] It is, therefore, necessary to reexamine the expression of the *hsrω* gene in the specific cell types in which the neurodegeneration is enhanced. It will also be interesting to study if other poly-Q expansion models in *Drosophila* (like for huntingtin) also display comparable enhancement due to *hsrω* mutant expression.

Functions of *hsrω* Transcripts

The above evidence suggest that the *hsrω* gene does have widespread and vital housekeeping functions in fly's life[31-33] so that this gene's complete absence results in extensive lethality and a variety of other phenotypes in the rare survivors. Recent studies have provided some insights into the novel housekeeping functions of these noncoding transcripts.

The Small Cytoplasmic *hsrω*–C Transcripts May Monitor the Translational Activities in Cell

Fini et al[62] showed that one or two ribosomes associate with the *hsrω*-c RNA and translate the small ORF, coding for 23-27 amino acids in different species (Figs. 2, 3), although the translated product is not detectable. Bendena et al[37] observed that inhibition of protein but not RNA synthesis results in quick stabilization of the *hsrω*-c transcript, which otherwise shows a high turnover. In view of these observations and the fact that the amino acid sequence encoded by the omega-ORF is not conserved (see Fig. 4), it has been suggested that the act of translation of the omega ORF is important rather than the translated product.[31-33,62] It appears that the act of translation of the omega-ORF is coupled with degradation of the template so that partial or complete inhibition of translation due to drugs or some other cellular perturbations results in proportionate accumulation of the *hsrω*-c transcripts. This may provide a mechanism for monitoring "health" of the translational machinery through smooth passage of a ribosome through the ORFω and the levels of *hsrω*-c transcripts in the cytoplasm.[32] An additional and/or alternative role for the *hsrω*-c RNA molecules may be to serve as docking sites for unengaged ribosomes. Every cell has a large number of ribosomes, all of which are unlikely be actively engaged with mRNAs all the times. Furthermore, each cell is also likely to experience significantly varying translational activities so that at times many ribosomes get unengaged. The unengaged ribosomes may need storage till later recruitment for active translation. It is possible that RNA molecules like *hsrω*-c provide the storage or docking sites for such unengaged ribosomes.

The *hsrω*-n Transcripts Organize "Omega" Speckles in Nucleoplasm to Regulate the Availability of hnRNPs for RNA Processing Activities

Antibodies against several nuclear nonhistone proteins, mostly belonging to the heterogeneous nuclear RNA-binding family of proteins (hnRNPs[88]), display more or less exclusive binding with the *hsrω* puff in chromosome spreads from heat shocked salivary glands of *Drosophila* larvae (see Table 1).[89-98] In squash preparations of unstressed polytene cells, these antibodies decorate a large number of transcriptionally active chromosome regions, including the *hsrω* site (Fig. 5). In nonpolytene interphase nuclei the binding of these proteins with active regions is seen as a diffuse staining of chromatin but in these cells also, heat shock causes accumulation of the hnRNPs and related proteins at the 93D site on the chromatin.[36] As the cells recover from heat shock, the hnRNPs move back from the *hsrω* site to the different chromosomal regions within one hour. This rather intriguing phenomenon of the binding of the various nuclear proteins more or less exclusively at the *hsrω* locus under heat shock conditions has now provided clues to possible functions of the large nuclear *hsrω*-n transcripts. Our studies[35,36,67,72] utilizing immuno-fluorescent localization of the various hnRNPs and related proteins in conjunction with in situ hybridization of *hsrω* specific riboprobes to cellular RNA in intact cells of *Drosophila* proved very informative. Among the known RNA-polymerase II dependent eukaryotic transcripts, the *hsrω*-n transcripts are unique in being present in the nucleus, besides at the site of transcription, as small granules or speckles distributed in nucleoplasm in close vicinity of chromosomes (Fig. 5).[35,36,67] These nucleoplasmic speckles of *hsrω*-n RNA also contain a variety of hnRNPs and related proteins (see Table 1). The speckles containing the *hsrω*-n RNA and hnRNPs and the related proteins were designated as "omega speckles".[36] The omega speckles are undetectable in squash preparations in which the nuclear envelope is disrupted, suggesting that they are not closely bound to chromatin. It is significant that the well-known nuclear speckles and inter-chromatin granule clusters with which the SR-family of RNA-binding nuclear proteins are associated,[99,100] are distinct from the omega speckles.[35,36,72] The omega speckles appear to be same as the large RNP-particles observed at the *hsrω* site and free in nucleoplasm in some of the earlier ultra structural studies on *Drosophila* cells.[91,101]

The omega speckles are completely absent in the *hsrω* nullisomic (*Df(3R)e*^{Gp4}*/Df(3R)GC14*) cells; in such cells the hnRNPs remain diffused through the nucleoplasm.[36] Furthermore, the immunoprecipitate obtained with antibodies against the hnRNPs contains *hsrω* RNA.[36] It is also noted that in different types of normal cells, there is a close correlation between the amount of *hsrω*-n RNA and the number of omega speckles in nucleus.[36,67] All these observations show that the *hsrω*-n transcript is essential for organizing the omega speckles.

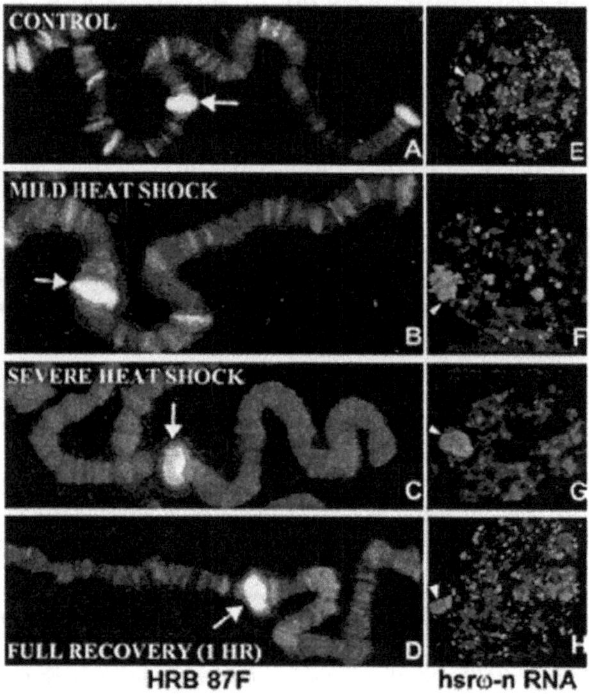

Figure 5. A-D. Heat shock causes all the hnRNPs (green/yellow fluorescence after immunostaining with antibody against HRB87F) to withdraw from the different active regions on polytene chromosomes (dull fluorescence) and finally accumulate at the 93D puff site (arrow) as in C; the hnRNPs return to their normal chromosomal locations within 1 hour of recovery frown heat shock (D). E-H. The *hsrω*-n RNA (localized by fluorescent in situ hybridization, dull fluorescence) in control cells (E) is present in speckles in the nucleoplsm close to chromatin (bright fluorescence) and at the 93D site (solid arrows) but following heat shock, the free speckles tend to coalesce with each other and accumulate at the 93D site (F) so that after about 40min of heat shock, all the *hsrw*-n transcripts are seen only at the 93D site (G). Although not shown here, the hnRNPs follow the same pattern as the *hsrω*-n RNA (see ref. 35,36). Like the hnRNPs (D), the *hsrw*-n transcripts also show the normal distribution within one hour recovery (H). Images adapted from ref 103.

The omega speckles are suggested to function as dynamic storage sites for the unengaged hnRNPs and the release of hnRNPs from them is coupled to degradation of the associated *hsrω*-n transcripts.[35,36] Depending upon the nuclear needs, the hnRNPs are either released from or sequestered in the omega speckles, and correspondingly the levels of *hsrω*-n transcripts go down or up.[35,36] The widely varying levels of *hsrω*-n transcripts in different cell types during normal development[66,67] reflect the varying needs of RNA processing in these cell types since the levels of hnRNPs in these cells also vary correspondingly. Heat shock severely inhibits general transcription and RNA processing activities in the nucleus.[82,102] Under conditions of reduced nuclear RNA processing, the *hsrω*-n level goes up, the omega speckles increase in size, possibly due to the greater sequestration of the hnRNPs, and coalesce with each other to form large aggregates. The aggregates finally get exclusively localized to the 93D chromosomal site itself

Table 1. Antibodies to the following proteins specifically bind with the* hsrω *puff in heat shocked polytene nuclei of* D. melanogaster *and other* Drosophila *species

Protein	Reference
hnRNPs: Hrp40 (hnRNP A, Squid), HRB87F (hnRNP A1/A2), Hrb57A (hnRNP K), S5 (hnRNP M)	89-95
Sxl	96,97
Nuclear non-histone proteins recognized by Q14, Q16, T29, P75 antibodies	89
Snf	97
Sera from auto-immune disorder patients	98

(see Fig. 5).[35,36] It is likely that the massive clustering of omega speckles in fully stressed cells ensures against illegitimate RNA processing activities under adverse conditions.[35,36] As the cells recover from heat stress and resume their normal transcriptional and processing activities, the hnRNPs are released from clusters of omega speckles and within one hour, the hnRNPs get restored to active chromosomal sites and typical omega speckles also reappear in the nucleoplasm. The level of *hsrω*-n transcripts also quickly goes down to the normal.

It is significant that in heat shocked *hsrω*-nullisomic cells, the hnRNPs do not completely move out of their chromosomal binding sites, do not form clusters and also do not regain their normal chromosomal distribution till about 2 hours of recovery from heat shock.[103] Thus in the absence of *hsrω*-n transcripts, the hnRNPs are not properly chaperoned and this may be the reason for the earlier noted thermo-sensitivity of *hsrω*-nullisomics.

The presence of Hsp83 chaperone protein also at the *hsrω* gene locus under heat shock condition[48] and the interaction of *hsrω* with *hsp83* gene mutation (see above and ref. 77) are significant. It is likely that the Hsp83 chaperone activity keeps the hnRNPs protected from thermal damage.[35] It has been reported that heterozygosity for *hsp83* null mutation in *Drosophila* causes epigenetic appearance of a variety of recessive mutant phenotypes, which normally remain masked.[104,105] It is likely that dysfunction of Hsp83/Hsp90 (e.g., reduced quantity in null mutant heterozygotes) affects the chaperoning of hnRNPs, which in turn may affect the desired splicing and/or other aspects of RNA processing. The *hsrω* transcripts may facilitate the chaperoning of hnRNPs by the Hsp83 by providing a common platform.[35]

Post-transcriptional processing of nascent transcripts involves a wide variety of protein and RNA-protein complexes. Among these, the hnRNPs constitute a large family of RNA-binding proteins with important roles in packaging, splicing, transport and degradation etc of RNA.[88,106-108] One of the important functions in which many of the hnRNPs and other proteins like the Sxl, Snf etc have a significant role, is the alternative splicing.[107-109] In recent years, a variety of other well defined nuclear speckles or granules/clusters have been identified and most of them are believed to function as storage sites for specific sub-sets of the various RNA processing factors.[99,100,109-119] It has been well established that the ratio of SR-proteins and hnRNPs in general, and the specific sub-sets of each of these families of proteins that are available for splicing, influence the selection of donor and acceptor splice sites in multi-intronic transcripts.[120,121] Therefore, it appears likely that a regulated release of hnRNPs from the omega speckles and of the SR proteins and other splicing factors by the other storage sites provides an efficient and self-organized machinery for modulation of alternative splicing of various transcripts. Disassembly of interchromatin granules affects the coordination of transcription and post-transcriptional processing[122] apparently due to disruption in the regulated release of RNA-processing factors. Likewise, since the *hsrω*-n transcripts have a pivotal role in organizing omega speckles as a mechanism for regulated availability of the hnRNPs and related proteins for RNA processing, any change in the levels of these transcripts in a cell may be expected to adversely affect splicing and/or transport of a variety of nuclear transcripts. This in turn would seriously compromise the cell's normal functioning. Therefore, the lethality of *hsrω*-nullisomic embryos[73,74,77] or the sterility of *hsrω*^{05241} mutant males[72] seems to result from widespread disruption of RNA processing due either to complete absence or over-abundance, respectively, of the *hsrω* transcripts. We are currently studying (Sengupta S, Lakhotia SC, unpublished) another *hsrω* mutant allele derived through local hopping of the P-transposon in the *hsrω*^{05241} chromosome; this derived mutant line displays late larval/pupal lethality and it is significant that the omega speckles show extensive clustering in different larval tissues, comparable to that seen[72] in cyst cells of the *hsrω*^{05241} mutant males. It is likely that a new mutation in the *hsrω* promoter in this line mis-regulates *hsrω*-n transcripts in most larval cells, causing over-sequestration of hnRNPs in large clusters of omega speckles. This in turn would cause a widespread mis-regulation of RNA processing and finally in larval/pupal lethality.

Future Prospects

The *hsrω*-n RNA is the first noncoding RNA to be distinctly shown to be responsible for organizing a well-defined nuclear domain, the omega speckles. Although it has been considered that some RNA species provide a structural role in sequestering the unengaged RNA processing factors,[115] no specific RNA has as yet been identified. The role played by the noncoding *hsrω*-n transcripts in organizing the omega speckles provides new paradigm for understanding the regulation of nuclear RNA processing activities and also for the novel roles that RNA can perform. Since the post-transcriptional processing events in the nucleus are highly conserved[88] and since every cell needs to fine tune the availability of RNA-processing factors in the nucleus to ensure that the RNA-processing events progress smoothly in a well coordinated manner as required by the specific and dynamically changing cellular needs, it is expected that noncoding RNAs comparable to the *hsrω* transcripts are present in all eukaryotic organisms. Furthermore, it is likely that all the different classes of nuclear domains/speckles are organized through one or more species of noncoding RNAs. A pro-active search for their identification will be rewarding. hsrω-like RNA species in humans may have clinical significance as well in view of the fact that this gene's dysfunction enhances poly-Q based neurodegeneration in *Drosophila* models.[69] Furthermore, since many of the fertility factors in man are RNA-binding proteins,[123] and since a dysfunction of the *hsrω* RNA in *Drosophila* testis causes sterility,[72] it will be interesting to examine if any of the infertility cases in human have a comparable etiology.

Functions of the small *hsrω*-c transcript have not been explored much and it will obviously be interesting to examine how the act of translation of the ORFω causes degradation of the template. Likewise, we know little about the spliced out intronic fragment, which is reported to be relatively stable.[37] Does this serve as a sink for some nuclear proteins or can it function as RNA regulator? These possibilities need to be examined. Detailed analysis of the secondary/tertiary structures of *hsrω* gene's various transcripts may provide a basis for search for similar transcripts in other eukaryotes.

Although amides have been instrumental in bringing this gene into focus, little is known about the action of amides on *Drosophila* cells that leads to the selective activation of the *hsrω* locus. The mechanism by which the *hsrω*-n transcripts are brought back to the *hsrω* locus after heat shock also remains intriguing. It will be interesting to examine if some components of nuclear lamins or other nuclear skeletal elements are involved in this movement.

It is interesting that a noncoding gene like the *hsrω* can have such varied functions in nucleus as well as cytoplasm, which help integrate cellular regulation. Such functions are of fundamental importance to cell's housekeeping activities and provide good examples of self-organizing systems, so characteristic of living organisms.[117] It is certain that as we explore deeper into the mysteries of genome with no preconceived notions, we will uncover many more instances of such noncoding RNA based self-organizing systems.

Acknowledgements

Research on the *hsrω* gene in my laboratory has been supported, at various times, by the University Grants Commission, New Delhi, Department of Atomic Energy (Govt. of India), Mumbai, Council of Scientific & Industrial Research, N. Delhi and the Department of Science & Technology (Govt. of India), N. Delhi. I also like to acknowledge the enthusiasm of my former and present Ph.D. students and post-doctoral fellows who chose to work on this gene and help make it ever more interesting to study.

References

1. Crick F. Central dogma of molecular biology. Nature 1970; 227:561-563.
2. Lakhotia SC. RNA polymerase II dependent genes that do not code for protein. Ind J Biochem Biophys 1996; 33:93-102.
3. Lakhotia SC. Noncoding RNAs: versatile roles in cell regulation. Curr Science 1999; 77:479-480.
4. Erdmann VA, Szymanski M, Hochberg A et al. Collection of mRNA-like noncoding RNAs. Nucleic Acids Res 1999; 27:192-195.
5. Erdmann VA, Barciszweska MZ, Szymanski M et al. The noncoding RNAs as riboregulators. Nucleic Acids Res 2001; 29:189-193.
6. Pesole G, Grillo G, Larizza A et al. The untranslated regions of eukaryotic mRNAs: structure, function evolution and bioinformatics tools for their analysis. Brief Bioinformatics 2000; 1:236-249.
7. MacIntosh GC, Wilkerson C, Green PJ. Identification and analysis of Arabidopsis expressed sequence tags characteristic of noncoding RNAs. Plant Physiol 2001; 127:765-776.
8. Beaton MJ, Cavalier-Smith T. Eukaryotic noncoding DNA is functional: evidence from the differential scaling of cryptomonad genomes. Proc R Soc Lond B 1999; 266:2053-2059.
9. Mattick JS. Noncoding RNAs: the architects of eukaryotic complexity. EMBO Rep 2001; 21:986-991.
10. Zamore PD. Ancient pathways programmed by small RNAs. Science 2002; 296:1265-1269.
11. Riddihough G. The other RNA world. Science 2002; 296:1259.
12. Schattner P. Searching for RNA genes using base-composition statistics. Nucl Acids Res 2002; 30:2076-2082.
13. Stotz G. An expanding universe of noncoding RNAs. Science 2002; 296:1260-1263.
14. Eddy SR. Noncoding RNA genes and the modern RNA world. Nature Rev Genet 2001; 2:919-929.
15. Eddy SR. Computational genomics of noncoding RNA genes. Cell 2002; 109:137-140.
16. Carter RJ, Dubchak I, Holbrook SR. A computational approach to identify genes for functional RNAs in genomic sequences. Nucl Acids Res 2001; 29:3928-3938.
17. Maison C, Bailly D, Peters AHFM et al. Higher-order structure in pericentric heterochromatin involves a distinct pattern of histone modification and an RNA component. Nat Genetics 2002; 30:329-334.
18. Mourelatos Z, Dostie J, Paushkin S et al. MiRNPs: a novel class of ribonucleoproteins containing numerous microRNAs. Genes Develop 2002; 720-728.
19. Moss EG. MicroRNAs: hidden in the genome. Curr Biol 2002; 12:R138-R140.
20. Blencowe BJ. Transcription: surprising role for an elusive small nuclear RNA. Curr Biol 2002; 12:R147-R149.
21. Szymanski M, Barciszweski J. Beyond the proteome: noncoding regulatory RNAs. Genome Biology 2002; 3:00051-00058.
22. Ritossa FM. A new puffing pattern induced by heat shock and DNP in *Drosophila*. Experientia 1962; 18:571-573.
23. Tissieres A, Mitchell HK, Tracy V. Protein synthesis in salivary glands of *D. melanogaster*: relation to chromosome puffs. J Mol Biol 1974; 84:389-398.
24. Ashburner M. Patterns of puffing activity in salivary gland chromosomes of *Drosophila*. V Responses to environmental treatment Chromosoma 1970; 31:356-376.

25. Lakhotia SC, Mukherjee AS. Activation of a specific puff by benzamide in *Drosophila melanogaster*. Dros Inf Serv 1970; 45:108.
26. Brown CJ, Hendrich BD, Rupert JL et al. The human *Xist* gene: analysis of a 17 kb inactive X-specific RNA that contains several conserved repeats and is highly localized within the nucleus. Cell 1992; 71:527-542.
27. Wutz A. *Xist* RNA associates with chromatin and causes gene silencing. In: Barciszewski J, Erdmann VA, eds. Nonprotein-coding RNAs. Landes Bioscience, 2002:.
28. Meller VH, Wu KH, Roma G et al. *rox*1 RNA paints the X-chromosome of male *Drosophila* and is regulated by the dosage compensation system. Cell 1997; 88:445-457.
29. Meller VH, Gordadze PR, Park Y et al. Ordered assembly of rox RNAs into MSL complexes on the dosage-compensated X chromosome in *Drosophila*. Curr Biol 2000; 10:136-43.
30. Kelly RL, Kuroda MI. The role of chromosomal RNAs in marking the X for dosage compensation. Curr Opin Genet Dev 2000; 10:555-561.
31. Lakhotia SC. The 93D heat shock locus in *Drosophila*- A review. J Genet 1987; 66:139-157.
32. Lakhotia SC. The 93D heat shock locus of *Drosophila melanogaster* modulation by genetic and developmental factors. Genome 1989; 31:677-683.
33. Lakhotia SC, Sharma A. The 93D *hsr*-omega locus of *Drosophila* noncoding gene with house keeping function. Genetica 1996; 97:339-348.
34. Bendena WG, Fini ME, Garbe JC et al. *hsrw*: A different sort of heat shock locus. In: Pardue ML, Ferimisco J, Lindquist S, eds. Stress-induced proteins. Alan R. Liss, Inc, 1989:3-14.
35. Lakhotia SC, Ray P, Rajendra TK et al. The noncoding transcripts of *hsrw* gene in *Drosophila*: do they regulate trafficking and availability of nuclear RNA-processing factors? Curr Science 1999; 77:553-563.
36. Prasanth KV, Rajendra TK, Lal AK et al. Omega speckles - a novel class of nuclear speckles containing hnRNPs associated with noncoding *hsr*-omega RNA in *Drosophila*. J Cell Sci 2000; 113:3485-3497.
37. Bendena WJ, Garbe JC, Traverse KL et al. Multiple inducers of the *Drosophila* heat shock locus 93D (*hsr* omega): inducer-specific patterns of the three transcripts. J Cell Biol 1989; 108:2017-2028.
38. Mukherjee T, Lakhotia SC 3H-uridine. incorporation in the puff 93D and in chromocentric heterochromatin of heat shocked salivary glands of *Drosophila melanogaster*. Chromosoma 1979; 74:75-82.
39. Ashburner M, Bonner JJ. The induction of gene activity in *Drosophila* by heat shock. Cell 1979; 17:241-254.
40. Lakhotia SC, Mukherjee T. Specific activation of puff 93D of *Drosophila melanogaster* by benzamide and the effect of benzamide treatment on the heat shock induced puffing activity. Chromosoma 1980; 81:125-136.
41. Lakhotia SC, Mukherjee T. Specific induction of the 93D puff in polytene nuclei of *Drosophila melanogaster* by colchicine. Ind J Exp Biol 1984; 22:67-70.
42. Tapadia MG, Lakhotia SC. Specific-induction of the *hsrw* locus of *Drosophila melanogaster* by amides. Chromosome Research 1997; 5:1-4.
43. Lakhotia SC, Singh AK. Conservation of the 93D puff of *Drosophila melanogaster* in different species of *Drosophila*. Chromosoma 1982; 86:265-268.
44. Burma PK, Lakhotia SC. Cytological identity of 93D-like and 87C-like heat shock loci in *Drosophila* pseudoobscura. Ind J Exp Biol 1984; 22:577-580.
45. Nath BB, Lakhotia SC. Search for a *Drosophila*- 93D-like locus in *Chironomus* and *Anopheles*. Cytobios 1991; 65:7-13.
46. Carmona MJ, Morcillo G, Galler R et al. Cloning and molecular characterization of a telomeric sequence from a temperature induced balbiani ring. Chromosoma 1985; 92:108-115.
47. Botella LM, Morcillo G, Barettino D et al. Heat-shock induction and cytoplasmic localization of transcripts from telomeric-associated sequences in *Chironomus* thummi. Exp Cell Res 1991; 196:206-209.
48. Morcillo G, Diez JL, Carbajal ME et al. HSP90 associates with specific heat shock puffs *hsrw* in polytene chromosomes of *Drosophila* and *Chironomus*. Chromosoma 1993; 102:648-656.
49. Spradling A, Pardue ML, Penman S. Messenger RNA in heat-shocked cells. J Mol Biol 1977; 109:559-587.

50. Langyel JA, Ransom LJ, Graham ML et al. Transcription and metabolism of RNA from the *Drosophila melanogaster* heat shock puff site 93D. Chromosoma 1980; 80:237-252.
51. Lakhotia SC, Mukherjee T. Absence of novel translational products in relation to induced activity of the 93d puff in *D. melanogaster*. Chromosoma 1982; 85:369-374.
52. Peters FPAMN, Lubsen NH, Walldorf U et al. The unusual structure of heat shock locus 2-48B in *Drosophila* hydei. Mol Gen Genet 1984; 197:392-398.
53. Walldorf U, Richter S, Ryseck R-P et al. Cloning of the heat-shock locus 93D from *Drosophila melanogaster*. EMBO J 1984; 3:2499-2504.
54. Garbe JC, Pardue ML. Heat shock locus 93D of *Drosophila melanogaster*, a spliced RNA most strongly conserved in the intron sequence. Proc Natl Acad Sci USA 1986; 83:1812-1816.
55. Garbe JC, Bendena WG, Pardue ML. Sequence evolution of the *Drosophila* heat shock locus *hsrw*. I. The nonrepeated portion of the gene Genetics 1989; 122:403-415.
56. Garbe JC, Bendena WG, Pardue ML. A *Drosophila* heat shock gene with a rapidly diverging sequence but a conserved structure. J Biol Chem 1986; 261:16889-16894.
57. Hovemann BT, Walldorf U,Ryseck R-P. Heat shock locus of 93D of *Drosophila melanogaster* an RNA with limiting coding capacity accumulates precursor transcripts after heat shock. Mol Gen Genet 1986; 204:334-340.
58. Hogan NC, Slot F, Traverse KL et al. Stability of tandem repeats in the *Drosophila melanogaster hsr*-omega nuclear RNA. Genetics 1995; 139:1611-1621.
59. Peters FPAMN, Lubsen NH, Sondermeijer PJA. Rapid sequence divergence in a heat shock locus of *Drosophila*. Chromosoma 1980; 81:271-280.
60. Ryseck R-P, Walldorf U, Hoffmann T. et al. Heat shock loci 93D of *Drosophila melanogaster* and 48B of *Drosophila* hydei exhibit a common structural and transcriptional pattern. Nucleic Acids Res 1987; 15:3317-3333.
61. Brockdorff N. X-chromosome inactivation: closing in on proteins that bind *Xist* RNA. Trends Genetics 2002; 18:352-358.
62. Fini ME, Bendena WJ, Pardue ML. Unusual behavior of the cytoplasmic transcript of *hsrw*. An abundant stress-inducible RNA that is translated, but which yields no detectable protein product. J Cell Biol 1989; 108:2045-2057.
63. Lakhotia SC, Sharma A. RNA metabolism in situ at the 93D heat shock locus in polytene nuclei of *Drosophila melanogaster* after various treatments. Chromosome Research 1995; 3:151-161.
64. Hogan NC, Traverse KL, Sullivan et al. The nucleus-limited *hsr*-omega-n transcript is a polyadenylated RNA with a regulated intranuclear turnover. J Cell Biol 994; 125:21-30.
65. Bendena WJ, Southgate AA, Garbe JC et al. Expression of heat shock locus *hsr* omega in nonstressed cells during development in *Drosophila melanogaster*. Dev Biol 1991; 144:65-77.
66. Mutsuddi M, Lakhotia SC. Spatial expression of the *hsr*-omega 93D gene in different tissues of *Drosophila melanogaster* and identification of promoter elements controlling its developmental expression. Dev Genet 1995; 17:303-311.
67. Lakhotia SC, Rajendra TK, Prasanth KV. Developmental regulation and complex organization of the promoter of the noncoding *hsrw* gene of *Drosophila melanogaster*. J Biosciences 2001; 26:25-38.
68. Posey KL, Jones LB, Cerda R et al. Survey of transcripts in the adult brain. Genome Biology 2001; 2:0008.1-0008.16.
69. Fernandez-Funez P, Nino-Rosales ML, de Gouyon B et al. Identification of genes that modify ataxin-1-induced neurodegeneration. Nature 2000; 408:101-106.
70. Lakhotia SC, Mutsuddi M. Heat shock but not benzamide and colchicine response elements are present within the -844bp upstream region of the *hsrw* gene of *Drosophila melanogaster*. J Biosciences 1996; 21:235-246.
71. Lakhotia SC, Tapadia MG. Genetic mapping of the amide response element/s of the *hsrw* locus of *Drosophila melanogaster*. Chromosoma 1998; 107:127-135.
72. Rajendra TK, Prasanth KV, Lakhotia SC. Male sterility associated with over-expression of the noncoding *hsrw* gene in cyst cells of testis of *Drosophila melanogaster*. J Genetics 2001; 80:97-110.
73. Mohler J, Pardue ML. Deficiency mapping of the 93D heat shock locus in *Drosophila*. Chromosoma 1982; 86:457-467.
74. Mohler J, Pardue ML. Mutational analysis of the region surrounding the 93D heat shock locus of *Drosophila melanogaster*. Genetics 1984; 106:249-265.

75. Burma PK, Lakhotia SC. Expression of the 93D heat shock puff of *Drosophila melanogaster* in deficiency genotypes and its influence on activity of the 87C puff. Chromosoma 1986; 94:273-278.
76. Ray P. Studies on interaction of the 93D (*hsr*-omega) locus with other genes during development of *D. melanogaster*. Ph.D. thesis,. Varanasi: Banaras Hindu University, 1997:.
77. Lakhotia SC, Ray P. HSP83 mutation is a dominant enhancer of lethality associated with absence of the nonprotein coding *hsrw* locus in *Drosophila melanogaster*. J Bioscience 1996; 21:207-219.
78. Pardue ML, Bendena WG, Fini ME et al. *hsrw*, a novel gene encoded by a *Drosophila* heat shock puff. Biol Bull 1990; 179:77-86.
79. McColl G, Hoffmann AA, McKechnie SW. Response of two heat shock genes to selection for knockdown heat resistance in *D. melanogaster*. Genetics 1996; 143:1615-1627.
80. McKechnie S W, Hafford M M, McColl G et al. Both allelic variation and expression of nuclear and cytoplasmic transcripts of *hsrw* are closely associated with thermal phenotype in *Drosophila*. Proc Natl Acad Sci USA; 95:2423-2428.
81. Fuller M. Spermatogenesis. In: Bate M, Martinez A, eds. The Development of *Drosophila melanogaster*. Cold.Spring Harbor: NY, Cold Spring Harbor Laboratory Press., 1993:71-147.
82. Lakhotia SC. Heat shock response – regulation and functions of coding and non genes in *Drosophila*. Proc Indian Natn Sci Acad (PINSA) 2001; B67:247-264.
83. Lakhotia SC, Prasanth KV. Tissue and development specific induction and turnover of hsp70 transcripts from 87A and 87C loci after heat shock and during recovery in *Drosophila melanogaster*. J Exp Biol 2002; 205:345-358.
84. Ray P, Lakhotia S C. Interaction of the nonprotein-coding developmental and stress-inducible *hsrw* gene with Ras of *Drosophila melanogaster*. J Biosci 19982; 3:377-386.
85. Cutforth T, Rubin GM. Mutations in Hsp83 and cdc37 impair signaling by Sevenless receptor tyrosine kinase in *Drosophila*. Cell 1994; 1027-1036.
86. Cummings CJ, Zoghbi HY. Trinucleotide repeats: mechanisms and pathophysiology. Annu Rev Genomics Hum Genet 2000; 1:281-328.
87. Kazemi-Esfarjani P, Benzer S. Genetic suppression of polyglutamine toxicity in *Drosophila*. Science 2000; 287:1837-1840.
88. Krecic AM, Swanson MS. hnRNP complexes: composition, structure and function. Curr Opin Cell Biol 1999; 11:363-371.
89. Saumweber H, Symmons P, Kabisch R et al. Monoclonal antibodies against chromosomal proteins of *Drosophila melanogaster*. Chromosoma 1980; 80:253-275.
90. Dangli A,Bautz EKF. Differential distribution of nonhistone proteins from polytene chromosomes of *Drosophila melanogaster* after heat shock. Chromosoma 1983; 88:201-207.
91. Dangli A, Grond C, Kloetzel P et al. Heat-shock puff 93D from *Drosophila melanogaster*: accumulation of a RNP-specific antigen associated with giant particles of possible storage function. EMBO J 1983; 2:1747-1751.
92. Risau W, Symmons P, Saumweber H et al. Nonpackaging and packaging proteins of hnRNA in *Drosophila melanogaster*. Cell 1983; 33:529-241.
93. Kabisch R, Bautz EKF. Differential distribution of RNA polymerase B and nonhistone chromosomal proteins in polytene chromosomes of *Drosophila melanogaster*. EMBO J 1983; 2:395-402.
94. Schuldt C, Kloetzel PM, Bautz EKF. Molecular organization of RNP complexes containing P11 antigen in heat-shocked and nonheat-shocked *Drosophila* cells. Eur J Biochem 1989; 181:135-142.
95. Hovemann BT, Dessen E, Mechler H et al. *Drosophila* snRNP associated protein P11 which specifically binds to heat shock puff 93D reveals strong homology with hnRNP core protein A1. Nucleic Acids Res 1991; 19:4909-4919.
96. Bopp D, Bell LR, Cline TW et al. Developmental distribution of female-specific Sex-lethal proteins in *Drosophila melanogaster*. Genes Dev 1991; 5:403-415.
97. Samuels ME, Bopp D, Colvin RA et al. RNA binding by Sxl proteins in vitro and in vivo. Mol Cell Biol 1994; 14:4975-4990.
98. Lakomek H-Z, Will H, Zech M et al. A new serologic marker in ankylosing spondylitis. Arthritis Rheum 1984; 27:961-967.
99. Spector DL. Macromolecular domains within the cell nucleus. Ann Rev Cill Biol 1993; 9:265-315.
100. Spector DL. Nuclear domains. J Cell Sci 2001; 114:2891-2893.

101. Derksen J, Berendes HD, Willart E. Production and release of a locus-specific ribonucleoprotein product in polytene nuclei of *Drosophila hydei.* J Cell Biol 1983; 59:661-668.
102. Lindquist S. The heat shock response. Annu Rev Biochem 1986; 55:1151-1191.
103. Prasanth KV. Studies on the regulation of intracellular localization and turnover of the heat shock gene products in different cell types of *Drosophila.* Ph.D. thesis . Varanasi : Banaras Hindu University, 2000.
104. Rutherford SL, Lindquist S. HSP90 as a capacitor for morphological evolution. Nature 1998; 396:336-342.
105. Rutherford SL. From genotype to phenotype: buffering mechanisms and the storage of genetic information. Bioessays 2000; 22:1095-1105.
106. Lorckovic ZJ, Dominika A, Wieczorek K et al. PremRNA splicing in higher plants. Trends Plant Sci 2000; 5:160-167.
107. Reed R, Hurst E. A conserved mRNA export machinery coupled to premRNA splicing. Cell 2002; 108:523-531.
108. Caputi M, Zahler AM. SR proteins and hnRNP H regulate the splicing of the HIV-1 tev-specific exon 6D. EMBO J 2002; 21:845-855.
109. Elefanty AG, Antoniou M, Custidio N et al. GATA transcription factors associate with a novel class of nuclear bodies in erythroblasts and megakaryocytes. EMBO J 1996; 15:319-333.
110. Saurin AJ, Shiels C, Williamson J et al. The human polycomb group complex associates with pericentromeric heterochromatin to form a novel nuclear domain. J Cell Biol 1998; 142:887-898.
111. Cotto J, Fox S, Morimoto R. HSF1 granules: a novel stress induced nuclear compartment of human cells. J Cell Sci 1997; 110:2925-2934.
112. Jolly C, Vourc'h C, Robert-Nicoud M et al. Intron-independent association of splicing factors with active genes. J Cell Biol 1999; 145:1133-43.
113. Weighardt F, Cobianchi F, Cartegni L et al. A novel hnRNP protein HAP/SAF-B enters a subset of hnRNP complexes and relocates in nuclear granules in response to heat shock. J Cell Sci 1999; 112:1465-1476.
114. Chiodi I, Bigglogera M, Denegri M et al. Structure and dynamics of hnRNP-labelled nuclear bodies induced by stress treatments. J Cell Sci 2000; 113:4043-4053.
115. Fardaei M, Larkin K, Brook JD et al. In vivo colocalisation of MBNL protein with *DMPK* expanded-repeat transcripts. Nucleic Acids res 2001; 29:2766-2771.
116. Fox AH, Lam YW, Leung AKL et al. Paraspeckles: a novel nuclear domain. Curr Biol 2002; 12:13-25.
117. Mistelli T. The concept of self-organization in cellular architecture. J Cell Biol 2001; 155:181-185.
118. Denegri M, Chiodi I, Corioni M et al. Stress-induced nuclear bodies are sites of accumulation of premRNA processing factors. Mol Biol Cell 2001; 12:3502-3514.
119. Jagtheesan G, Thanumalayan H, Muralikrishna B et al. Colocalization of intranuclear lamin foci with RNA splicing factors. J Cell Sci 1999; 112:4651-4661.
120. Caceres JF, Kornblihtt AR. Alternative splicing: multiple control mechanisms and involvement in human disease. Trends Genet 2002; 18:186-193.
121. Mayeda A, Krainer AR. Regulation of alternative premRNA splicing by hnRNP A1 and splicing factor SF2. Cell 1992; 68:365-375.
122. SaccoBubulya P, Spector DL. Disassembly of interachromatin granule clusters alters coordination of transcription and premRNA splicing. J Cell Biol 2002; 156:425-436.
123. Venables JP, Eperon IC. The roles of RNA-binding proteins in spermatogenesis and male infertility. Curr Opin Genet Dev 1999; 346-354.

Chapter 15

Adapt Gene RNA Transcripts as Riboregulators

Dana R. Crawford and Kelvin J. A. Davies

Abstract

There is growing interest in the study of so-called riboregulator or noncoding RNAs. These spliced and polyadenylated RNAs contain either a very short or no apparent open reading frame or translational product. However, they are associated with a wide range of biological activities that suggest that they represent important cellular regulators. We initially identified the *adapt15* and *adapt33* mRNAs by their induction following oxidative stress under conditions where a protective "adaptive response" occurred. This adaptive response involved the induction of RNAs in response to a modest concentration of hydrogen peroxide that protected the cells against further oxidative damage. Sequence and translational analysis revealed that both RNAs are polyadenylated and spliced but neither contain any obvious open reading frame nor generate a detectable protein product. They therefore qualify as riboregulators. *adapt15*, also known as *gadd7*, is involved in growth suppression and may associate with intracellular v-Src protein. The intracellular function of *adapt33* is not clear although it may play a role in the regulation of protein synthesis. A new classification acronym, SIR, is proposed for those riboregulators that are induced by stress.

Background

Riboregulator RNAs are spliced and polyadenylated RNAs that contain no apparent open reading frame or translational product. Due to their infrequent identification and lack of a functional protein product, their importance and characterization has been largely ignored. It has been increasingly clear, however, that these RNAs are associated with a wide range of biological activities which suggests that they represent important cellular regulators. These riboregulators, also referred to as noncoding RNAs (ncRNAs) in eukaryotes, have been implicated in transcriptional regulation, RNA processing, tumor suppression, prenatal lethality, chromosome condensation, developmental timing, regulation of protein synthesis, and growth arrest.[1-5] An increasing number of them have been reported in organisms ranging from bacteria and plants to humans. A recent review listed twenty-eight such riboregulators.[6]

Oxidative stress is known to induce the expression of a number of mammalian genes.[7,8] Some of these inducted genes are thought to be part of protective responses by cells.[7-9] We have studied the induction of genes in cells undergoing a so-called protective "adaptive response". Adaptive response refers to the ability of cells or organisms to better resist the damaging effects of a toxic agent when first preexposed to a lower dose of the same or similar agent. It is a widespread phenomena that has been observed in prokaryotes, yeast, mammals and plants in

Noncoding RNAs: Molecular Biology and Molecular Medicine, edited by Jan Barciszewski and Volker A. Erdmann.

response to many different types of stress.[8-10] The induction of stress response-genes is an integral part of this protection. The identification of such genes has led to important insights into basic cellular function and response.[9,11-13] Such stress genes are able to respond to a various combinations of stress including ionizing radiation, UV radiation, peroxides, heat, aldehydes, heavy metals, and alkylating agents, and these have been extensively studied in bacteria and to a lesser extent in mammalian cells. Mammalian stress genes implicated in adaptive responses include heat shock, glucose-regulated protein, heme oxygenase, and ferritin.[11-13]

It has previously been shown that HA-1 hamster fibroblasts undergo an adaptive response when exposed to a minimally cytotoxic ("pretreatment") dose of hydrogen peroxide that confers resistance against a subsequent higher, lethal ("challenge") dose of peroxide.[10,14] Importantly, this adaptive response is dependent upon the de novo production of protective RNAs and proteins since inhibitors of protein and RNA synthesis dramatically inhibit the adaptive response.[10] Identifying RNAs that are induced by the pretreatment dose of hydrogen peroxide thus allows for the identification of genes involved in protective cellular responses to oxidative stress and potentially other stress.

Identification of *adapt15* and *adapt33*

Since it was previously demonstrated that inhibition of RNA and protein synthesis also strongly inhibits the adaptive response of HA-1 cells to hydrogen peroxide[10], we decided to use the pretreatment dose to identify oxidant-inducible mRNAs that may also protect cells against oxidative damage. A significant inhibition of the adaptive response by RNA and protein synthesis inhibitors was known to occur throughout the 18 hours that followed exposure to peroxide, so the cells should be a source of newly synthesized protective sequences at all time points within this span. We extracted total RNA from HA-1 cells at multiple time points after exposure to a pretreatment dose of hydrogen peroxide, and using the Differential display technique,[14] identified and confirmed five genes, designated "*adapts*", whose RNA products are induced in response to a pretreatment concentration of hydrogen peroxide. The expression of all the *adapts* characterized to date, (*adapt15, 33, 66, 73,* and *78)* have been found to be induced by multiple stresses including methyl methanesulphonate, cis(II)-platinum, hydrogen peroxide, 2-deoxyglucose, heavy metals, heat, and calcium elevation by calcium ionophore A23187 and thapsigargin. Two of these, *adapt15* and *adapt33*, qualify as riboregulators as discussed below.

Adapt15

For *adapt15*, a cDNA library from HA-1 cells treated for 5 hours with a pretreatment concentration of peroxide was screened and 30 positives were found after secondary screening. Two of these clones were sequenced in their entirety. Twenty eight other clones were mostly-but not entirely-sequenced. The two entirely sequenced clones were respectively 753 bases, plus 110 bases of polyadenylate (designated P8); and 746 bases, plus 38 bases of polyadenylate (designated P9). When these two cDNAs were used to probe a Northern blot, the presence of an inducible RNA species was again confirmed.[14] We observed maximum induction (average 10.2-fold) and minimal basal levels when we stressed cells growing in log phase (Fig. 1). The size of the inducible RNA species was 950 bases, and it was designated *adapt15* RNA. Since Northern blot analysis had revealed that *adapt15* had a size of 950 bases, these two clones were close to full length when taking into account the normal-length of a polyadenylation tail (150-200 bases). We also assessed the possible contribution of calcium to *adapt15* RNA induction by H_2O_2 (Fig. 2). Preincubation of cells with the acetoxymethyl ester form of the intracellular calcium chelator bis(aminophenoxy)ethane *N,N'*-tetraacetic acid (BAPTA-AM) totally

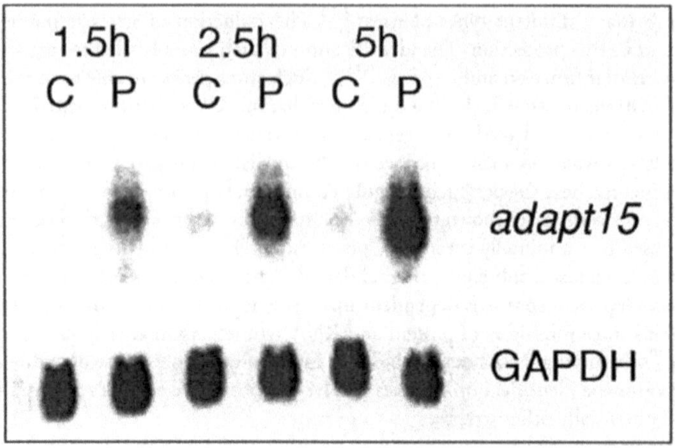

Figure 1. Induction of *adapt15* mRNA by a pretreatment dose of hydrogen peroxide in HA-1 hamster fibroblasts. Cells were exposed to a pretreatment concentration of hydrogen peroxide (4.1 μmoles/10^7 cells) for the times indicated. RNA was then extracted, electrophoresed and Northern blotted, and probed with *adapt15* cDNA. C-control; P-peroxide treated. The Northern blot was also probed with GAPDH cDNA as a loading control. (Reproduced from Crawford et al[25] with permission).

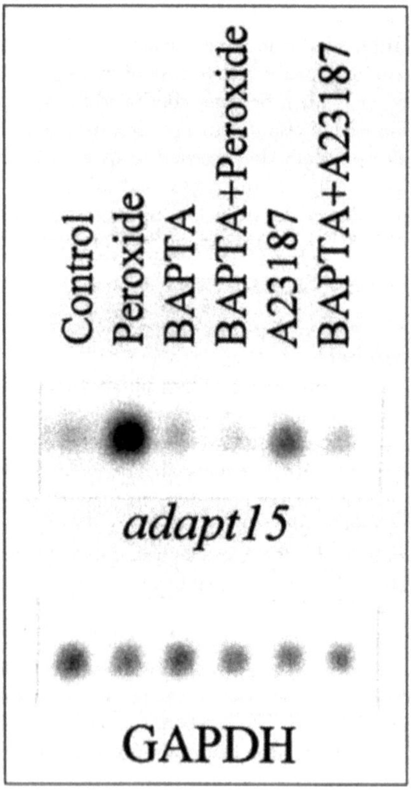

Figure 2. The involvement of calcium in *adapt15* induction by hydrogen peroxide. HA-1 cells were pretreated for 1 hour with 20 μM of the intracellular calcium chelator BAPTA-AM dissolved in DMSO, followed by hydrogen peroxide (4.1 μmoles H_2O_2 pretreatment dose per 10^7 cells) or A23187 (1 μg/ml) for 5 hours. HA-1 cells treated only with BAPTA-AM served as a control for these analyses. Hydrogen peroxide and A23187 were also added to separate cultures without BAPTA pretreatment. All cultures, including control, contained equal amounts of DMSO (during pretreatment) and ethanol (at time 0). Total RNA was then extracted, electrophoresed, blotted, and probed with *adapt15* cDNA. GAPDH was used as a loading control. (Reproduced from Crawford et al[15] with permission).

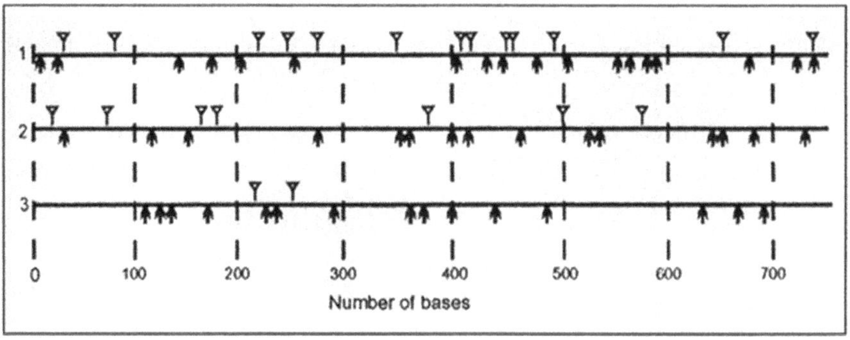

Figure 3. Possible initiation and termination sites of *adapt15* cDNA. Predicted initiation (triangles) and termination (arrows) sites of *adapt15* mRNA, excluding the polyadenylation tail, are indicated. Possible open reading frames are indicated as 1, 2 and 3. (Reproduced from Crawford et al[14] with permission).

prevented *adapt15* RNA induction by H_2O_2. In addition, the calcium ionophore A23187 by itself induced *adapt15* RNA, to a level 30% and 100% of the induction observed with H_2O_2, in two separate experiments.[15] These inductions were also inhibited by BAPTA-AM. Thus, calcium is integrally involved in the induction of *adapt15* RNA by H_2O_2.

Sequence analysis revealed that neither *adapt15* cDNA form had a long open reading frame as shown in Figure 3.[14] The predicted protein (peptide) size of the longest open reading frame was only about 5,000 daltons. A consensus Kozak box-like sequence was also found starting at base 76 and consists of GCCCAGATGG. Initiation of translation at this site would generate a protein (peptide) of 2,371 daltons. We also found that P8 and P9 were identical in sequence with the exception of an extra three bases (GCA) in the P9 sequence that was not present in P8. Overall, we found 25 P8 and P9 homologous clones, all with polyadenylation tails. No translation product of 5000 daltons or greater was observed in our in vitro transcription-translation analysis of *adapt15*.[14] However, we were not technically able to analyze for smaller translation products. It has recently been reported that a translated product (reported to be only 8 amino acids in length) of an *adapt15* homologue from hamster binds to the C-terminus of v-Src protein.[16] If true, it would indicate that a small portion of the *adapt15* mRNA is translated. However, it is also possible that the transcribed product is an artifact of the two-hybrid system from which it was identified as there has been no subsequent report confirming or extending this observation.

Using cell fractionation, we determined that the majority of *adapt15* RNA is located in the cytoplasm (Fig. 4). Subfractionation of the cytoplasm into actively-translating fraction and nonactive fraction is a way of identifying RNAs thought to be important to the cell. We determined that *adapt15*·RNA is abundantly associated with the monosome/polysome area of the cytoplasm.[14]. It is mostly concentrated right at the monosome peak, and extends well into the polysome area (the region that extends to the right of the monosome). Importantly, a dramatic shift in *adapt15* distribution is observed following disruption of the monosomes/polysomes with EDTA, an indication that a large percentage of *adapt*15 RNA is associated with active translation. A similar distribution and shift is observed for hydrogen peroxide treated HA-1 cells (Fig. 4B).

We have also isolated a genomic clone for *adapt15*. Sequencing analysis has revealed that *adapt15* has multiple introns, consistent with its designation as a small cytoplasmic RNA or riboregulator.

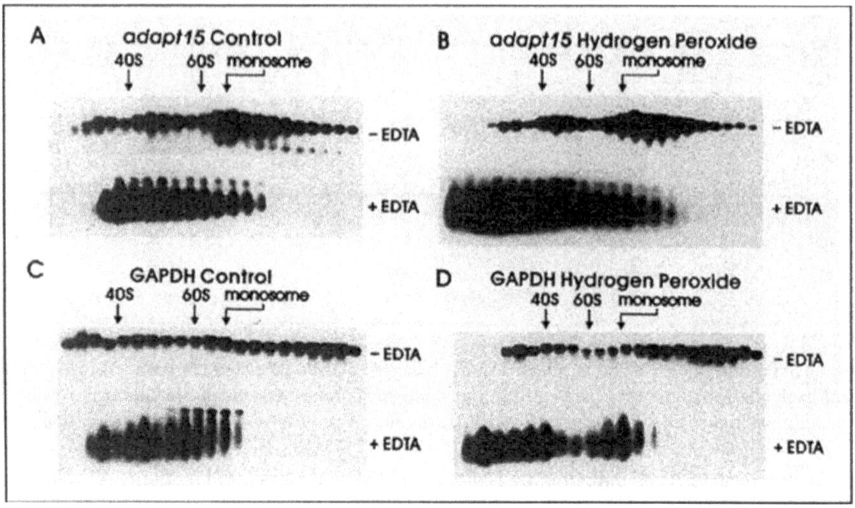

Figure 4. Cytoplasmic fractionation of *adapt15* mRNA. Cytoplasmic fractionation of *adapt15* RNA. Cytoplasmic preparations of both peroxide treated- and nontreated-cells were incubated on ice in the presence or absence of EDTA. The samples were then layered onto 10-50% (w/v) sucrose density gradients with a 60% (w/v) cushion and spun 5 hours at 288,000 x g. Fractions were collected, extracted with phenol / chloroform, and the RNA run out on a formaldehyde gel. After transfer, Northern blots were probed with *adapt15* and GAPDH cDNAs. Blots were then exposed to X-ray film for varying amounts of time in order to best detect the full range of hybridization signal in each fraction (i.e., these blots are not quantitatively comparable) Panels A and C—cytoplasmic fractions from untreated cells probed with *adapt15* or GAPDH; Panels B and D—cytoplasmic fractions from peroxide-treated cells probed with *adapt15* or GAPDH. (Reproduced from Crawford et al[14] with permission).

Adapt33

Subsequent to *adapt15*, another RNA was detected by differential display whose RNA was also strongly induced by H_2O_2 during adaptive response in a calcium-dependent manner.[17] Confirmation of *adapt33* induction by Northern blot analysis and cloning were carried out as previously described for *adapt15*. Northern blot analysis identified two inducible *adapt33* RNA bands of 990 and 1460 bases, respectively, both of which were strongly induced by H_2O_2 (Fig. 5). *adapt33* RNA was found to be significantly less abundant than is *adapt15*. The previously mentioned 5 hour cDNA library from peroxide-treated cells was then screened for homologous *adapt33* clones, and five secondary screen positives were obtained after initially screening approximately 80,000 recombinants. These cDNA clones were confirmed by Northern Blot analysis. All five clones were at least partially sequenced and some were fully sequenced. They were all found to contain significant regions of overlapping sequence and, as expected, polyadenylation tails. The longest *adapt33* cDNA (clone 7) was fully sequenced but revealed a relatively small putative truncated open reading frame of 210 bases at the 5' end (Fig. 6). However, in vitro transcription-translation of *adapt33* did not generate any observable protein product.

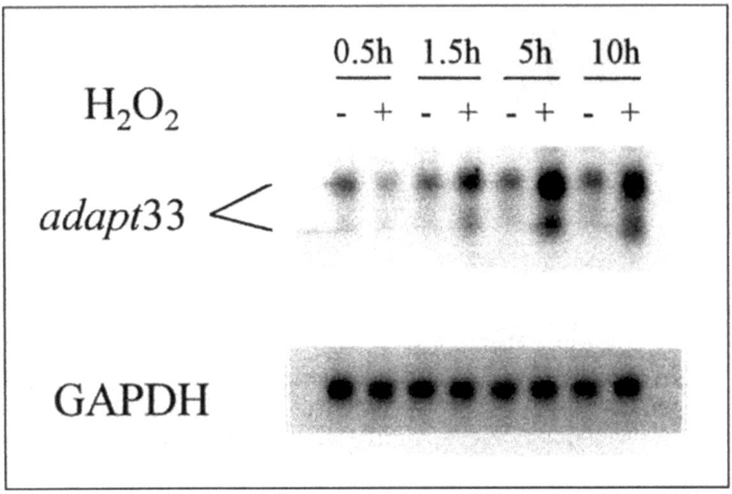

Figure 5. Induction of *adapt33* by hydrogen peroxide. A Northern blot containing RNA from HA-1 cells treated with 4.1 μmoles of hydrogen peroxide per 10^7 cells for 30 minutes. 90 minutes, 5 hours and 10 hours was probed with labeled *adapt33* cDNA. C, control; P, peroxide treated. The Northern blot was also probed with GAPDH cDNA as a loading control. (Reproduced from Crawford et al[17] with permission).

Mechanism of Action of *adapt15* and *adapt33*

Adapt15

Two other known sequences, *gadd*45 and *gadd*153, were also found to be induced under our pretreatment conditions.[15] These genes are known to be inducible by growth arrest and DNA damage. Therefore, we assessed the possible association of *adapt15* with growth arrest and DNA damage. We found that, like *gadd45* and *gadd153*, the levels of *adapt15* RNA were low during proliferation, but high during density saturation- and low serum (G_0)- growth arrests. *adapt15* RNA levels were assessed under different growth and growth arrest conditions. As shown in Figure 7 (left panel), HA-1 cells that were trypsinized and plated exhibited typical lag, log and plateau phases of growth. Growth saturation (144 hours for this experiment) and 0.1% fetal calf serum-containing cultures are known to induce growth arrest in G_0 phase. Aphidicolin arrests cells at the G_1/S phase interface. Analysis of Northern blots (right panel) containing RNA extracted at these different stages of growth and probed with *adapt15* cDNA revealed a low level of *adapt15* RNA during proliferation (especially 51 hours) and a much higher level during G_0 growth arrest (144 hours and 0.1% fetal calf serum) (Fig. 7—right panel). Aphidicolin (designated lane "A") had only a modest effect. Furthermore, the *adapt15* expression profile was very similar to that observed for *gadd45* and *gadd153*. Thus, *adapt15* is a growth arrest-associated sequence that appears to have strong specificity for G_0. Importantly, a subsequent report demonstrated that the overexpression of *adapt15/gadd7* leads to growth suppression in both hamster and human cells.[4]

Exposure of HA-1 cells to DNA damaging agents revealed significant induction of *adapt15* RNA by methylmethanesulphonate and cis-platinum but not X-irradiation. As for the growth arrest studies, near identical responses were also observed for *gadd45* and *gadd153* RNAs, suggesting coordinate regulation of *adapt15*, *gadd45* and *gadd153*. All three RNAs were also increased, relative to controls, following heat shock, a nongenotoxic treatment. Finally, the

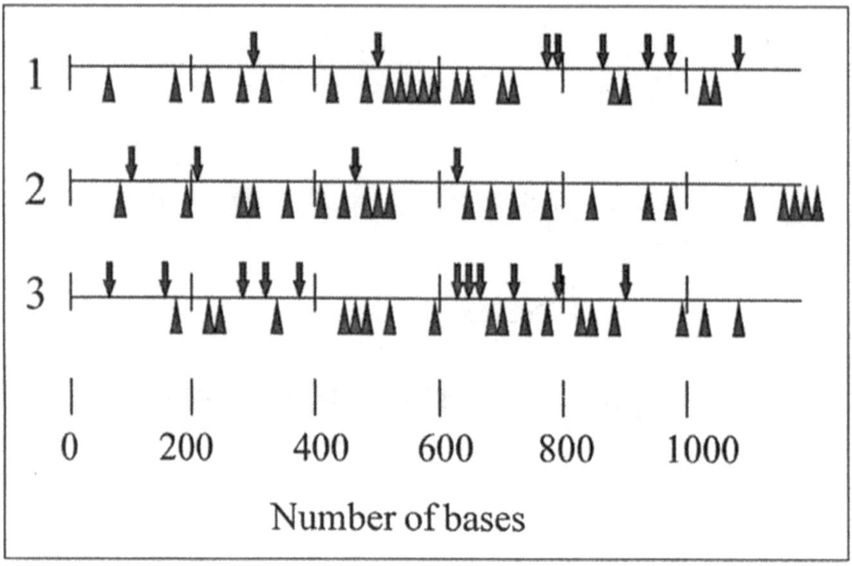

Figure 6. Initiation and termination sites of *adapt33* mRNAs. The predicted initiation (arrows) and termination (triangles) sites of the 2 mRNA species of *adapt33* are shown, excluding polyadenylation tails. The three possible reading frames are indicated. (Reproduced from Crawford et al[17] with permission).

induction of *adapt15* by hydrogen peroxide was strongly dependent upon calcium, a hallmark of *gadd153* induction. The coordinate inductions of *adapt15, gadd45* and *gadd153* by multiple agents; and by growth arrest; and their induction by an adaptive-response level of hydrogen peroxide (but not by a nonadaptive); suggest these RNAs may act in concert to protect cells against the damaging effects of oxidative stress.

The determination that *adapt15* RNA is abundantly associated with the monosome/polysome area of the cytoplasm suggests a role for the RNA in protein translation. Other riboregulators have also been reported with translational effects including baterial OxyS and *C. elegans lin-4*, which inhibit protein synthesis[18-20] and another small RNA in bacteria (DsrA) that activates protein synthesis by interfering with secondary RNA structure.[19,20]

As discussed above, one report determined that a small, eight amino acid peptide can be translated from a hamster homologue to *adapt15* RNA, and that this peptide binds v-Src protein. Since there have not been any subsequent reports confirming or extending this observation, it is difficult to conclude its importance at this time. Nonetheless, even if this is true in mammalian cells, there would still be a lot of untranslated *adapt15* RNA remaining in the cell that may very well account for the observed growth suppressive effects of *adapt15* RNA. We also theorize that *adapt15* RNA may act directly as a regulatory RNA at the site of actively translating ribosomes during translation to protect cells against the damaging effects of oxidative stress.

Adapt33

The intracellular function of *adapt33* is not yet known. Cell fractionation studies have revealed that a significant proportion of *adapt33* RNA is associated with the actively translating polysome region. This is demonstrated in Figure 8 where *adapt33* RNA as well as the

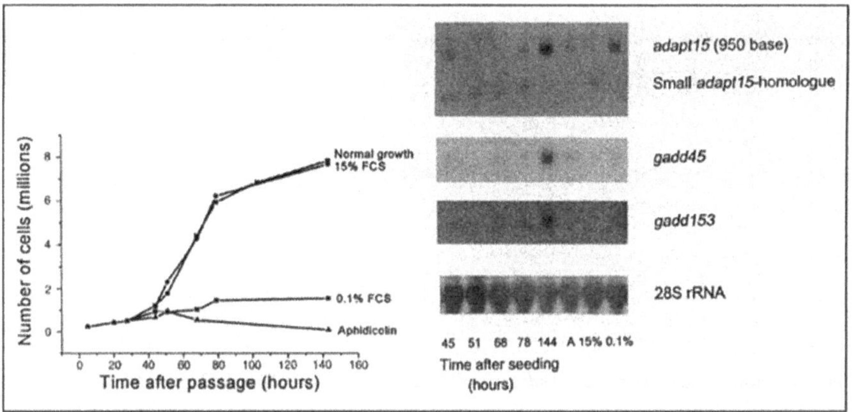

Figure 7. The effect of HA-1 cell growth and growth arrest on the levels of *adapt15* RNA. Cells were trypsinized from confluent cultures and replated at 200,000 per 60 mm plate, a density that is well below confluence. Cell growth was then followed over 6 days by cell counting and trypan blue inclusion analyses (left panel). RNA was extracted at each time point, as well as 24 hours after treating early log phase cultures with aphidicolin, or after changing media and replacing with either 0.1% or the normal 15% fetal calf serum. Northern blots containing this RNA were then probed with *adapt15* cDNA (right panel). The top band in the right panel, marked as *adapt15* (950 base), represents the oxidant-inducible 950 base *adapt15* RNA species; the bottom band is the relatively nonoxidant-modulated, but homologous, smaller species. RNA from aphidicolin- treated cultures is designated "A" at the bottom of the gels. (Reproduced from Crawford et al[15] with permission).

known translated mRNAs for GAPDH and *gadd153* fractionate to the actively translating polysomal region located to the right of the monosome peak. Like *adapt15/gadd7*, *adapt33* RNA may act directly as a regulatory RNA at the site of actively translating ribosomes during translation. As has been previously speculated for *adapt15/gadd7*, it is also possible that *adapt33* RNA acts both via RNA structure and a small peptide translated from one of the many small open reading frames. Just like *adapt15*, *adapt33* is strongly induced by hydrogen peroxide under adaptive response conditions where cellular protection is observed. This suggests that, whatever the mechanism, these RNAs are both involved in protecting cells against the damaging effects of oxidative stress and perhaps other stresses.

Conclusions

Our identification and characterization of the expression of *adapt15* and *33* suggest a novel mammalian stress response system that combines the actions of untranslated riboregulator *adapt* RNAs with other known protein stress-response mediators. If true, it would resemble the bacterial OxyR system where both noncoding oxyS RNA and other protein-encoding RNAs are induced following exposure to hydrogen peroxide; and the *Drosophila* heat shock response where both hsr[omega] noncoding RNAs are induced along with other RNAs encoding heat shock proteins, although the latter are induced independently.[6,21] Clearly, there are a number of similarities between *adapt33* and *adapt15/gadd7* including inducibility by hydrogen peroxide under conditions of adaptive response; their similar peroxide-induction kinetics; the role of calcium in these inductions; and their lack of large open reading frames. It is therefore possible that these two *adapts* belong to the same family. We have identified *adapt15/gadd7* as a growth arrest- and DNA damage-inducible sequence.[15] Induction of *adapt15* by growth arrest, methyl

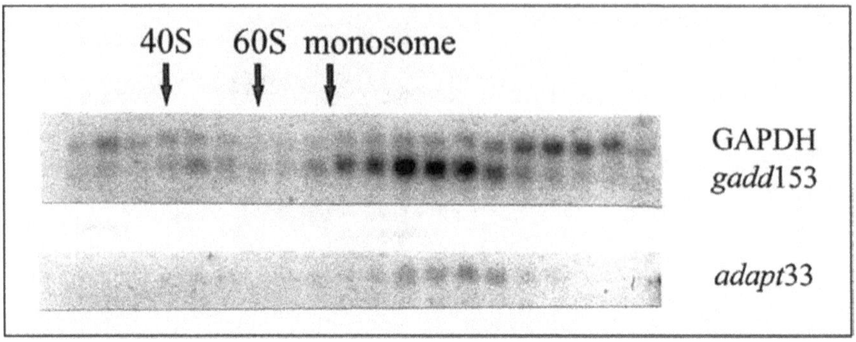

Figure 8. Cytoplasmic fractionation of *adapt33* mRNA. A cytoplasmic extract from peroxide-treated HA-1 cells was layered onto 10-50% (w/v) sucrose density gradients with a 60% (w/v) cushion and spun 5 hours at 288,000 x g. Fractions were collected, extracted with phenol /chloroform, and the RNA run out on a formaldehyde gel. After transfer, the Northern blot was probed, sequentially, with *adapt33, gadd153,* and GAPDH cDNAs. Monosome, 60S ribosomal subunit, and 40S ribosomal subunit, peak fractions are indicated. (Reproduced from Crawford et al[17] with permission).

methane sulfonate, cis-platinum, hydrogen peroxide, and heat shock is very similar to that observed for the well-known *gadd45* and *gadd153* RNAs, both of which encode for protein. Thus, *adapt15* and possibly *adapt33* appear to belong to the same family as *gadd45* and *gadd153*. This suggests that *adapt15* and perhaps *adapt33* RNAs act as riboregulators in concert with Gadd45 and Gadd153 proteins to bring about growth arrest and perhaps protect against DNA damage.

We propose a new additional classification for those riboregulators that are induced by stress: SIR, short for stress-inducible riboregulators. Thus, *adapt15, adapt33,* OxyS, and *hsr[omega]* would be part of this SIR gene family. Comparison of the actions of these RNAs will be interesting as more is learned about these regulators and combined, may provide valuable insight into the intracellular function(s) of SIR.

The relatively new grouping of certain RNAs as noncoding RNAs/riboregulators requires qualification or perhaps even modified designations as more and more examples are reported. While in the strict sense it refers to nontranslated RNAs, there is always the possibility that translation of a small proteins occurs from these RNAs that are undetectable by current analytical methods, or that may only be synthesized under certain conditions or stages of growth. We would argue that even if true, any such RNA, perhaps including *adapt15 or adapt33,* should still be characterized as a riboregulator because a large majority of its sequence is not translated. Furthermore, there have been several reports of RNAs with significant open reading frames also containing noncoding sequence with important functions. These include tumor suppression activity in the 3' untranslated region of alpha-tropomyosin, prohibitin, and mel-18.[22-24] These sequences act as important regulators at the RNA level despite the fact that a different portion of the molecule is translated. Thus they should also be considered as riboregulators in the broader sense despite a significant open reading frame. Along the same lines, it is possible that some riboregulators identified to date act both via RNA structure and a small peptide translated from one of the many small open reading frames.

References

1. Brockdorff N, Ashworth A, Kay GF et al. The product of the mouse Xist gene is a 15 kb inactive X- specific transcript containing no conserved ORF and located in the nucleus. Cell 1992; 71:515-526.
2. Brunkow ME, Tilghman SM. Ectopic expression of the H19 gene in mice causes prenatal lethality. Genes Dev 1991; 5:1092-1101.
3. Wickens M, Takayama K. RNA. Deviants—or emissaries [news]. Nature 1994; 367:17-18.
4. Hollander MC, Alamo I, Fornace AJ Jr. A novel DNA damage-inducible transcript, gadd7, inhibits cell growth, but lacks a protein product. Nucleic Acids Res 1996; 24:1589-1593.
5. Storz G. An expanding universe of noncoding RNAs. Science 2002; 296:1260-1263.
6. Erdmann VA, Barciszewska MZ, Szymanski M et al The noncoding RNAs as riboregulators. Nucleic Acids Res 2001; 29:189-193.
7. Crawford DR. Regulation of gene expression by reactive oxygen species. In: Gilbert DL, Colton CA, eds. Reactive oxygen species in biological systems: an interdisciplinary approach. New York: Plenum publishing, 1999:155-171.
8. Crawford DR, Suzuki T, Davies KJA. Oxidant-modulated gene expression. In: Sen CK, Sies H, Baeuerle, PA, eds. Antioxidant and redox regulation of genes. San Diego: Academic Press, 2000:21-45.
9. Crawford DR, Davies KJA. Adaptive response and Oxidative stress. Environ Health Perspect 1994; 102:25-28.
10. Wiese AG, Pacifici RE, Davies KJA. Transient Adaptation to Oxidative stress in Mammalian Cells. Arch Biochem Biophys 1995; 18:231-240.
11. Vile GF, Basu-Modak S, Waltner C et al. Heme oxygenase 1 mediates an adaptive response to oxidative stress in human skin fibroblasts. Proc Natl Acad Sci USA 1994; 91:2607-2610.
12. Riabowol KT, Mizzen LA, Welch WJ. Heat shock is lethal to fibroblasts microinjected with antibodies against hsp70. Science 1988; 242:433-436.
13. Gomer CJ, Ferrario A, Rucker N et al. Glucose regulated protein induction and cellular resistance to oxidative stress mediated by porphyrin photosensitization. Cancer Res 1991; 51:6574-6579
14. Crawford DR, Schools GP, Salmon SL et al. Hydrogen peroxide induces the expression of *adapt15*, a novel RNA associated with polysomes in hamster HA-1 cells. Arch Biochem Biophys 1996; 325:256-264.
15. Crawford DR, Schools GP, Davies KJA. Oxidant-inducible adapt15 RNA is associated with growth arrest- and DNA damage-inducible gadd153 and gadd45. Arch Biochem Biophys 1996; 329:137-144.
16. Mizenina O, Yanushevich Y, Musatkina E et al. C-terminal end of v-src protein interacts with peptide coded by gadd7/adapt15-like RNA in two-hybrid system. FEBS Lett 1998; 422:79-84.
17. Wang Y, Crawford DR, Davies KJA. *adapt33*, a novel oxidant-inducible RNA from hamster HA-1 cells. Arch Biochem Biophys 1996; 332:255-260.
18. Ruvkun G. Molecular biology—Glimpses of a tiny RNA world. Science 2001; 294:797-799.
19. Wassarman KM, Zhang AX, Storz G. Small RNAs in *Escherichia coli*. Trends Microbiol 2001; 7:37-45.
20. Altuvia S, Wagner EGH. Switching on and off with RNA. Proc Natl Acad Sci USA 2000; 97:9824-9826.
21. Erdmann VA, Szymanski M., Hochberg A et al. Collection of mRNA-like noncoding RNAs. Nucleic Acids Res 1999; 27:192-195.
22. Rastinejad F, Conboy MJ, Rando TA. Tumor Suppression by RNA from the 3' Untranslated Region of Alpha-Tropomyosin. Cell 1993; 75:1107-1117.
23. Jupe ER, Liu XT, Kiehlbauch JL et al. The 3' untranslated region of prohibitin and cellular immortalization. Exper Cell Res 1996; 224:128-135.
24. Ishiwatari H, Nakanishi K, Kondoh G et al. Suppression of tumor growth by the 3' untranslated region of mel-18 in 3Y1 cells transformed by the E6 and E7 genes of human papillomavirus type 18. Cancer Lett 1997; 117:57-65.
25. Crawford DR, Kochheiser JC, Schools GP et al. Differential display: A critical analysis. Gene Express 2002; 10:101-107.

Chapter 16

RNA Pathogenesis in Dominant Noncoding Microsatellite Expansion Disorders

Laura P.W. Ranum and John W. Day

Abstract

Microsatellite expansions cause 18 inherited human neurodegenerative diseases, including Huntington's disease (HD), Friedreich's ataxia (FA), nine forms of spinocerebellar ataxia (SCA), and myotonic dystrophy (DM). Most of these disorders are caused by trinucleotide expansions, although recently pathogenic tetranucleotide (DM2) and pentanucleotide (SCA10) expansions have been described.

Disease-related repeat expansions appear to be divided into the following three mechanistic categories: a) recessive or X-linked diseases caused by triplet expansions that interfere with transcription of the mutated gene resulting in a loss of the protein product; b) dominantly inherited diseases caused by expansions located in protein coding regions resulting in altered proteins; c) a third pathogenic mechanism involves dominantly inherited diseases caused by noncoding expansions such as occur in DM1 and SCA8. The discovery that DM2 is caused by a CCTG expansion within an intron, along with other discoveries on DM1 pathogenesis, have led to the understanding that the mutations in DM1 and DM2 are pathogenic at the RNA level. The genetic causes of myotonic dystrophy and the identification of other microsatellite disorders caused by noncoding microsatellite expansions indicate the existence of a new category of disease wherein repeat expansions in RNA alter cellular function resulting in specific disease phenotypes. The role of RNA pathogenesis in myotonic dystrophy type 1 and type 2 (DM1 and DM2) will be discussed along with the potential role for similar dominant RNA mechanisms in spinocerebellar ataxia type 8, 10 and 12 (SCA8, SCA10 and SCA12) and Huntington's disease like 2 (HDL2).

Introduction

Microsatellite repeats involving 2, 3, 4 or more nucleotides in a set pattern that are repeated several times in succession are common in normal human genome. During the past decade 18 inherited neurodegenerative disorders have been found to be caused by microsatellite repeats that abnormally expand and thus become pathogenic.[1-3]

The vast majority of pathogenic expansions found to date have been trinucleotide repeats. Triplet repeats cause several different classes of dominantly inherited disease, including eight types of spinocerebellar ataxia (SCA),[2] two forms of Huntington's disease,[4,5] and myotonic dystrophy type 1.[6,7] Trinucleotide expansions also underlie the recessively inherited Friedreich's ataxia (FA)[8] and fragile-X mental retardation (FMR).[9] More recently, a pentanucleotide

---

Noncoding RNAs: Molecular Biology and Molecular Medicine, edited by Jan Barciszewski and Volker Erdmann.

expansion was identified as the cause of spinocerebellar ataxia type 10 (SCA10),[10] and a tetranucleotide expansion as the cause of myotonic dystrophy type 2 (DM2).[11]

The recessive inheritance of FA is associated with decreased transcription of the *frataxin* gene.[12] Similarly, transcription of the *FMR-1* gene is diminished in affected boys due to the CGG trinucleotide expansion associated with Fragile X syndrome.[13] If an individual carries one of these expansions and also a normal allele (as is the case in both FA parents, and mothers of FMR boys), the function of the normal gene is sufficient to preserve cell function and the carriers are either unaffected or minimally affected. In individuals with two large FA expansions, or in boys with a large FMR expansion on their only copy of the X chromosome, the triplet repeat results in loss of gene product, which leads to the clinical syndrome.

The dominantly inherited diseases caused by microsatellite expansions in coding regions of genes cause pathogenic effects due to changes in the protein encoded by the mutated gene. Many of these diseases are caused by expansion of CAG trinucleotide repeats, resulting in expanded polyglutamine tracts within the mutant proteins.[2] Pathogenic models include: 1) the accumulation of misfolded proteins that overload proteasome degradation pathways and result in a buildup of many types of proteins that are normally degraded;[14,15] 2) misfolded stretches of expanded polyglutamines can sequester other proteins involved in transcriptional regulation and thus may have global effects on the expression of many downstream genes.[16-19] It remains to be seen whether the many polyglutamine diseases cause their pathogenic effects through a common pathway, or whether different diseases involve different molecular mechanisms.

The molecular mechanisms responsible for the dominantly inherited disorders caused by noncoding microsatellite expansions, which are the focus of this review, have been a source of confusion and controversy since 1992, when the first such mutation was identified for DM1.[6,7] Several mechanisms were proposed to explain the effects of the DM1 microsatellite expansion, which is located in the 3' untranslated portion of the *myotonica-protein kinase (DMPK)* gene. The discovery of similar noncoding mutations in SCA8 (1999) and SCA10 (2000) provided additional examples of noncoding mutations, but given the clinical and molecular distinctions between these diseases and DM1 these discoveries did not allow a direct comparison and evaluation of potential pathogenic mechanisms. Recent descriptions of deleterious effects of CUG expansions in RNA, the development of a transgenic model of DM1, and most recently the determination that DM2 is caused by a CCTG expansion in intron 1 of the zinc finger protein 9 gene, have led to the understanding that CUG and CCUG expansions in RNA cause the multisystemic clinical features common to DM1 and DM2. The growing body of evidence that microsatellite repeat expansions expressed at the RNA level play an important role in disease pathogenesis will be discussed.

Myotonic Dystrophy Type 1: Gene Discovery and Proposed Pathogenic Models

Myotonic dystrophy (DM), which was first described nearly 100 years ago, is a dominantly-inherited, multisystemic disease[20,21] with a consistent group of rare clinical features including myotonia, muscular dystrophy, cardiac conduction defects, posterior subcapsular iridescent cataracts, and a peculiar and specific set of endocrine and serologic changes.[22] In 1992, several groups reported that the chromosome-19 form of DM is caused by a CTG expansion, located in the 3' untranslated region (UTR) of the *dystrophia myotonica-protein kinase (DMPK)* gene.[23-27] Because most dominant disorders are caused by the expression of an abnormal protein product with an altered function, it has not been clear how the multisystemic clinical features of dominantly inherited DM1 could be caused by a trinucleotide repeat that did not affect the protein coding portion of a gene.[6]

Initially it was thought that altered kinase activity may cause the multisystemic features of DM, and that the CTG expansion in the 3' untranslated region might decrease or otherwise alter *DMPK* expression from the mutant allele. Since individuals affected by DM1 carry a second completely normal *DMPK* allele, a single pathogenic DM1 expansion might result in haploinsufficiency, reducing overall gene expression by about half. Most of the early expression studies were consistent with this hypothesis indicating that *DMPK* mRNA and protein levels were reduced in muscle and cell cultures derived from DM1 subjects.[23,28,29] However, *DMPK* knockout mice, developed to test whether loss of the kinase was actually pathogenic, did not develop the typical multisystemic features of DM1. Initial descriptions of these mice showed only a very mild, late-onset myopathy that is not typical of DM1.[30,31] More recently, both hemizygous and homozygous *DMPK* knockout mice have been reported to have cardiac conduction abnormalities.[32,33] Taken together, these results suggested that although *DMPK* may contribute to the cardiac features of DM1, haploinsufficiency of *DMPK* does not cause the multisystemic clinical features of DM1. A further indication that the multisystemic features of DM1 are not simply caused by *DMPK* haploinsufficiency comes from the fact that there are no reported cases of DM caused by point mutations in the *DMPK* gene.

A second mechanism proposed for DM1 pathogenesis is that the expanded repeat affects the expression of multiple genes in the vicinity of the expansion. Support for this hypothesis comes from the observation that the CTG expansion is a strong nucleosome binding site and that subsequent alterations in chromatin structure could have regional effects on the expression of multiple genes.[34,35] Another possible cause for the involvement of multiple genes stems from the discovery that in addition to being at the 3' end of *DMPK* the CTG expansion is also located in the 5' promoter region of the immediately downstream homeodomain gene *SIX5*. Since *SIX5* is involved in eye and distal limb muscle development, and because cataracts and distal muscle wasting are common in DM1, haploinsufficiency of *SIX5* was suggested as a possible contributor to DM1 pathogenesis.[36-38] In addition to *DMPK* and *SIX5*, other neighboring genes that have been suggested to be involved in aspects of DM1 pathogenesis. *DMWD*, located immediately upstream of *DMPK*, is expressed in the testis and proposed to play a role in the male infertility characteristic in DM.[39] *FCGRT*, an IgG receptor gene located 4 Mb from the CTG expansion has been proposed to underlie the low IgG levels in DM.[40] In this model, the multisystemic features of DM1 would be explained by haploinsufficiency of a number of neighboring genes, with expression levels and hence disease severity, dependent on repeat length. The most compelling support for neighboring genes comes from the fact that *Six5* knockout mice develop cataracts.[41,42] Although cataracts are a prominent feature of DM, the cataracts in the *Six5* knockouts do not have the distinctive iridescent opacities or the posterior location that are typical of the cataracts in the human disease.

A third hypothesis of DM1 pathogenesis is that enlarged CUG containing transcripts, which accumulate as nuclear foci in cultured cells and biopsied tissue from affected patients,[43,44] exert a dominant effect that disrupts splicing and possibly other cellular functions.[45-49] Initial attempts to develop animal models to study the effects of the expression of elongated CUG expansions in transgenic mice were hampered by infertility of the mice, which prevented successful breeding.[50] In 2000 Mankodi et al, avoided this infertility problem by developing a mouse model in which the expression of a pathogenic CTG expansion (~250 repeats) was restricted to skeletal muscle by inserting the expansion into the 3' end of the human skeletal actin gene, a gene not thought to be involved in DM1 but whose expression is limited to skeletal muscle.[51] This mouse model, which expressed a CUG expansion as part of the 3' UTR of the actin mRNA, was the first to develop the myotonia and the myopathic features characteristic of DM1.[51] Because the expression of the CUG containing transcripts was limited to skeletal muscle the role of the CUG expansion in the multisystemic features of DM was not addressed.

An additive model was then proposed in which each of the above mechanisms contributes to DM1 pathogenesis,[6,52-54] with some aspects of the disease caused by haploinsufficiency of *DMPK, SIX5* and other neighboring genes, and other clinical features resulting from effects of the CUG expansion in RNA. An inconsistency with the additive model of DM1 pathogenesis was that the genetic locus for the second form of myotonic dystrophy (DM2) maps to an entirely different chromosome. Because there is no synteny between the DM1 region on chromosome 19 and the DM2 region on chromosome 3, it would be unlikely that these two chromosomal regions would contain similar sets of genes.[55]

Spinocerebellar Ataxia Type 8

In 1999 we demonstrated that an untranslated CTG expansion causes a novel, dominantly inherited form of spinocerebellar ataxia (SCA8). In contrast to other SCA mutations, the repeat tract is transcribed in the CTG orientation, as in DM1, but not the CAG orientation, as demonstrated for all of the other SCAs that had been previously described. SCA8 has the dominant inheritance pattern and clinical features that are typical of the spinocerebellar ataxias, but it is caused by an untranslated CTG expansion, which had only previously been seen in DM1. As opposed to *DMPK,* which is quite widely expressed, SCA8 expression is limited predominantly to the central nervous system, which may underlie the clinically different diseases caused by these two similar mutations. In contrast to other dominant ataxias, repeat sizes found in affected individuals have a much broader size range and at all repeat lengths the mutation shows a high degree of reduced penetrance.[56-61]

The SCA8 CTG repeat is preceded by a polymorphic but stable CTA tract, with the configuration $(CTA)_{1\text{-}21}(CTG)_n$.[56,58] The CTG portion of the repeat is elongated on pathogenic alleles, and nearly always changes in size when transmitted from generation to generation. In contrast to other triplet repeat diseases, expanded alleles found in affected SCA8 individuals can have either a pure uninterrupted CTG repeat tracts or an allele with one or more CCG, CTA, CTC, CCA or CTT interruptions.[61] Surprisingly, we found six different sequence configurations of the CTG repeat on expanded alleles in a seven-generation family. In two instances duplication of CCG interruptions occurred over a single generation and in other instances duplications that had occurred in different branches of the family could be inferred.[61] SCA8 instability in sperm samples from individuals with expansions ranging in size from 80 to 800 repeats in blood also show a surprising trend in that all, or nearly all, of the expanded alleles undergo contractions *en masse* to repeat sizes $\leq$110 CTG repeats.[61] These *enmasse* repeat contractions in sperm may underlie the reduced penetrance associated with paternal transmissions that is seen in some families.[56,57,61]

We further showed that the 5' end of the SCA8 gene overlaps the 5' end of the actin-binding gene, *Kelch-like 1 (KLHL1),* transcribed in the opposite direction.[56,62] Although no functional relationship between the two transcripts has been demonstrated, the genomic organization of *SCA8* and *KLHL1* suggests the possibility that the normal function of the SCA8 transcripts may be as an antisense regulator of *KLHL1*.[56,62] Although a shorter version of the transcript has been found in mouse, the 3' end of the human SCA8 transcripts containing the CTG repeat is not conserved.[63] Although potential pathogenic mechanisms of SCA8 include dysregulation of *KLHL1*, pathogenic effects of the CUG containing transcripts may be responsible for both DM1 and SCA8. Because SCA8 and DM1 have no clinical similarities, however, identification of SCA8 did not help resolve the controversy about the pathogenic effects of the DM1 mutation.

Spinocerebellar Ataxia Type 10

In 2000, Matsuura et al reported that another dominant spinocerebellar ataxia (SCA10), characterized by ataxia and seizures,[64] is also caused by a noncoding microsatellite expansion. For SCA10 the pathogenic repeat motif involves an ATTCT pentanucleotide repeat expansion. This expansion is located in an intron of a novel gene and can contain up to 4,500 repeats.[10] The *SCA10* gene encodes a novel protein (475 amino acids) with no recognizable motifs or predicted structures.[10] Although it is possible that this enormous expansion may cause haploinsufficiency of *SCA10* and possibly neighboring genes, the dominant inheritance pattern and the fact the expression of the *SCA10* transcript does not appear to be reduced in affected individuals[10] suggests this potential mechanism is unlikely. Alternatively it has been proposed that SCA10 may involve a gain of function RNA mechanism.[10] Although SCA10 involves an untranslated microsatellite expansion, the repeat motif is distinct from DM1 and SCA8, and so the identification of the SCA10 mutation did little to resolve the controversy of DM1 pathogenesis or to suggest that a common pathogenic mechanism may be involved in these three diseases.

Clinical and Molecular Distinctions between DM1 SCA8 and SCA10

By 2001, three dominantly inherited disorders had been identified as being caused by untranslated microsatellite expansions. DM1 was known to be caused by a CTG expansion in the 3' UTR of a broadly expressed gene, SCA8 had been shown to be caused by a CTG expansion in the 3' UTR of a gene expressed at very low steady state levels almost exclusively in the central nervous system,[65,56] and SCA10 had been shown to be caused by a pentanucleotide (ATTCT) repeat expansion within an intron of a broadly expressed gene.[66] Because of the clinical and molecular differences between these diseases, there were insufficient common elements to suggest a common disease mechanism. The prevailing opinion was that noncoding microsatellite expansions somehow caused disease by affecting the expression of genes in their vicinity, possibly in combination with some direct deleterious effects of the mutations as expressed in RNA.

As an independent approach toward determining the pathogenic cause of myotonic dystrophy, we studied families with a second genetic form of the disease. After genetic testing became available for DM1, we identified several families that had previously been diagnosed as having DM, but did not have a DM1 CTG expansion. Furthermore these families were not linked to the chromosome 19 DM1 locus. We hypothesized that identification of a second human mutation that causes the multisystemic effects of DM and a determination of what is common to these diseases at the molecular level provided an independent means of evaluating the pathogenic pathway(s) of DM. If DM2 were caused by a gene involved in the *DMPK* or *SIX5* pathways, it would substantiate the role of either of those genes in DM pathogenesis. Alternatively, if DM2 was caused by an untranslated microsatellite expansion, it would substantiate the role of RNA pathogenesis in DM.

Myotonic Dystrophy Type 2

Although myotonic dystrophy was first described in the early 1900s,[20] the existence of a second genetic locus was only recognized after genetic testing became available for DM1.[67,68] In 1998 we mapped the DM2 locus to chromosome 3q21[55] and Ricker subsequently reported that disease locus in many families with proximal myotonic myopathy (PROMM) also mapped to the same region 3q21.[69] Individuals affected by DM2 have a complex clinical presentation that is strikingly similar to DM1 and includes: i) myotonia; ii) a characteristic pattern of muscle weakness and histological changes; iii) posterior iridescent cataracts; iv) cardiac arrhythmias or progressive cardiomyopathy; v) hypotestosteronism and oligospermia in males; vi) insulin insensitivity and vii) a specific set of serological changes including low gamma-globulin, elevated creatine kinase and elevated follicle stimulating hormone in males.[70-72]

We recently reported that DM2 is caused by the expansion of a CCTG repeat located in intron 1 of the *zinc finger protein 9 (ZNF9)* gene.[11] The DM2 CCTG repeat tract is part of a complex repeat motif with the overall configuration $(TG)_n(TCTG)_n(CCTG)_n$.[11] The longest normal allele observed had 26 CCTG repeats, whereas the expanded range is extremely broad (~75-11,000 CCTGs, mean ~5,000) with a high degree of somatic mosaicism. The smallest pathogenic size is not yet clear because uncommon shorter expansions were found in individuals with multiple allele sizes in blood.[11] Similar to what has been reported for SCA1[73] and FMR[74] all eight of the normal alleles examined had interruptions within the CCTG repeat tract that were not present on expanded alleles.[11] Loss of sequence interruptions on normal SCA1 and FMR alleles is thought to predispose alleles to further expansion.[73,74] Dramatic somatic instability of the DM2 expansion is illustrated by a pair of identical twins with repeat sizes that differ by ~3000 repeats (at 31 yrs). Further evidence of somatic instability is indicated by the multiple expansion sizes often found in blood.[11] The DM2 expansion can increase with age (500 repeats over 3 yrs in one individual), which may explain the unexpected tendency for offspring to have smaller repeat tracts than their affected parents.[11] Although we did not observe a correlation between age of onset in DM2 and expansion size, the somatic instability of the repeat complicates this and other clinical correlations between repeat length and disease.[11]

Similar to DM1, the DM2 expansion is located in a transcribed but untranslated portion of a gene but this time in intron 1 of the *zinc finger protein 9 (ZNF9)* gene.[11] The normal function of *ZNF9* as a nucleic acid binding protein[75,76] appears unrelated to any of the proteins encoded in the DM1 region of chromosome 19. Similarly, genes in the DM2 region (*KIAA1160, Rab 11B, glycoprotein IX, FLJ11631*, and *FLJ12057*) bear no obvious relationship to the genes at the DM1 locus.[11] Even if the DM2 expansion alters the regulation of *ZNF9* and other genes in the DM2 region, it would be unlikely that alterations in the regulation of different sets of proteins at the DM1 and DM2 loci would result in diseases with such strikingly similar multisystemic clinical features.

Fluorescent in-situ hybridization has been used to detect nuclear RNA foci containing the CUG expansion in DM1 cells.[43] Evidence for a common pathogenic mechanism at the RNA level comes from experiments showing that similar CCUG-containing RNA foci are found in DM2 muscle.[11] These results demonstrate that the CCTG expansion is expressed at the RNA level, but additional experiments are needed to determine if the RNA foci contain the entire unprocessed *ZNF9* transcript or if the transcript is normally processed but the intron or the repeat tract alone resists degradation forming the RNA foci.

RNA Pathogenesis in the Myotonic Dystrophies

As mentioned above, substantial evidence has accumulated supporting the model that a gain of function alteration at the RNA level plays a role in DM1 pathogenesis.[6] Evidence supporting this hypothesis includes: i) the expression of mutant *DMPK* mRNA containing an expanded repeat in the 3' UTR inhibits differentiation of cultured myoblasts[77] and ii) transgenic models in which >250 CTG repeats expressed at RNA level causes myotonia and muscular dystrophy.[51,78]

A generalized mechanism of RNA pathogenesis has been proposed in which CUG repeat-containing *DMPK* transcripts that accumulate as nuclear RNA foci alter the regulation of CUG binding proteins, including CUG-BP[45] and three different forms of muscleblind,[48,79] altering their normal functions.[6] There is direct evidence that CUG repeat induced alterations in CUG-BP result in the altered RNA splicing of several genes, including cardiac troponin T

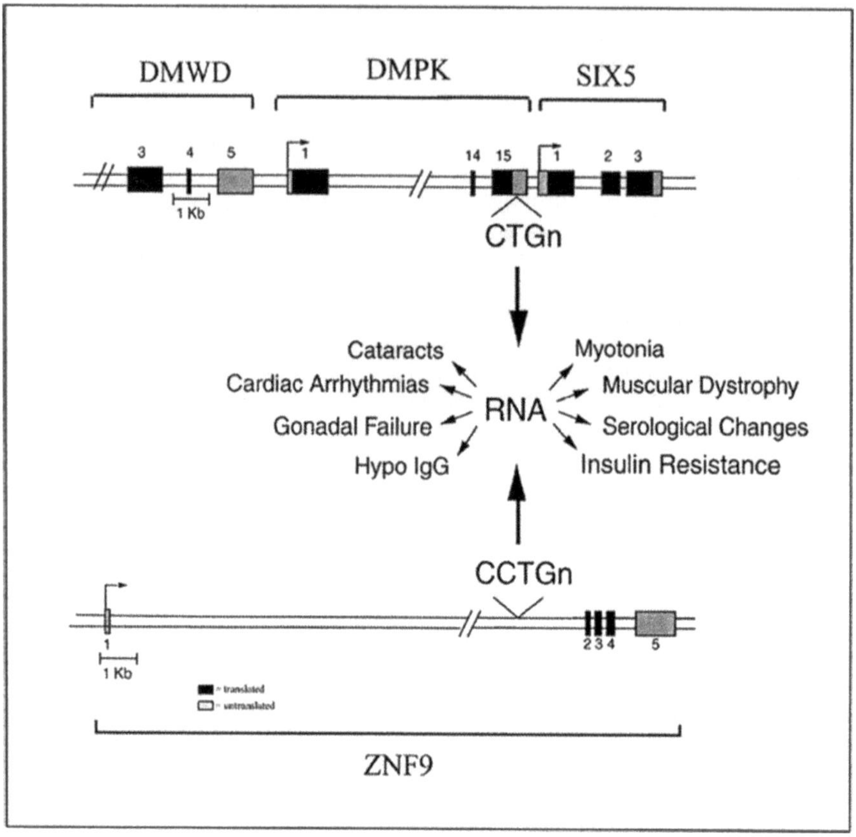

Figure 1. RNA model of DM pathogenesis. Schematic representations of genes in the DM1 region on chromosome 19, and of *ZNF9* gene at the DM2 locus on chromosome 3 are shown. The relative spacing of exons is depicted, with translated regions colored black and untranslated regions colored gray. The location of the causative repeat expansions is shown, in the 3' untranslated region of *DMPK* for DM1, and within intron 1 of *ZNF9* for DM2. The two diseases share the clinical features listed, and both diseases have demonstrable nuclear accumulation of untranslated RNA repeat expansions. These clinical and molecular parallels between DM1 and DM2 simplify previous models of DM pathogenesis and demonstrate that the multisystemic features shared by DM1 and DM2 are caused by an RNA mechanism mediated by transcripts containing CUG and CCUG repeat expansions. Figure reprinted with permission of *Current Opinion in Genetics and Development.*

(cTNT),[46] the insulin receptor (IR)[49] and the skeletal muscle chloride channel (CLCN1)[80] which may contribute to various features of DM, including cardiac involvement, insulin insensitivity and myotonia, respectively. Further evidence of downstream molecular parallels resulting from a common RNA mechanism comes from the fact that alterations in the splicing of the IR leading to a predominance of the insulin-resistant splice form is seen in both DM1 and DM2.[49,81] This splicing alteration is thought to predispose both DM1 and DM2 patients to diabetes. Additional evidence that RNA foci containing the DM1 and DM2 repeat motifs behave in a similar manner is that several forms of the RNA binding protein muscleblind (MBNL, MBLL, MBXL) colocalize to the repeat containing foci in both diseases.[82,83]

Although the additive model of DM1 suggested that CUG repeats in RNA cause the myotonia and muscular dystrophy of DM1, the causes of other DM features including cardiac conduction defects and cataracts were ascribed to haploinsufficiency of genes in the DM1 region. The clinical and molecular parallels between DM1 and DM2 suggest a simpler model of DM pathogenesis (Fig. 1) in which the clinical features common to both diseases including myotonia, muscular dystrophy, cataracts, cardiac arrythmias, insulin insensitivity and diabetes, hypogammaglobulinemia, and testicular failure are caused by the pathogenic effects of RNA containing the CUG and CCUG expansions.[11,71]

The clinical similarities between DM1 and DM2 have helped to clarify the extensive role that RNA containing a CUG or CCUG expansion plays in DM pathogenesis. On the other hand, although DM1 and DM2 phenotypes are strikingly similar, they are not identical. DM2 does not show a congenital form or the severe central nervous system involvement seen in DM1.[11,72] Defining the molecular differences between DM1 and DM2, which could involve temporal and spatial expression patterns of the DM1 and DM2 mutations, or altered regulation of locus specific genes such as *DMPK*, *SIX5*, or *ZNF9*, or different downstream effects of the CUG or CCUG expansions, will be important for understanding the clinical distinctions between these diseases.[11] Fillipova et al, have recently suggested that a possible mechanism for the congenital form of DM1 may involve methylation at the DM1 locus which could disrupt normal CTCF insulator binding sites and lead to increased expression of *DMPK* and hence higher levels of CUG containing transcripts.[54]

Other Noncoding Expansion Disorders

The pathogenic mechanism of the DM1 and DM2 expansions are evident only because the two diseases share such a distinctive phenotype. Although the striking clinical and molecular parallels between DM1 and DM2 demonstrate that RNA pathogenesis plays a much broader role in the multisystemic features of myotonic dystrophy than was previously suspected, noncoding microsatellite expansions for SCA8[56] and SCA10,[10] were identified before the DM2 expansion. Molecular parallels between the SCA8 CTG repeat and the DM1 CTG and DM2 CCTG repeats, and the known toxic properties of transcripts with elongated CUGs/CCUGs[51,77,84] suggest that a similar gain of function RNA mechanism may underlie the cerebellar degeneration in SCA8 at the cellular level. Consistent with the multisystemic features of DM1 and DM2, the *DMPK* and *ZNF9* transcripts are broadly expressed.[11,30] Likewise, SCA8 transcripts are almost exclusively expressed in the brain, which is consistent with its central nervous system involvement.[56,65] Although in SCA10 it is possible that the enormous ATTCT expansion in an intron of a gene may cause disease through haploinsufficiency, the dominant inheritance pattern and the fact the expression of the SCA10 transcript does not appear to be reduced in affected individuals[10] make a gain of function RNA mechanism the more attractive hypothesis. Unlike SCA8, the gene that harbors the mutation for SCA10 is ubiquitously expressed, indicating that if a toxic RNA mechanism is involved, secondary proteins that interact with the ATTCT repeat motif may confer organ specific pathogenicity.[10]

Two other dominantly inherited diseases that involve noncoding repeat expansions are SCA12 and HD-like2 (HDL2). SCA12 is caused by a CAG repeat expansion (55-78 triplets) located in the 5' region of the *PPP2R2B* gene encoding a brain-specific regulatory subunit of the protein phosphatase PP2A,[85] SCA12 is a degenerative disease characterized by ataxia, tremor and dementia.[86] Because the repeat expansion may be located within the transcript (5' UTR), possible pathogenic mechanisms include a toxic effect at the RNA level or altered expression of the *PPP2R2B* gene.[87] HDL2 is a dominantly inherited disease similar to HD that causes a movement disorder, dementia and psychiatric abnormalities.[88] HDL2 is caused by a moderately sized CTG expansion (51-57 repeats) in the *junctophilin-3* gene.[5] Several different alternatively spliced forms of the transcript have been found. In two splice forms the CTG expansion is predicted to encode an elongated polyleucine or polyalanine tract, but in the other two forms the CTG repeat would be transcribed but untranslated either as part of intron 1 or as part of the 3' UTR suggesting the possibility of a dominant RNA mechanism.

Conclusions

Clinical and molecular parallels between DM1 and DM2 simplify previous models of DM1 pathogenesis and establish that CUG and CCUG expansions in RNA are responsible for the multisystemic features common to both diseases. One of downstream effects of the CUG and CCUG containing transcripts is that they mediate changes in splicing of other genes. Although to date the splicing regulation of several different genes (cTNT, IR, CLCN1) has been shown to be altered in DM1 and DM2, there are potentially a huge number of additional genes whose expression may be altered at different times during development and in different tissues by microsatellite expansions that are expressed at the RNA level. It will be important to determine if a dominant RNA mechanism proves to be a general pathogenic pathway that in addition to causing the multisystemic clinical features of DM1 and DM2 is also involved in SCA8 and SCA10 and possibly SCA12 and HDL2.

References

1. Brice A. Unstable mutations and neurodegenerative disorders. J Neurol 1998; 245:505-510.
2. Zoghbi HY, Orr HT. Glutamine repeats and neurodegeneration. Annu Rev Neurosci 2000; 23:217-247.
3. Margolis RL, Ross CA. Expansion explosion: new clues to the pathogenesis of repeat expansion neurodegenerative diseases. Trends Mol Med 2001; 7:479-482.
4. The Huntington's disease collaborative research group. A novel gene containing a trinucleotide repeat that is expanded and unstable on Huntington's disease chromosomes. Cell 1993; 72:971-983.
5. Holmes SE, O'Hearn E, Rosenblatt A et al. A repeat expansion in the gene encoding *junctophilin-3* is associated with Huntington disease-like 2. Nat Genet 2001; 29:377-378.
6. Tapscott SJ. Deconstructing myotonic dystrophy. Science 2000; 289:1701-1702.
7. Tapscott SJ, Thornton CA. Reconstructing myotonic dystrophy. Science 2001; 293:816-817.
8. Campuzano V, Montermini L, Molto MD et al. Friedreich's ataxia: autosomal recessive disease caused by an intronic GAA triplet repeat expansion. Science 1996; 271:1423-1426.
9. Fu Y-H, Kuhl DPA, Pizzuti A et al. Variation of the CGG repeat at the fragile X site results in genetic instability: resolution of the Sherman paradox. Cell 1991; 67:1047-1058.
10. Matsuura T, Yamagata T, Burgess DL et al. Large expansion of the ATTCT pentanucleotide repeat in spinocerebellar ataxia type 10. Nat Genet 2000; 26:191-194.
11. Liquori C, Ricker K, Moseley ML et al. Myotonic dystrophy type 2 caused by a CCTG expansion in intron 1 of *ZNF9*. Science 2001; 293:864-867.
12. Sakamoto N, Ohshima K, Montermini L et al. Sticky DNA, a self-associated complex formed at long GAA*TTC repeats in intron 1 of the frataxin gene, inhibits transcription. J Biol Chem 2001; 276:27171-27177.
13. Jin P, Warren ST. Understanding the molecular basis of fragile X syndrome. Hum Mol Genet 2000; 9:901-908.
14. Jana NR, Zemskov EA, Wang G et al. Altered proteasomal function due to the expression of polyglutamine-expanded truncated N-terminal huntingtin induces apoptosis by caspase activation through mitochondrial cytochrome c release. Hum Mol Genet 2001; 10:1049-1059.
15. Bence NF, Sampat RM, Kopito RR. Impairment of the ubiquitin-proteasome system by protein aggregation. Science 2001; 292:1552-1555.
16. Nucifora Jr FC, Sasaki M, Peters MF et al. Interference by huntingtin and atrophin-1 with cbp-mediated transcription leading to cellular toxicity. Science 2001; 291:2423-2428.
17. Shimohata T, Nakajima T, Yamada M et al. Expanded polyglutamine stretches interact with TAFII130, interfering with CREB-dependent transcription. Nat Genet 2000; 26:29-36.

18. Dunah A, Jeong H, Griffin A et al. Sp1 and TAFII130 transciptional activity disrupted in early Huntington's disease. Science 2002; 296:2238-2243.
19. Okazawa H, Rich T, Chang A et al. Interaction between mutant ataxin-1 and POBP-1 affects transcription and cell death. Neuron 2002; 34:701-713.
20. Steinert H. Myopathologische Beitrage 1. Uber das klinischeund anatomische Bild des muskelschwunds der myotoniker. Dtsch Z Nervenheilkd 1909; 37:58-104.
21. Batten F, Gibb H. Myotonia atrophica. Brain 1909; 32:187-205.
22. In Harper PS, ed. Myotonic dystrophy. 2 ed. London: WB Saunders, 1989.
23. Fu Y-H, Pizzuti A, Fenwick RGJ et al. An unstable triplet repeat in a gene related to myotonic muscular dystrophy. Science 1992; 255:1256-1258.
24. Buxton J, Shelbourne P, Davies J et al. Detection of an unstable fragment of DNA specific to individuals with myotonic dystrophy. Nature 1992; 355:547-548.
25. Harley HG, Brook JD, Rundle SA et al. Expansion of an unstable DNA region and phenotypic variation in myotonic dystrophy. Nature 1992; 355:545-546.
26. Brook JD, McCurrah ME, Harley HG et al. Molecular basis of myotonic dystrophy: Expansion of a trinucleotide (CTG) repeat at the 3' end of a transcript encoding a protein kinase family member. Cell 1992; 68:799-808.
27. Mahadevan M, Tsilfidis C, Sabourin L et al. Myotonic dystrophy mutation: an unstable CTG repeat in the 3' untranslated region of the gene. Science 1992; 255:1253-1255.
28. Hoffmann-Radvanyi H, Lavedan C, Rabes JP et al. Myotonic dystrophy: absence of CTG enlarged transcript in congenital forms, and low expression of the normal allele. Hum Mol Genet 1993; 2:1263-1266.
29. Novelli G, Gennarelli M, Zelano G et al. Failure in detecting mRNA transcripts from the mutated allele in myotonic dystrophy muscle. Biochem Mol Bio Int 1993; 29:291-297.
30. Jansen G, Groenen PJTA, Bachner D et al. Abnormal myotonic dystrophy protein kinase levels produce only mild myopathy in mice. Nat Genet 1996; 13:316-324.
31. Reddy S, Smith DB, Rich MM et al. Mice lacking the myotonic dystrophy protein kinase develop a late onset progressive myopathy. Nat Genet 1996; 13:325-335.
32. Berul CI, Maguire CT, Aronovitz MJ et al. DMPK dosage alterations result in atrioventricular conduction abnormalities in a mouse myotonic dystrophy model. J Clin Invest 1999; 103:R1-R7.
33. Berul CI, Maguire CT, Gehrmann J et al. Progressive atrioventricular conduction block in a mouse myotonic dystrophy model. J Interv Card Electrophysiol 2000; 4:351-358.
34. Otten AD, Tapscott SJ. Triplet repeat expansion in myotonic dystrophy alters the adjacent chromatin structure. Proc Natl Acad Sci U S A 1995; 92:5465-5469.
35. Wang Y-H, Amirhaeri S, Kang S et al. Preferential nucleosome assembly at DNA triplet repeats from the myotonic dystrophy gene. Science 1994; 265:669-671.
36. Boucher CA, King SK, Carey N et al. A novel homeodomain-encoding gene is associated with a large CpG island interrupted by the myotonic dystrophy unstable (CTG)n repeat. Hum Mol Genet 1995; 4:1919-1925.
37. Jansen G, Bachner D, Coerwinkel M et al. Structural organization and developmental expression pattern of the mouse WD-repeat gene DMR-N9 immediately upstream of the myotonic dystrophy locus. Hum Mol Genet 1995; 4:843-852.
38. Shaw DJ, McCurrach M, Rundle SA et al. Genomic organisation and transcriptional units at the myotonic dystrophy locus. Genomics 1993; 18:673-679.
39. Alwazzan M, Newman E, Hamshere MG et al. Myotonic dystrophy is associated with a reduced level of RNA from the DMWD allele adjacent to the expanded repeat. Hum Mol Genet 1999; 8:1491-1497.
40. Junghans RP, Ebralidze A, Tiwari B. Does (CUG)n repeat in DMPK mRNA 'paint' chromosome 19 to suppress distant genes to create the diverse phenotype of myotonic dystrophy?: A new hypothesis of long-range cis autosomal inactivation. Neurogenetics 2001; 3:59-67.
41. Klesert TR, Cho DH, Clark JI et al. Mice deficient in Six5 develop cataracts: implications for myotonic dystrophy. Nat Genet 2000; 25:105-109.
42. Sarkar PS, Appukuttan B, Han J et al. Heterozygous loss of Six5 in mice is sufficient to cause ocular cataracts. Nat Genet 2000; 25:110-114.

43. Taneja KL, McCurrach M, Schalling M et al. Foci of trinucleotide repeat transcripts in nuclei of myotonic dystrophy cells and tissues. J Cell Bio 1995; 128:995-1002.
44. Davis BM, McCurrach ME, Taneja KL et al. Expansion of a CUG trinucleotide repeat in the 3' untranslated region of myotonic dystrophy protein kinase transcripts results in nuclear retention of transcripts. Proc Natl Acad Sci U S A 1997; 94:7388-7393.
45. Timchenko LT, Miller J , Timchenko NA et al. Identification of a (CUG)n triplet repeat RNA-binding protein and its expression in myotonic dystrophy. Nucleic Acids Res 1996; 24:4407-4414.
46. Philips AV, Timchenko LT, Cooper TA. Disruption of splicing regulated by a CUG-binding protein in myotonic dystrophy. Science 1998; 280:737-741.
47. Lu X, Timchenko NA, Timchenko LT. Cardiac elav-type RNA-binding protein (ETR-3) binds to RNA CUG repeats expanded in myotonic dystrophy. Hum Mol Genet 1999; 8:53-60.
48. Miller JW, Urbinati CR, Teng-Umnuay P et al. Recruitment of human muscleblind proteins to (CUG)(n) expansions associated with myotonic dystrophy. Embo J 2000; 19:4439-4448.
49. Savkur RS, Philips AV, Cooper TA. Aberrant regulation of insulin receptor alternative splicing is associated with insulin resistance in myotonic dystrophy. Nat Genet 2001; 29:40-47.
50. Monckton DG, Ashizawa T, Siciliano MJ. Murine models for myotonic dystrophy, in Genetic instabilities and hereditary neurological diseases. In: Wells RD, Warren ST, eds. San Diego: Academic Press, 1998:181-193.
51. Mankodi A, Logigian E, Callahan L et al. Myotonic dystrophy in transgenic mice expressing an expanded CUG repeat. Science 2000; 289:1769-1773.
52. Groenen P, Wieringa B. Expanding complexity in myotonic dystrophy. Bioessays 1998; 20:901-912.
53. Larkin K, Fardaei M. Myotonic dystrophy-a multigene disorder. Brain Res Bull 2001; 56:389-395.
54. Filippova GN, Thienes CP, Penn BH et al. CTCF-binding sites flank CTG/CAG repeats and form a methylation-sensitive insulator at the DM1 locus. Nat Genet 2001; 28:335-343.
55. Ranum L, Rasmussen P, Benzow K et al. Genetic mapping of a second myotonic dystrophy locus. Nat Genet 1998; 19:196-198.
56. Koob MD, Moseley ML, Schut LJ et al. An untranslated CTG expansion causes a novel form of spinocerebellar ataxia (SCA8). Nat Genet 1999; 21:379-384.
57. Day JW, Schut LJ, Moseley ML et al. Spinocerebellar ataxia type 8: clinical features in a large family. Neurology 2000; 55:649-657.
58. Stevanin G, Herman A, Durr A et al. Are (CTG)n expansions at the SCA8 locus rare polymorphisms? Nat Genet 2000; 24:213.
59. Worth PF, Houlden H, Giunti P et al. Large, expanded repeats in SCA8 are not confined to patients with cerebellar ataxia. Nat Genet 2000; 24:214-215.
60. Moseley ML, Schut LJ, Bird TD et al. Reply. Nat Genet 2000; 24:215.
61. Moseley ML, Schut LJ, Bird TD et al. SCA8 CTG repeat: *enmasse* contractions in sperm and intergenerational sequence changes may play a role in reduced penetrance. Hum Mol Genet 2000; 9:2125-2130.
62. Nemes JP, Benzow KA, Moseley ML et al. The SCA8 transcript is an antisense RNA to a brain-specific transcript encoding a novel actin-binding protein (KLHL1). Hum Mol Genet 2000; 9:1543-1551. [Correction/Addition Hum Mol Genet 9:2777].
63. Benzow KA, Koob MD. The KLHL1-antisense transcript (KLHL1AS) is evolutionarily conserved. Mamm Genome 2002; 13:134-141.
64. Rasmussen A, Matsuura T, Ruano L et al. Clinical and genetic analysis of four Mexican families with spinocerebellar ataxia type 10. Ann Neurol 2001; 50:234-239.
65. Janzen MA, Moseley ML, Benzow K A et al. Limited expression of SCA8 is consistent with cerebellar pathogenesis and toxic gain of function RNA model. Am J Hum Genet 1999, 65:A267.
66. Matsuura T, Ashizawa T. Polymerase chain reaction amplification of expanded ATTCT repeat in spinocerebellar ataxia type 10. Ann Neurol 2002; 51:271-272.
67. Ricker K, Koch MC, Lehmann-Horn F et al. Proximal myotonic myopathy: a new dominant disorder with myotonia, muscle weakness, and cataracts. Neurology 1994; 44:1448-1452.
68. Thornton CA, Griggs RC, and Moxley RT. Myotonic dystrophy with no trinucleotide repeat expansion. Ann Neurol 1994; 35:269-272.

69. Ricker K, Grimm T, Koch M C et al. Linkage of proximal myotonic myopathy to chromosome 3q. Neurology 1999; 52:170-171.
70. Ricker K, Koch M, Lehmann-Horn F et al. Proximal myotonic myopathy: clinical features of a multisystem disorder similar to myotonic dystrophy. Arch Neurol 1995; 52:25-31.
71. Day JW, Roelofs R, Leroy B et al. Clinical and genetic characteristics of a five-generation family with a novel form of myotonic dystrophy (DM2). Neuromuscul Disord 1999; 9:19-27.
72. Day J, Liquori C, Johnson C et al. Myotonic dystrophy Type 2 (DM2) in Minnesota. Am J Hum Genet 2000; 67:1850.
73. Chung M-y, Ranum LPW, Duvick LA et al. Evidence for a mechanism predisposing to intergenerational CAG repeat instability in spinocerebellar ataxia type 1. Nat Genet 1993; 5:254-258.
74. Kunst CB, Warren ST. Cryptic and polar variation of the fragile X repeat could result in predisposing normal alleles. Cell 1994; 77:853-861.
75. Pellizzoni L, Lotti F, Maras B et al. Cellular nucleic acid binding protein binds a conserved region of the 5' UTR of Xenopus laevis ribosomal protein mRNAs. J Mol Biol 1997; 267:264-275.
76. Pellizzoni L, Lotti F, Rutjes SA et al. Involvement of the Xenopus laevis Ro60 autoantigen in the alternative interaction of La and CNBP proteins with the 5'UTR of L4 ribosomal protein mRNA. J Mol Biol 1998; 281:593-608.
77. Amack JD, Paguio AP, Mahadevan MS. Cis and trans effects of the myotonic dystrophy (DM) mutation in a cell culture model. Hum Mol Genet 1999; 8:1975-1984.
78. Seznec H, Agbulut O, Sergeant N et al. Mice transgenic for the human myotonic dystrophy region with expanded CTG repeats display muscular and brain abnormalities. Hum Mol Genet 2001; 10:2717-2726.
79. Fardaei M, Larkin K, Brook JD et al. In vivo colocalisation of MBNL protein with DMPK expanded-repeat transcripts. Nucleic Acids Res 2001; 29:2766-2771.
80. Mankodi A, Takahashi M, Beck C et al. Myotonia is associated with loss of transmembrane chloride conductance and aberrant splicing of Clcn1, the skeletal muscle chloride channel, in a transgenic model of myotonic dystrophy (DM1). Am J Hum Gen 2001; 69:A211.
81. Ranum LPW, Liquori C, Moseley ML et al. Myotonic dystrophy type 2 is caused by a CCTG expansion in intron 1 of *ZNF9*. Am J Hum Genet 2001; 69:A211.
82. Mankodi A, Urbinati CR, Yuan QP et al. Muscleblind localizes to nuclear foci of aberrant RNA in myotonic dystrophy types 1 and 2. Hum Mol Genet 2001; 10:2165-2170.
83. Fardaei M, Rogers MT, Thorpe HM et al. Three proteins, MBNL, MBLL and MBXL, colocalize in vivo with nuclear foci of expanded-repeat transcripts in DM1 and DM2 cells. Hum Mol Genet 2002; 11:805-814.
84. Amack JD, Mahadevan MS. The myotonic dystrophy expanded CUG repeat tract is necessary but not sufficient to disrupt C2C12 myoblast differentiation. Hum Mol Genet 2001; 10:1879-1887.
85. Holmes SE, O'Hearn EE, McInnis MG et al. Expansion of a novel CAG trinucleotide repeat in the 5' region of PPP2R2B is associated with SCA12. Nat Genet 1999; 23:391-392.
86. O'Hearn E, Holmes SE, Calvert PC et al. SCA-12: Tremor with cerebellar and cortical atrophy is associated with a CAG repeat expansion. Neurology 2001; 56:299-303.
87. Holmes SE, Hearn EO, Ross CA et al. SCA12: an unusual mutation leads to an unusual spinocerebellar ataxia. Brain Res Bull 2001; 56:397-403.
88. Margolis RL, O'Hearn E, Rosenblatt A et al. A disorder similar to Huntington's disease is associated with a novel CAG repeat expansion. Ann Neurol 2001; 50:373-380.

CHAPTER 17

Noncoding RNAs Encoded by Bacterial Chromosomes

E. Gerhart H. Wagner and Jörg Vogel

Abstract

Small noncoding RNAs are common to bacterial plasmids, phages and transposons, in which they regulate biological processes by acting as antisense RNAs. By contrast, only few small, noncoding RNAs had been shown to be encoded by bacterial chromosomes until very recently. Some of these provided house-keeping functions, while others were regulators of gene expression. Since most riboregulators had been discovered serendipitously, several research groups embarked on systematic searches for chromosomally encoded noncoding RNAs, based on the conviction that such RNAs previously had escaped attention. In the last year more than 30 novel small RNAs were identified, and many more were predicted by bioinformatics. Even though the cumbersome work of assigning biological functions to these new RNAs is still ongoing, it appears likely that many of them are stress response regulators. Thus—in line with similarly interesting discoveries of many more small RNAs in eukaryotic systems and in archea—the role of RNA in regulation of gene expression must be re-evaluated.

Preface

This chapter aims at giving an educated, non-specialist reader an overview over an area of research that has recently been revolutionized. Noncoding RNAs have previously been regarded as a minor addition to the bulk of the known functional RNAs: mRNA, tRNA, rRNA. Here, we review the recent discovery of many new small, noncoding RNAs encoded by bacterial chromosomes, primarily of *Escherichia coli.* We choose to illustrate mechanistic concepts, peculiarities in function, and various biological roles, using examples from the recent literature, rather than providing an encyclopedic list of bacterial small RNAs. Other chapters in this book cover similar classes of RNAs in the other kingdoms of life. Except for comparisons, this review omits an abundant and well-characterized class of RNA-based regulators, the antisense RNAs encoded by plasmids, transposons, and bacteriophages (see refs. 1, 2, for reviews).

The RNAs addressed here will be denoted sRNAs (small RNAs) or ncRNAs (noncoding RNAs), though many other terms have been used, according to taste and style of various authors: snmRNA (small non-messenger RNA), fRNA (functional RNA), regulatory RNA, riboregulator, etc. None of these denotations is without problems, since each one is inappropriate for at least some known RNAs, or subsumes additional RNAs (mRNA, tRNA, etc.).

Noncoding RNAs: Molecular Biology and Molecular Medicine, edited by Jan Barciszewski and Volker A. Erdmann.

Introduction

The year 2001 saw a startling change in the way scientists perceived the role of RNA in biology. Traditionally, the two main roles of RNA have been in genetic information transfer—as mRNA, and as the genetic material of some viruses—and in protein synthesis—as ribosomal RNAs and tRNA adaptors. Catalytic activities of RNAs (ribozymes) added to that, and small, untranslated RNAs were known to be important for post-transcriptional processes such as splicing (snRNAs) and rRNA modification (snoRNAs) in eukaryotes. By contrast, the regulatory potential of RNA appeared to be mainly confined to small RNAs that control biological functions in bacterial accessory elements.[1,2] These RNAs act on target RNAs by base-pairing, and are denoted antisense RNAs. Their versatility, efficiency, and specificity suggested that genes encoded by bacterial chromosomes could be regulated by interaction between complementary RNAs as well. In addition, other sRNA-based functions were plausible, e.g., modulation of the activities of proteins. Yet, a survey of the data available a few years ago listed only ten small ncRNAs (disregarding tRNAs and rRNAs) in the well-characterized enterobacterium *E. coli*.[3] A few of these were involved in housekeeping functions (M1 RNA of RNase P, tmRNA, 4.5S), two were—at the time—of unknown function (6S, Spot 42), and less than a handful were bona fide regulators of gene expression (MicF, OxyS, DsrA, DicF, CsrB). Even fewer sRNAs had been discovered in other bacteria.

Since these RNAs were discovered fortuitously, by genetic screens or by radiolabeling of total RNA and isolation from gels,[4,5] one could expect that a feasible search protocol might identify many more small, noncoding RNAs. In 2001, three groups published the experimental identification of 31 new sRNAs in *E. coli*.[6-8] In all these three genome-wide approaches, intergenic regions (more precisely, inter-ORF regions) were searched for the presence of putative sRNA-encoding genes, by bioinformatics and "wet" experiments. Conservation of noncoding regions between related bacterial species proved to be a strong indicator of sRNA genes. Some, but not all RNAs were identified in more than one lab, yet it is clear that the search is not saturated. In fact, many more sRNA-encoding genes were predicted and await experimental testing,[6,9] and a few additional RNAs have been discovered recently (Hüttenhofer A, Vogel J, Tang TH et al, unpublished).

Thus, the list of sRNAs in *E. coli* has grown, and current work is dedicated to assigning biological roles. While at this point very little is known about these RNAs, expression analyses show that most of them are induced during stationary phase, suggesting functions associated with adaptive responses during stress conditions. Below, we will put emphasis on a discussion of features that have emerged in studies of previously and more recently discovered sRNAs. A number of recent reviews have covered various aspects of noncoding RNAs in bacteria and elsewhere.[2,3,10-17]

Housekeeping sRNAs

The chromosome of *E. coli* encodes three sRNAs that play housekeeping roles.

RNaseP

All tRNAs of bacteria are transcribed as precursors and require processing to generate mature 5'- and 3'-ends. The endoribonuclease responsible for tRNA 5'-end formation is called RNase P, and is indispensable.[18,19] In *E. coli,* the enzyme is composed of two subunits, M1 RNA and C5 protein. All bacteria (and, for that matter, all other organisms) encode homologues of this essential RNA. From an extensive body of in vitro work it is clear that M1 RNA alone can carry out correct cleavage of tRNA precursors, though the C5 protein is required in vivo. Thus, M1 RNA is a classical ribozyme (catalytic RNA).

tmRNA

Occasionally, truncated or otherwise damaged mRNAs stall ribosomes during translation. This creates a serious problem: stalled ribosomes must be recycled in order to maintain overall translation rates. Incompletely synthesized polypeptides add to this problem, and must be degraded. A second housekeeping RNA, the ~363 nt tmRNA (=transfer-messenger mRNA; or SsrA or 10Sa RNA), deals with these problems.[20,21] This highly structured RNA[22,23] has remarkable properties. It carries one structural domain that mimics a tRNA (tRNA domain), and a second domain that encodes a short amino acid sequence (mRNA domain; Fig. 1A). The tRNA-like domain is recognized and aminoacylated by alanine tRNA synthetase, bound by elongation factor Tu, and enters the free A-site of a ribosome stalled on a problematic mRNA. These steps are aided by a protein, SmpB, that binds tmRNA.[24] Through transpeptidation, the nascent peptide becomes transferred to tmRNA. After translocation and release of the problematic mRNA, translation resumes—in *trans*—on the mRNA segment within tmRNA, resulting in the addition of a short peptide tag that targets the aberrant protein for degradation by proteases. At the same time, the ribosome is freed and recycled.

4.5S RNA

A third sRNA has its role in protein secretion. Bacteria contain a signal recognition particle which appears to be a simplified version of its eukaryotic counterpart.[25] It consists of a protein, Ffh, and the 114 nt 4.5S RNA. This complex recognizes signal sequences on nascent peptide chains emerging from the ribosome. 4.5S RNA is essential but it is as yet unclear whether this is entirely attributable to its function in secretion, or might indicate a second role in translation. A recent paper demonstrated Ffh-independent binding of 4.5S to the 23S rRNA by cross-linking experiments,[26] and other reports suggested an interaction with elongation factor G.[27]

Regulatory Antisense RNAs Come in Two Flavors

In contrast to the few examples of housekeeping RNAs, most or all of the other known sRNAs appear to have regulatory roles. In principle, regulation can occur by a number of mechanisms, but considering the nature of RNAs, it will most likely involve sequence-specific binding to a target RNA. This implies complementarity and base-pairing between the regulator, i.e., the antisense RNA, and its target RNA.

Regulatory antisense RNAs were first discovered in 1981.[28,29] In two plasmids, copy number control was exerted by small, untranslated RNAs. RNAI (of plasmid ColE1) inhibits primer maturation, and CopA (of plasmid R1) inhibits translation of a replication protein. These so-called *cis*-encoded RNAs are fully complementary to their respective target RNAs, since they are transcribed from the same locus, but in an opposite orientation (Fig. 2). By contrast, antisense and target RNAs can also be encoded by separate genetic loci. In this case, an antisense RNA is said to be *trans*-encoded,[30,31] and regions of complementarity are generally short and often non-contiguous (Fig. 2). It seems that most of the chromosomally encoded antisense RNAs fall into this second category.

The Trans-Encoded Antisense RNA MicF Is a Stress Response Regulator

The first trans-encoded antisense RNA identified was MicF of *E. coli*.[32] This RNA (Fig. 1B) is induced by a variety of environmental stresses (changes in osmolarity, temperature and others (e.g., refs. 33, 34). Binding of MicF to its target, the translation initiation region (TIR) of *ompF* mRNA, blocks translation of the OmpF protein. As a consequence, the ratio between OmpF and OmpC—both porin proteins—is offset, and membrane properties are changed.

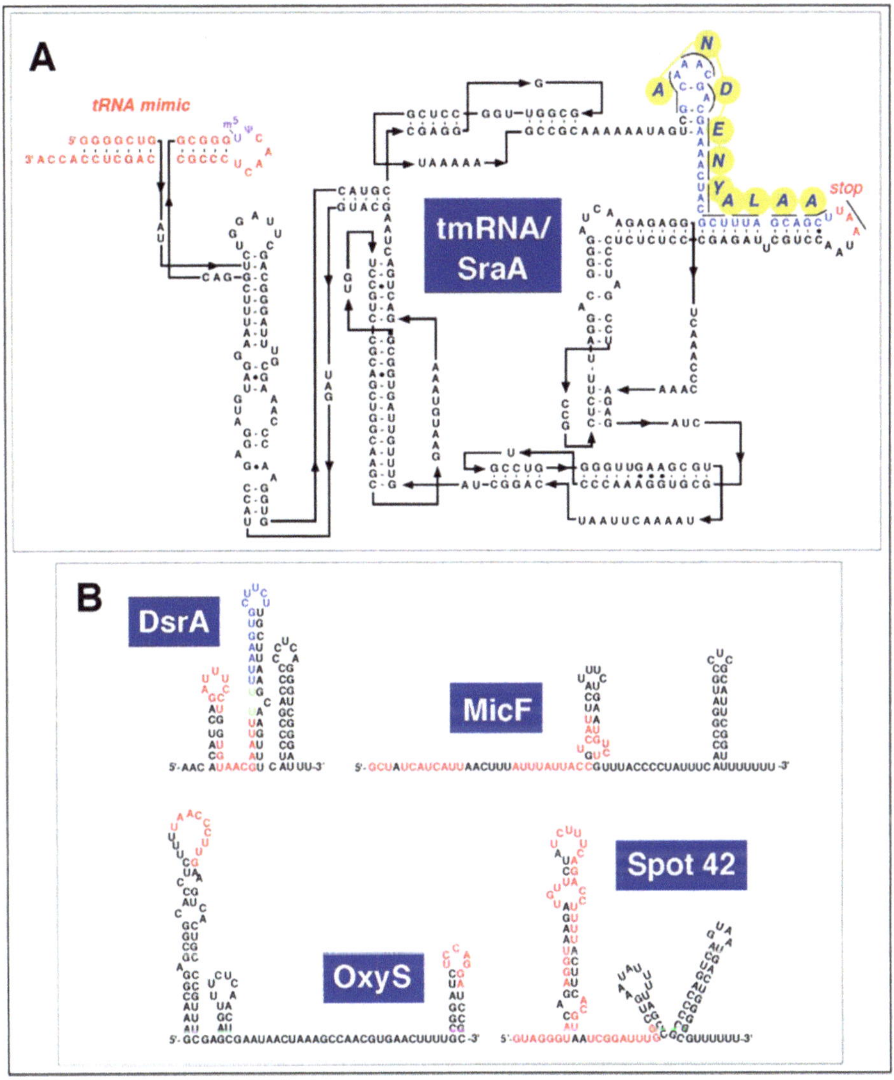

Figure 1. Structures of non-coding RNAs of *E. coli*. A: Example of a housekeeping sRNA, tmRNA. The tertiary structure is based on enzymatic and chemical probing in vitro, phylogenetic comparisons, and in vivo probing (see text for refs.). The tRNA-like domain is shown in red, and the short reading frame of the peptide tag is indicated; filled circles—amino acids encoded. Modification sites are shown in purple.[81] Arrows follow the backbone of the nucleotide chain. B: Four regulatory sRNAs (see text for details). Secondary structures are based on probing and/or folding algorithm predictions. Colored nucleotides base-pair with target RNA sequences. In the case of DsrA, red indicates interaction with *rpoS* mRNA, blue with *hns* mRNA, and green with both (overlap).

The complex formed between MicF and its target RNA illustrates a recurrent theme in all trans-encoded antisense RNA systems: incomplete complementarity between the RNAs results in partial duplexes, i.e., non-contiguous stretches of base-paired regions. Given the separate location of the genes encoding the two interacting RNAs, a sufficient extent of complementarity

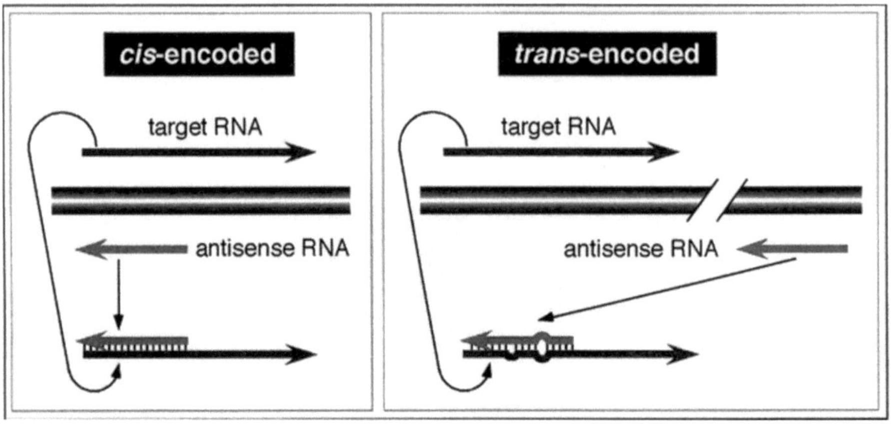

Figure 2. Schematics of *cis*- and *trans*-encoded antisense RNAs. Complexes formed by interaction between antisense and target RNAs are simplified (often, cis-encoded antisense/target RNA complexes are not fully paired[82]). Non-contiguous helices are indicated in the right hand panel.

appears to be selected for, yet full duplex formation is not. MicF remained an orphan example of a *trans*-encoded antisense RNA for several years, and the only one encoded by a bacterial chromosome.

sRNAs Can Regulate Multiple Genes

Novick and co-workers[35] identified a 514 nt RNA, RNAIII, which was encoded within the virulence locus *agr* of *Staphylococcus aureus*. RNAIII is a key regulator of virulence in this pathogen, and displays several novel features. Firstly, it is unusually long for a bacterial riboregulator (see also below). Secondly, it acts as a multi-gene regulator, mediating repression of cell-surface protein genes as well as activation of toxin genes. Thirdly, in addition to its RNA-based regulatory function, it also acts as an mRNA encoding ∂-hemolysin; recent results indicate that different domains of RNAIII are responsible for these activities[36] (Fig. 3).

The mechanism of action on one target, the *hla* mRNA (encoding α-hemolysin), has been elucidated. Surprisingly, RNAIII is an antisense RNA that activates translation of its target RNA rather than causing inhibition (in contrast to all examples of characterized antisense RNAs in plasmids). The mechanism of activation involves partial pairing of RNAIII with a segment of the *hla* mRNA, such that a self-complementary, inhibitory structure encompassing the *hla* TIR is prevented from forming[37] (Fig. 3).

OxyS RNA of *E. coli* may serve as a second example of a *trans*-encoded multi-gene regulator. This 109 nt RNA (Fig. 1B) regulates the expression of ≈40 genes, as part of a major defense against oxidative damage; when cells are treated with hydrogen peroxide, OxyS rapidly accumulates to a high level.[38] The effects on two target genes, *fhlA* (a transcriptional activator protein) and *rpoS* (stress/ stationary phase σ subunit of RNA polymerase) have been investigated. Translation of the *fhlA* mRNA is inhibited by the formation of two loop-loop ("kissing") complexes between the two major loops of OxyS and complementary structures near the TIR of the *fhlA* mRNA[39,40] (Fig. 4). Negative control by OxyS is also observed with *rpoS* mRNA, though the molecular mechanism appears to be different. One hypothesis invokes titration of Hfq (a.k.a. host factor I, a protein to which we will return in more detail below), a small, abundant protein that is required for efficient *rpoS* translation. Formation of OxyS-Hfq com-

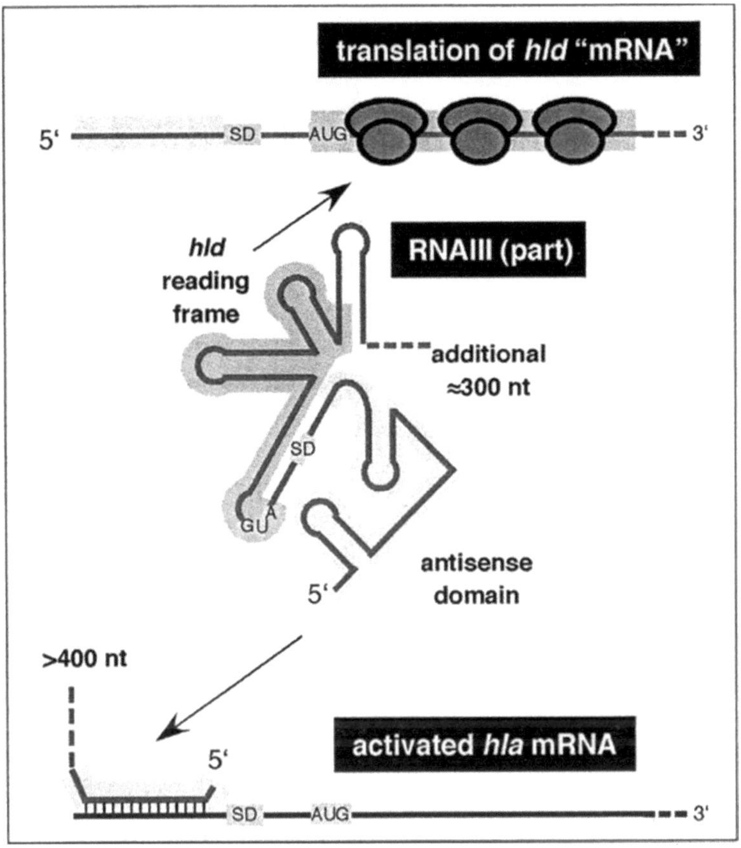

Figure 3. Dual function of RNAIII of the *agr* locus in *Staphylococcus aureus*. The 5'-segment of the proposed folded structure RNAIII, lacking approximately 300 nt towards the 3'-end, is shown schematically (middle). The two functional domains are: antisense domain (light shaded), reading frame encoding δ-hemolysin (darker shaded). Approximate positions of the *hld* Shine-Dalgarno sequence and the AUG start codon are indicated. Upper panel: under appropriate conditions, the *hld* mRNA segment is translated. Lower panel: binding of the 5' antisense domain of RNAIII to a 5' segment of *hla* mRNA sequesters the anti-Shine-Dalgarno sequence and thus activates Hla translation.[35,37]

plexes has been shown in vitro, and post-transcriptional inhibition of *rpoS* expression by OxyS in vivo.[41]

RNAIII and OxyS both affect the expression of many genes. However, it is worth noting that such effects could be indirect. In principle, a regulatory RNA could, by inhibition (or activation) of a single target RNA, decrease (or increase) the synthesis of a single regulatory protein which in turn might affect a great number of downstream genes. RNAIII so far has one identified target RNA, and so has OxyS. A priori, it is unlikely that a single RNA species can base-pair to and directly regulate many different mRNAs. Thus, as a rule, regulation of a great number of genes by an sRNA can be expected to be mediated through the effect on single targets. Does this imply that acting on several targets is impossible?

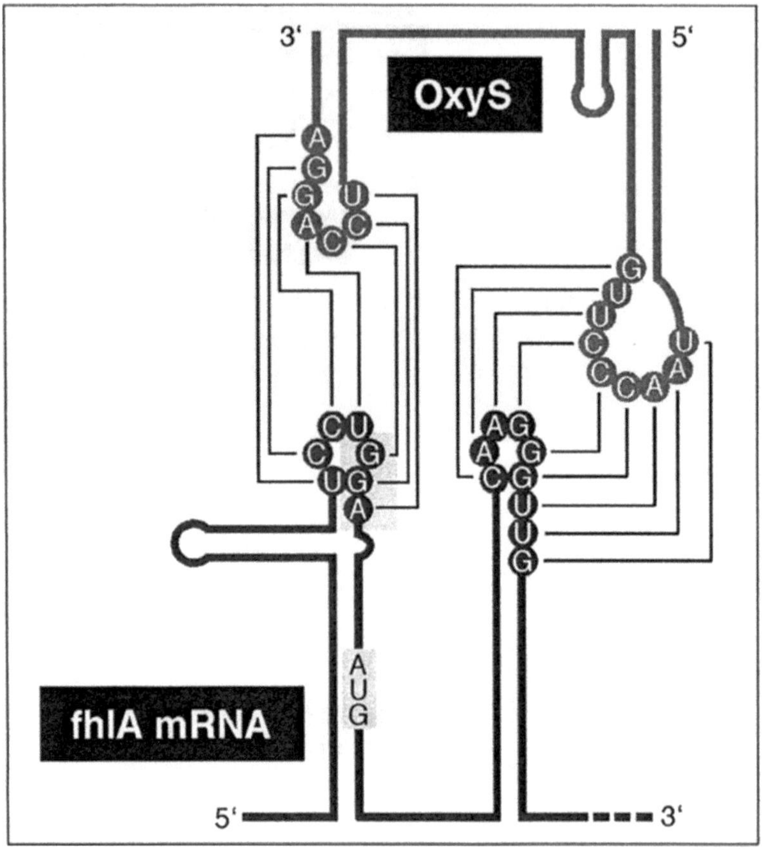

Figure 4. Bipartite kissing interactions between OxyS and *fhlA* mRNA. Secondary structures of OxyS and the 5'-segment of *fhlA* mRNA are shown. Nucleotides that base-pair are shown as filled circles connected by lines. The positions of the *fhlA* Shine-Dalgarno element and start codon are indicated by shaded boxes.

One Regulatory RNA—Two (or More) Target RNAs

The first and to date only RNA that shows proof of principle is DsrA. This 87 nt RNA (Fig. 1B) was identified by its effect on polysaccharide capsule synthesis, and is both induced and stabilized at low temperature.[42] Two mRNAs, *rpoS* and *hns*, are currently known to be targeted by DsrA. Notably, both of them encode global transcriptional regulator proteins that in turn control many genes.

How can one RNA bind two targets that show no apparent sequence similarity? A dissection of DsrA indicated that two different sequence segments are responsible for base-pairing with either mRNA. One stretch of nucleotides binds the TIR of *hns* mRNA, resulting in inhibition of HN-S translation. Possibly, this complex is additionally stabilized by interactions with the 3' portion of the *hns* mRNA.[43] The other sequence element in DsrA can base-pair to a stretch of nucleotides near the 5' end of the *rpoS* mRNA, which otherwise sequesters the *rpoS* TIR. Thus, pairing between DsrA and the *rpoS* mRNA results in activated translation (Fig. 5A), similar to the case of RNAIII (see above). Lease and Belfort[43] proposed, based on sequence gazing, that at least two additional mRNAs may be DsrA targets, but experimental support is so far lacking. Nevertheless, it is clear that short RNA sequences can carry information that permits functional interactions with more than one target.[44]

More Than One RNA Acting on the Same Target

The opposite reasoning applies to this paragraph: is it possible that the same target RNA is regulated by more than one sRNA? We have already briefly addressed *rpoS* regulation. Its gene product is a key factor in adaptation to various stresses, and is responsible for regulation of many stress-related genes. For bacterial cells in trouble, it seems important to integrate several, sometimes even conflicting, signals in order to mount proper responses. Recent work indicates that one point at which such signals may converge is regulation of *rpoS*. Both OxyS and DsrA affect *rpoS* translation, though in opposite directions. Yet another sRNA was recently shown to increase *rpoS* translation[45]. RprA (RpoS regulator RNA; 106 nt RNA,) targets the same region of the *rpoS* mRNA as does DsrA[46] (Fig. 5A). Notably, both sRNAs use non-contiguous stretches of nucleotides for base-pairing, which might explain how two sRNAs that differ in sequence can bind the same target region (Fig. 5B). Why does *E. coli* employ three sRNAs for regulation of the same target, in particular when simultaneous target access should be impossible? The reason for this could be the different expression patterns of DsrA, RprA and OxyS. DsrA accumulates at low temperature, RprA is required under osmotic stress and is under regulation

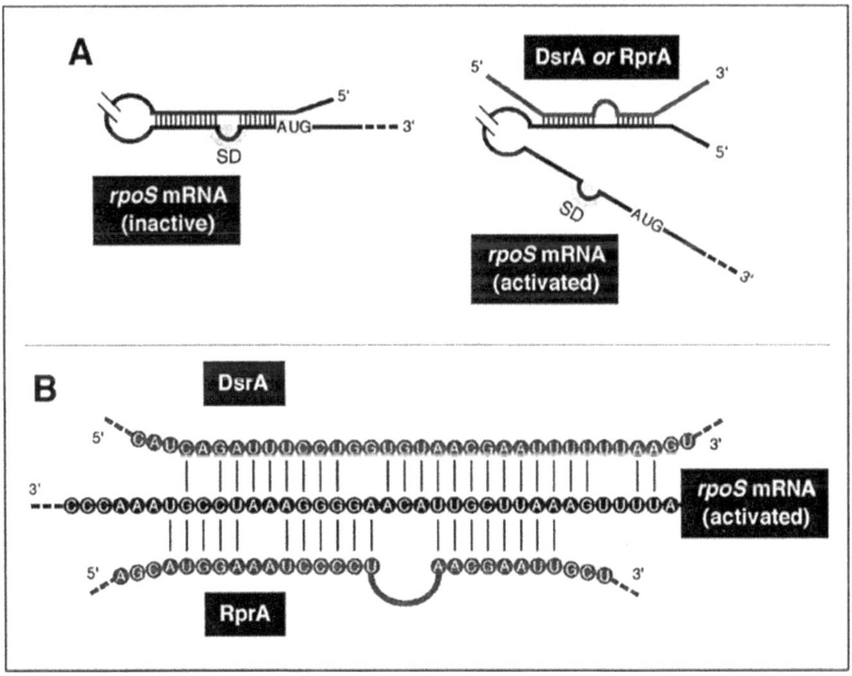

Figure 5. Two sRNAs targeting the same region of an mRNA. A: The inactive (sequestered) conformation of *rpoS* mRNA is shown. Shine-Dalgarno and AUG start codon are indicated. Upon binding of either DsrA or RprA, the TIR is freed, and translation is activated. B: Proposed base-pairing between either DsrA, or RprA, to the open conformation of *rpoS* mRNA. Lines indicate base-pairing. Connecting, non-paired nucleotides in RprA are indicated by a curved line.

of the RcsC/B phosphorelay system (regulating capsule synthesis), whereas OxyS is induced by oxidative stress. Hence, these three sRNAs will exert their effects—activation or inhibition of *rpoS* translation—under different conditions, and may serve to integrate specific stress responses with global stress.

Discoordinate Expression of Genes within Operons—Another Role for sRNAs

The aforementioned sRNAs post-transcriptionally regulate single genes, or at least the first genes within an operon. If a regulatory RNA could target a TIR of a downstream reading frame within a polycistronic mRNA, it would break co-ordinate expression of the (often functionally related) proteins encoded by the same operon. Two sRNAs of this kind have surfaced recently. One is the 109 nt Spot 42 (Fig. 1B), an RNA discovered as early as in 1973.[4] Although abundant, its function had remained obscure until recently. Deletion of its gene, *spf* (spot forty-two), showed no phenotype, and overexpression affected growth, but no mechanism (or target) was found. Recently, Spot 42 was demonstrated to be responsible for discoordinate expression of cistrons within the *galETKM* (galactose) operon.[47] By elegant in vitro experiments, the 5'-portion of the sRNA was shown to base-pair with the *galK* (third cistron of the operon) TIR, thereby preventing ribosome access. Since translation of the upstream reading frames is not affected, the GalK/GalET ratio decreases (Fig. 6). The physiological significance of this regulation so far remains unclear, but fine-tuning of carbon source utilization is a possible scenario. Such a speculation is also based on the fact that transcription of Spot 42 is regulated by cAMP levels. A second target, suggested based on sequence complementarity, might be *sucC* in the *sucABCD* operon, encoding TCA cycle enzymes.

The ~90 nt RhyB RNA[8] (in parallel identified as SraI[7]) represents another regulator of discoordinate operon expression. Its transcription is regulated by iron (Fe^{2+}), being repressed by the ferric uptake protein, Fur, when iron concentration is high. The discovery of RhyB solves a puzzle: Fur is known foremost as a repressor of genes involved in iron uptake and metabolism, but in addition upregulates a number of genes under repression conditions. Masse and Gottesman[48] identified a region of complementarity between RhyB and the TIR of the second cistron within the *sdhCDAB* operon. The *sdh* genes are required for growth on succinate. Upon overexpression of *rhyB*, or when *fur* was mutated, cells failed to grow on this carbon source. Thus, Fur-dependent activation of *sdhD* occurs by repression of RhyB transcription (=repression of a repressor). Although base-pairing interactions remain to be experimentally demonstrated, it is plausible that down-regulation of *sdhD* is mediated by an antisense mechanism, i.e., RhyB-induced *sdhC* mRNA decay, but this might also be a consequence of translational inhibition.

These two cases illustrate that sRNAs can modulate the relative output of proteins encoded within a single operon in response to environmental signals such as iron or cAMP concentration.

Some sRNAs Require Helper Proteins

In a cellular environment, RNAs cannot be expected to be naked. Instead, they form complexes with neutralizing cations, and frequently also with proteins. A number of RNA (and DNA) binding proteins are known, e.g., Hfq (or HF-I; a protein identified as required for phage Qβ replication; see below), the RNA chaperone StpA,[49,50] the histone-like protein HU,[51] and the transcriptional regulator H-NS.[50] Since these proteins are very abundant, ranging from ~20-30 000 (StpA, H-NS) to ~50 000 (Hfq, HU) molecules per cell in logarithmic growth,[52] it is reasonable to assume that they bind sRNAs and possibly affect sRNA-mediated regulation, either directly or indirectly.

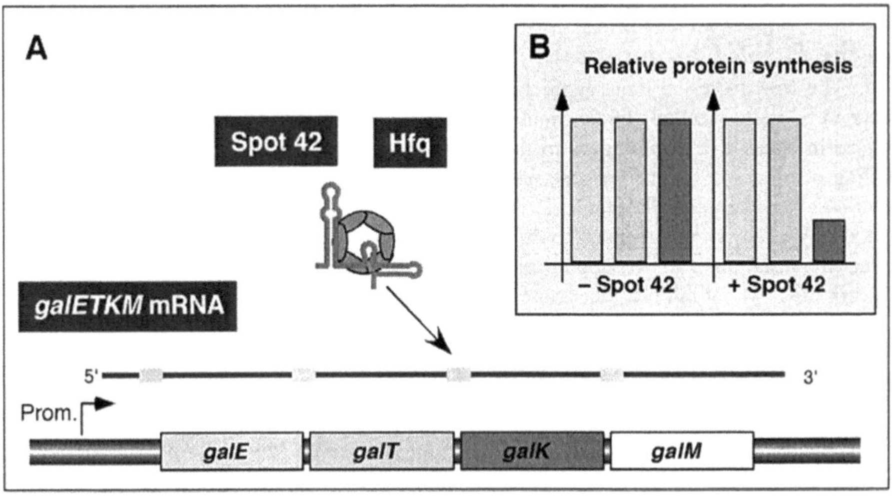

Figure 6. Discoordinate regulation of the *galETKM* operon mRNA by Spot 42. A: Location of the genes within the operon is shown by boxes. Rightward arrow: transcription initiation site. The line above represents the entire mRNA, and TIRs are indicated by shaded boxes. The structure of Spot 42 is shown schematically, in complex with the hexameric Hfq protein. Targeting of the *galK* TIR is indicated by an arrow. B: Relative output of the GalE, GalT (both grey) and GalK (dark) proteins, in the presence or absence of Spot 42 in vivo (based on ref. 47).

Two mechanisms have indeed received experimental support. One affects the abundance of sRNA, and thereby indirectly sRNA-mediated regulation. For example, StpA destabilizes MicF, resulting in rapid decay of this RNA.[53] Conversely, Hfq binding increases the half-life of Spot 42;[54] in an *hfq* mutant strain, Spot 42 is rapidly degraded. The other mechanism is illustrated by OxyS and, again, by Spot 42. In both cases, sRNA-mediated regulation of *fhlA* and *galK*, respectively, is strictly dependent on the presence of a functional *hfq* gene. In vitro experiments demonstrated that the binding between sRNA and target is greatly enhanced by added Hfq protein.[54,55] This protein may facilitate interaction by partial unfolding of structures within the sRNA, or stabilize the complex formed—either way the efficiency of regulation is increased. HU also binds at least one sRNA, DsrA, in vitro with a nanomolar Kd,[51] but functional implications have yet to be investigated. StpA and its close relative, H-NS, promote *td* intron splicing in vivo and in vitro,[49] indicating that these proteins can also stabilize RNA. In addition, StpA, but not H-NS, has strong RNA-RNA annealing activity;[50] however, neither of these two proteins has been analyzed for sRNA-mediated effects.

Protein Hfq deserves further attention. Firstly, seven of the recently identified sRNAs could be efficiently co-immunoprecipitated by anti-Hfq-antibodies.[50] Since RprA, DsrA, RhyB, OxyS, and Spot 42 are dependent on this protein for regulatory activity,[45,48,54-56] Hfq may turn out to be a major player in (trans-encoded) antisense RNA mechanisms. Interestingly, Hfq is an Sm-like protein. Sm-proteins of eukaryotes are involved in splicing and other RNA transactions. Like the Sm proteins, Hfq binds RNA with a preference for unstructured A/U-rich regions. A recent crystallographic study showed that *Staphylococcus aureus* Hfq forms a hexameric ring structure and binds a model substrate, 5'AUUUUUG, in an unfolded conformation within its cavity.[57] Perhaps, interactions between partially complementary RNAs are aided by simultaneous binding and unfolding of the interactant RNAs.

The Exception (?): *cis*-Encoded Antisense RNAs in the *E. coli* Chromosome

The cases described above argue that antisense RNAs encoded by the *E. coli* chromosome tend to be trans-encoded. Two examples represent exceptions. A family of killer genes, conserved in bacteria, is homologous to the well-characterized *hok/sok* locus (post-segregational killing of plasmids[58]). The *gef* gene encodes a toxin which rapidly kills cells. Uncontrolled synthesis would entail cellular suicide. A short RNA, Sof, is *cis*-encoded from the same locus; the sRNA is fully complementary to the leader of the *gef* mRNA.[59] Translation of *gef* is dependent on translation of an overlapping reading frame (ORF69). Since the Sof target overlaps the start of ORF69, Sof inhibits killer gene expression indirectly by blocking ribosome loading at the ORF69 TIR.

A recent report addressed the significance of four so-called long direct repeat (LDR) sequences in *E. coli*.[60] Surprisingly, LDR-D, one of these repeats, harbors a toxin-antitoxin gene pair which may be analogous (but not homologous) to *gef/sof*. A stable mRNA encodes the LdrD toxin. Killing is prevented by the metabolically unstable ≈60 nt RdlD antisense RNA which acts as an inhibitor of translation. Though such systems are abundantly present in bacteria, the biological role of conditional suicide genes remains enigmatic.

Regulation of Gene Expression by Interaction between sRNAs and Proteins

Some sRNAs do not act by antisense mechanisms, but instead through interaction with proteins. The CsrA protein is a key regulator of carbon storage, adhesion, cell surface properties,[61] and virulence (in *Erwinia*[62]). CsrA interacts with target mRNAs and either facilitates their rapid degradation or, in some cases, activates translation. CsrB, a 363 nt long ncRNA, is a CsrA antagonist. This RNA is predicted to fold into a phylogenetically conserved structure with 18 CsrB binding sites, repeated mostly in short stem-loops. When CsrB is at high concentration, CsrA is sequestered in a globular complex (≈18 CsrA monomers bound per CsrB). Thus, CsrB appears to determine the activity of CsrA by controlling the availability of free CsrA. An additional twist in this system is a feedback loop: CsrA indirectly regulates transcription of the antagonistic CsrB RNA.[63]

The 184 nt 6S RNA of *E. coli* was among the early discovered sRNAs.[64] In spite of deletion and overexpression analyses, no detectable phenotype could be established. Recently, Wassarman and Storz showed that 6S RNA co-fractionates with RNA polymerase holo-enzyme from a cellular extract.[65] Binding is specific, and probably involves σ^{70} (housekeeping σ factor). Whether 6S RNA can be regarded a bona fide regulator is still unclear and awaits further experiments. However, effects of the presence or absence of the sRNA indicated that, in stationary phase, stable σ^{70}/core RNA polymerase association requires 6S RNA. Furthermore, 6S RNA accumulates from ≈1000 molecules per cell in logarithmic growth to ≈10 000 in stationary phase. Therefore, modulation of RNA polymerase (state or activity) could promote the switch from σ^{70}- to σ^{S}-dependent promoters, which is required for survival when cells cease to grow after reaching a high density.

Structures of sRNAs

As far as secondary or tertiary structures of sRNAs are concerned, little can be generalized. Most of the hitherto identified regulatory sRNAs are small (60-200 nt). From enzymatic and chemical probing, or based on predictions by folding algorithms, several sRNA secondary structures have been proposed (Fig. 1). Similar to the cis-encoded antisense RNAs, sRNAs most often carry stem-loops, usually connected by regions of low secondary structure. Regions important for base-pairing to target RNAs are generally located in the single-stranded loops, or in connectors between stems. As an exception, DsrA carries target complementary sequences in two of its three stable stem-loop regions[43] (Fig. 1B). In such a case, unfolding by Hfq or other proteins may be needed to present nucleotides for interaction.

Structure/ function relationships in sRNAs have so far not been adequately addressed. From the predicted structural conservation (compensatory base changes etc.) that parallels phylogenetic conservation,[6,9] it could be argued that short target complementarity only works if presented in an appropriate structural context.

More structural information is available for the three housekeeping RNAs of *E. coli*. Based on phylogenetic conservation as well as in vitro and in vivo structure probing, a secondary and tertiary structure model has been proposed for tmRNA[22,23] (Fig. 1A). A three-dimensional model of RNase P has been generated by integration of phylogenetic, mutational, and structure probing data,[66] and the crystal structure of the signal recognition particle core (4.5S RNA and Ffh protein) has been published.[67]

Stable or Unstable in Cells?

The often compact, defined secondary structures might suggest that sRNAs are metabolically stable, i.e., are turned over slowly. Indeed, many of the sRNAs discussed in this study were found to be stable (when induced), exhibiting half-lives of 20-60 min. For several of them, stabilization requires protein binding. E.g., the absence of Hfq renders Spot 42 unstable[54] (but leaves OxyS entirely unaffected and stable[55]). DsrA displays a whole range of stabilities. When functioning as a regulator at low temperature, it is stable (half-life 23 min), whereas at 37 °C this value decreases to 4 min,[68] and further decreases to 1 min in an *hfq* mutant background.[56]

In contrast to the unstable antisense RNAs of plasmids, whose decay pathways have been elucidated,[69-73] next to nothing is known about how sRNAs are turned over. Probably, the presence of accessible sites for endonuclease RNase E (the main RNA decay enzyme in *E. coli*) is a major determinant. The stabilization of sRNAs by Hfq may then be tentatively explained by sequestration of such sites; RNase E and Hfq share a preference for single-stranded A/U-rich regions.[57,74] Another determinant of intracellular stability is the 3'-end of an RNA, where exoribonucleases like PNPase and RNase II act. Often, 3'-exonucleolytic activity is modulated by PcnB/PAPI-dependent addition of short oligo-A tails, as exemplified by RNAI, CopA, and Sok from plasmids.[69-72] Interestingly, some of the recently identified sRNAs carry A-tails.[7]

Primary or Processed Transcripts

Many sRNAs accumulate as unprocessed, primary transcripts (e.g., DsrA, Spot42, MicF, OxyS). Several others, though, are generated from longer precursors, and are trimmed to final size. This is true for 4.5S RNA, tmRNA, M1 (RNase P) RNA, 6S RNA, DicF, and probably a few more. RNases implicated in the maturation of 5'- or 3'-ends vary, from RNase P (for tmRNA and 4.5S) over the double-strand-specific endonuclease RNase III (tmRNA, DicF) to RNase E (M1) (see references in 3).

Processing is often important to create an active RNA species. When an sRNA is observed to be present in more than one size, further analysis is required as to which species carries out a given function. Of the newly discovered sRNAs, several occur as multiple size species, and some undergo growth rate-dependent changes in size, consistent with growth condition-dependent processing.[7,8] Not knowing their roles, it remains unclear whether the different RNA variants are functional, or whether conversion of an inactive precursor form to a mature form is part of the regulatory circuit.

What Is the Job of the New sRNAs?

Though a subject of intensive research, most of the new sRNAs have not yet been assigned biological roles. So far, almost all sRNA gene knock-outs are viable, and have failed to show characteristic phenotypes under standard laboratory conditions (ref. 8; Argaman L, Altuvia S, Vogel J et al, unpublished). Based on what is known about the previously discovered sRNAs, one might guess that many of the newcomers also are stress response regulators;[12] their expression profiles under different growth conditions suggest this to be likely.[7,8] In the absence of striking physiological effects upon under- or overexpression, several lines of research are currently being followed. Proteomics can be used to monitor sRNA overexpression- or knockout-dependent changes in the abundance of proteins, e.g., suggesting involvement in a certain metabolic pathway. Microarray analyses are probably less promising to identify regulatory targets if sRNAs act post-transcriptionally, though effects on target mRNA abundance have been observed.[43,48] It has been argued that most sRNAs may be antisense RNAs.[10] If so, bioinformatics could prove useful in searching for putative target sequences.

And by What Mechanisms?

It would be exciting if new mechanisms would emerge in studies of novel sRNAs. However, mechanisms relying on antisense-target interactions, and in some cases on sRNAs interacting with proteins, will most likely dominate. For trans-encoded antisense RNAs, proteins such as Hfq or even StpA will frequently be involved (as discussed above). For cis-encoded antisense RNAs, e.g., CopA of plasmid R1, help appears not to be required (Slagter-Jäger C, Wagner EGH, unpublished). Most sRNAs will prove to be post-transcriptional regulators, affecting the translation or stability of target mRNAs.

How Many Are There?

The simple answer, of course, is: we don't know. The list of experimentally supported sRNAs—old and new ones—has by now grown to ~50 (ref 6-8, 75; Hüttenhofer A, Vogel J, Tang TH et al, unpublished), whereas the number of predicted sRNA genes is as high as 275[6] or 370.[9] Since few known sRNAs are constitutively expressed, and appropriate conditions for the induction of all others are difficult to predict, many candidates might have escaped detection. Most sRNAs are "non-essential", i.e., their absence (or mutation) does not cause lethality. However, essentiality is usually defined under rich medium laboratory conditions and may simply reflect our ignorance about conditions under which an sRNA-related activity might be required for growth.

It is already clear that sRNAs are much more prevalent than anticipated but, apart from that, an estimate of how many are present depends on which criteria we choose as qualifyers. The presence of an sRNA of distinct size visualized, e.g., by Northern analysis, does not automatically ascertain that the RNA has a biological role. On the one hand, some sRNAs may be merely 5'-leader or 3'-trailer segments of mRNAs, accumulated inadvertantly after endonucleolytic processing, and lack functional significance altogether. On the other hand, even such an RNA may turn out to be functional on its own, e.g., being a leader when part of an mRNA, but having a separate role when accumulating as a distinct sRNA species. The location of many of the predicted, and some experimentally supported sRNAs genes, indicates that they belong to this uncertain category. Again some other sRNA-candidates turned out to be, upon closer inspection, putative mRNAs encoding short, not previously annotated ORFs.[8,75]

Conclusions

The last few years have changed our view concerning the roles RNAs play in organisms. The significance of RNAs as regulators of various biological processes has been beautifully illustrated by several small, ncRNAs in *E. coli* and other bacteria. Add to that the experimental verification of many new sRNAs still awaiting a functional characterization, and an even greater number of predicted sRNA genes, and we begin to see that RNAs play important roles in addition to genetic information transfer. The theme of sRNAs as regulators is likely a universal one. The discovery of numerous snoRNAs[76] and microRNAs in eukaryotes[77-79] (many of which may be developmental regulators), as well as recent reports on novel sRNAs in archea[80] suggests that challenging questions about ncRNAs will keep scientists busy in the years to come.

Acknowledgements

The authors gratefully acknowledge funding from HFSP, VR, Wallenberg Foundation (all to E.G.H. Wagner) and from EMBO (long-term fellowship, J. Vogel).

References

1. Wagner EG, Simons RW. Antisense RNA control in bacteria, phages, and plasmids. Annu Rev Microbiol 1994; 48:713-742.
2. Wagner EGH, Altuvia S, Romby P. Antisense RNAs in bacteria and their genetic elements. Adv Genet 2002; 46:361-398.
3. Wassarman KM, Zhang A, Storz G. Small RNAs in *Escherichia coli*. Trends Microbiol 1999; 7:37-45.
4. Ikemura T, Dahlberg JE. Small ribonucleic acids of *Escherichia coli*. I. Characterization by polyacrylamide gel electrophoresis and fingerprint analysis. J Biol Chem 1973; 248:5024-5032.
5. Griffin BE. Separation of 32P-labelled ribonucleic acid components. The use of polyethylenimine-cellulose (TLC) as a second dimension in separating oligoribonucleotides of '4.5 S' and 5 S from *E. coli*. Febs Lett 1971; 15:165-168.
6. Rivas E, Klein RJ, Jones TA et al. Computational identification of noncoding RNAs in *E. coli* by comparative genomics. Curr Biol 2001; 11:1369-1373.
7. Argaman L, Hershberg R, Vogel J et al. Novel small RNA-encoding genes in the intergenic regions of *Escherichia coli*. Curr Biol 2001; 11:941-950.
8. Wassarman KM, Repoila F, Rosenow C et al. Identification of novel small RNAs using comparative genomics and microarrays. Genes Dev 2001; 15:1637-1651.
9. Carter RJ, Dubchak I, Holbrook SR. A computational approach to identify genes for functional RNAs in genomic sequences. Nucleic Acids Res 2001; 29:3928-3938.
10. Wagner EGH, Flärdh K. Antisense RNAs everywhere? Trends Genet 2002; 18:223-226.
11. Storz G. An expanding universe of noncoding RNAs. Science 2002; 296:1260-1263.
12. Wassarman KM. Small RNAs in bacteria: diverse regulators of gene expression in response to environmental changes. Cell 2002; 109:141-144.
13. Eddy SR. Computational genomics of noncoding RNA genes. Cell 2002; 109:137-140.
14. Eddy SR. Noncoding RNA genes and the modern RNA world. Nature Rev Genet 2001; 2:919-929.
15. Erdmann VA, Barciszewska MZ, Hochberg A et al. Regulatory RNAs. Cell Mol Life Sci 2001; 58:960-977.
16. Erdmann VA, Barciszewska MZ, Szymanski M et al. The noncoding RNAs as riboregulators. Nucleic Acids Res 2001; 29:189-193.
17. Zeiler B, Simons RW. Antisense RNA structure and function. In: Simons RW, Grunberg-Manago M, eds. RNA Structure and Function. Cold Spring Harbor: Cold Spring Harbor Laboratory Press, 1998:437-64.
18. Altman S, Kirsebom LA. Ribonuclease P. In: Gesteland RF, Cech TR, Atkins JF, eds. The RNA world. Cold Spring Harbor: Cold Spring Harbor Laboratory Press, 1999:351-80.
19. Frank DN, Pace NR. Ribonuclease P: unity and diversity in a tRNA processing ribozyme. Annu Rev Biochem 1998; 67:153-180.
20. Keiler KC, Waller PR, Sauer RT. Role of a peptide tagging system in degradation of proteins synthesized from damaged messenger RNA. Science 1996; 271:990-993.
21. Gillet R, Felden B. Emerging views on tmRNA-mediated protein tagging and ribosome rescue. Mol Microbiol 2001; 42:879-885.

22. Felden B, Himeno H, Muto A et al. Probing the structure of the Escherichia coli 10Sa RNA (tmRNA). RNA 1997; 3:89-103.
23. Lindell M, Romby P, Wagner EGH. Lead(II) as a probe for investigating RNA structure in vivo. RNA 2002; 8:534-41.
24. Wower J, Zwieb CW, Hoffman DW et al. SmpB: a protein that binds to double-stranded segments in tmRNA and tRNA. Biochemistry 2002; 41:8826-8836.
25. Keenan RJ, Freymann DM, Stroud RM et al. The signal recognition particle. Annu Rev Biochem 2001; 70:755-775.
26. Rinke-Appel J, Osswald M, von Knoblauch K et al. Crosslinking of 4.5S RNA to the *Escherichia coli* ribosome in the presence or absence of the protein Ffh. RNA 2002; 8:612-625.
27. Nakamura K, Miyamoto H, Suzuma S et al. Minimal functional structure of Escherichia coli 4.5 S RNA required for binding to elongation factor G. J Biol Chem 2001; 276:22844-22849.
28. Stougaard P, Molin S, Nordström K. RNAs involved in copy-number control and incompatibility of plasmid R1. Proc Natl Acad Sci USA 1981; 78:6008-6012.
29. Tomizawa J, Itoh T, Selzer G et al. Inhibition of ColE1 RNA primer formation by a plasmid-specified small RNA. Proc Natl Acad Sci USA 1981; 78:1421-1425.
30. Altuvia S, Wagner EGH. Switching on and off with RNA. Proc Natl Acad Sci USA 2000; 97:9824-9826.
31. Delihas N. Regulation of gene expression by trans-encoded antisense RNAs. Mol Microbiol 1995; 15:411-414.
32. Andersen J, Delihas N, Ikenaka K et al. The isolation and characterization of RNA coded by the micF gene in *Escherichia coli.* Nucleic Acids Res 1987; 15:2089-2101.
33. Andersen J, Forst SA, Zhao K et al. The function of micF RNA. micF RNA is a major factor in the thermal regulation of OmpF protein in *Escherichia coli.* J Biol Chem 1989, 264:17961-17970.
34. Delihas N, Forst S. MicF: an antisense RNA gene involved in response of *Escherichia coli* to global stress factors. J Mol Biol 2001; 313:1-12.
35. Novick RP, Ross HF, Projan SJ et al. Synthesis of staphylococcal virulence factors is controlled by a regulatory RNA molecule. EMBO J 1993; 12:3967-3975.
36. Benito Y, Kolb FA, Romby P et al. Probing the structure of RNAIII, the Staphylococcus aureus agr regulatory RNA, and identification of the RNA domain involved in repression of protein A expression. RNA 2000; 6:668-679.
37. Morfeldt E, Taylor D, von Gabain A et al. Activation of alpha-toxin translation in *Staphylococcus aureus* by the trans-encoded antisense RNA, RNAIII. EMBO J 1995; 14:4569-4577.
38. Altuvia S, Weinstein-Fischer D, Zhang A et al. A small, stable RNA induced by oxidative stress: role as a pleiotropic regulator and antimutator. Cell 1997; 90:43-53.
39. Altuvia S, Zhang A, Argaman L et al. The *Escherichia* coli OxyS regulatory RNA represses fhlA translation by blocking ribosome binding. EMBO J 1998; 17:6069-6075.
40. Argaman L, Altuvia S. fhlA repression by OxyS RNA: kissing complex formation at two sites results in a stable antisense-target RNA complex. J Mol Biol 2000; 300:1101-1112.
41. Zhang A, Altuvia S, Tiwari A et al. The OxyS regulatory RNA represses rpoS translation and binds the Hfq (HF-I) protein. EMBO J 1998; 17:6061-6068.
42. Sledjeski D, Gottesman S. A small RNA acts as an antisilencer of the H-NS-silenced rcsA gene of *Escherichia coli.* Proc Natl Acad Sci USA 1995; 92:2003-2007.
43. Lease RA, Belfort M. A trans-acting RNA as a control switch in *Escherichia coli*: DsrA modulates function by forming alternative structures. Proc Natl Acad Sci USA 2000; 97:9919-9924.
44. Lease RA, Belfort M. Riboregulation by DsrA RNA: trans-actions for global economy. Mol Microbiol 2000; 38:667-672.
45. Majdalani N, Chen SA, Murrow J et al. Regulation of RpoS by a novel small RNA: the characterization of RprA. Mol Microbiol 2001; 39:1382-1394.
46. Majdalani N, Hernandez D, Gottesman S. Regulation and mode of action of a second small RNA activator of RpoS translation, RprA. Mol Microbiol 2002; in press.
47. Møller T, Franch T, Udesen C et al. Spot 42 RNA mediates discoordinate expression of the *E. coli* galactose operon. Genes Dev 2002; 16:1696-1706.
48. Masse E, Gottesman S. A small RNA regulates the expression of genes involved in iron metabolism in *Escherichia coli.* Proc Natl Acad Sci USA 2002; 99:4620-4625.

49. Waldsich C, Grossberger R, Schroeder R. RNA chaperone StpA loosens interactions of the tertiary structure in the td group I intron in vivo. Genes Dev 2002; 16:2300-2312.
50. Cusick ME, Belfort M. Domain structure and RNA annealing activity of the *Escherichia* coli regulatory protein StpA. Mol Microbiol 1998; 28:847-857.
51. Balandina A, Kamashev D, Rouviere-Yaniv J. The bacterial histone-like protein HU specifically recognizes similar structures in all nucleic acids. DNA, RNA, and their hybrids. J Biol Chem 2002; 277:27622-2768.
52. Ali Azam T, Iwata A, Nishimura A et al. Growth phase-dependent variation in protein composition of the *Escherichia coli* nucleoid. J Bacteriol 1999; 181:6361-6370.
53. Deighan P, Free A, Dorman CJ. A role for the Escherichia coli H-NS-like protein StpA in OmpF porin expression through modulation of micF RNA stability. Mol Microbiol 2000; 38:126-139.
54. Møller T, Franch T, Højrup P et al. Hfq: A bacterial Sm-like protein that mediates RNA-RNA interaction. Mol Cell 2002; 9:23-30.
55. Zhang A, Wassarman KM, Ortega J et al. The Sm-like Hfq protein increases OxyS RNA interaction with target mRNAs. Mol Cell 2002; 9:11-22.
56. Sledjeski DD, Whitman C, Zhang A. Hfq is necessary for regulation by the untranslated RNA DsrA. J Bacteriol 2001; 183:1997-2005.
57. Schumacher MA, Pearson RF, Møller T et al. Structures of the pleiotropic translational regulator Hfq and an Hfq-RNA complex: a bacterial Sm-like protein. EMBO J 2002; 21:3546-3556.
58. Gerdes K, Poulsen LK, Thisted T et al. The hok killer gene family in gram-negative bacteria. New Biologist 1990; 2:946-956.
59. Poulsen LK, Refn A, Molin S et al. The gef gene from *Escherichia coli* is regulated at the level of translation. Mol Microbiol 1991; 5:1639-1648.
60. Kawano M, Oshima T, Kasai H et al. Molecular characterization of long direct repeat (LDR) sequences expressing a stable mRNA encoding for a 35-amino-acid cell-killing peptide and a cis-encoded small antisense RNA in *Escherichia coli.* Mol Microbiol 2002; 45:333-349.
61. Liu MY, Gui G, Wei B et al. The RNA molecule CsrB binds to the global regulatory protein CsrA and antagonizes its activity in *Escherichia coli.* J Biol Chem 1997; 272:17502-17510.
62. Liu Y, Cui Y, Mukherjee A, Chatterjee AK. Characterization of a novel RNA regulator of *Erwinia carotovora* ssp. carotovora that controls production of extracellular enzymes and secondary metabolites. Mol Microbiol 1998; 29:219-234.
63. Gudapaty S, Suzuki K, Wang X et al. Regulatory interactions of Csr components: the RNA binding protein CsrA activates csrB transcription in *Escherichia coli.* J Bacteriol 2001; 183:6017-6027.
64. Hindley J. Fractionation of 32P-labelled ribonucleic acids on polyacrylamide gels and their characterization by fingerprinting. J Mol Biol 1967; 30:125-136.
65. Wassarman KM, Storz G. 6S RNA regulates *E. coli* RNA polymerase activity. Cell 2000; 101:613-623.
66. Massire C, Jaeger L, Westhof E. Derivation of the three-dimensional architecture of bacterial ribonuclease P RNAs from comparative sequence analysis. J Mol Biol 1998; 279:773-793.
67. Batey RT, Rambo RP, Lucast L et al. Crystal structure of the ribonucleoprotein core of the signal recognition particle. Science 2000; 287:1232-1239.
68. Repoila F, Gottesman S. Signal transduction cascade for regulation of RpoS: temperature regulation of DsrA. J Bacteriol 2001; 183:4012-4023.
69. Dam Mikkelsen N, Gerdes K. Sok antisense RNA from plasmid R1 is functionally inactivated by RNase E and polyadenylated by poly(A) polymerase I. Mol Microbiol 1997; 26:311-320.
70. Söderbom F, Binnie U, Masters M et al. Regulation of plasmid R1 replication: PcnB and RNase E expedite the decay of the antisense RNA, CopA. Mol Microbiol 1997; 26:493-504.
71. Söderbom F, Wagner EGH. Degradation pathway of CopA, the antisense RNA that controls replication of plasmid R1. Microbiology 1998; 144:1907-1917.
72. Xu F, Lin-Chao S, Cohen SN. The Escherichia coli pcnB gene promotes adenylylation of antisense RNAI of ColE1-type plasmids in vivo and degradation of RNAI decay intermediates. Proc Natl Acad Sci USA 1993; 90:6756-6760.
73. Lin-Chao S, Cohen SN. The rate of processing and degradation of antisense RNAI regulates the replication of ColE1-type plasmids in vivo. Cell 1991; 65:1233-1242.

74. McDowall KJ, Lin-Chao S, Cohen SN. A+U content rather than a particular nucleotide order determines the specificity of RNase E cleavage. J Biol Chem 1994; 269:10790-10796.
75. Chen S, Lesnik EA, Hall TA et al. A bioinformatics based approach to discover small RNA genes in the *Escherichia coli* genome. Bio Systems 2002; 65:157-177.
76. Hüttenhofer A, Kiefmann M, Meier-Ewert S et al. RNomics: an experimental approach that identifies 201 candidates for novel, small, non-messenger RNAs in mouse. EMBO J 2001; 20:2943-2953.
77. Lee RC, Ambros V. An extensive class of small RNAs in *Caenorhabditis elegans*. Science 2001; 294:862-864.
78. Lau NC, Lim LP, Weinstein EG et al. An abundant class of tiny RNAs with probable regulatory roles in *Caenorhabditis elegans*. Science 2001; 294:858-862.
79. Lagos-Quintana M, Rauhut R, Lendeckel W et al. Identification of novel genes coding for small expressed RNAs. Science 2001; 294:853-858.
80. Tang TH, Bachellerie JP, Rozhdestvensky T et al. Identification of 86 candidates for small non-messenger RNAs from the archaeon *Archaeoglobus fulgidus*. Proc Natl Acad Sci USA 2002; 99:7536-7541.
81. Felden B, Hanawa K, Atkins JF et al. Presence and location of modified nucleotides in *Escherichia coli* tmRNA: structural mimicry with tRNA acceptor branches. EMBO J 1998; 17:188-196.
82. Kolb F, Malmgren C, Westhof E et al. An unusual structure formed by antisense-target RNA binding involves an extended kissing complex with a four-way junction and a side-by-side helical alignment. RNA 2000; 6:311-324.

Chapter 18

We Are Legion:

Noncoding Regulatory RNAs and Hfq

Cristin C. Brescia and Darren D. Sledjeski

Summary

Regulation by small, noncoding RNAs (ncRNA) has a long history in bacteria. Plasmid copy number control via the antisense interaction of two RNAs (RNAI and RNAII) was first described over 30 years ago in *Escherichia coli*. One of the first trans-acting ncRNAs discovered was the *E. coli* MicF RNA. It is now apparent that MicF expression is involved in the global stress response of *E. coli*. More recently, 3 RNAs have been described that regulate Rpos translation, DsrA, OxyS and RprA. RNAs have also been identified in *E. coli* that effect RNA polymerase activity (6S), carbon storage (CsrB), and oligopeptide transport (GcvB). In addition, ncRNAs have been identified in *Staphylococcus aureus* (Agr RNAIII) and *Streptococcus pyogenes* (FasX). The regulation of the expression of many of these ncRNAs is complex. A common theme among many of the newly identified ncRNAs is their interaction with the broadly conserved RNA binding protein Hfq. Recent work, including the elucidation of the structure of an Hfq:RNA co crystal, has shed new light on the functions of Hfq and ncRNAs.

Although the history of ncRNA-mediated regulation in bacteria stretches back decades, it is only recently that the use of bioinformatics and DNA micro arrays has led to the identification of dozens of novel putative ncRNAs in a variety of bacteria. It is clear that the much still needs to be learned about these formerly hidden regulatory networks.

Noncoding RNAs

Small, noncoding regulatory RNAs (ncRNAs) have a long history. The earliest examples were the RNAI and RNAII mediated control of ColE1 plasmid replication[1-5] and the regulation of *ompF* translation by *micF* RNA.[6] It is amazing that after 30 years of intensive research only a few ncRNAs have been identified in *Escherichia coli* and these were fortuitous.[7-9] The major obstacles to identification were the small size (85-300 bases) of the genes that encoded the RNA molecules, making them hard to discern with genetics, and the lack of mathematical algorithms allowing their identification from DNA sequence. It is only in recent years that the abundance, importance and ubiquity of ncRNAs has become clear.[10-12]

It is estimated that *E. coli* contains up to 50 ncRNAs.[11-13] A computational genomics approach has also identified 30 potential ncRNAs in two hyperthermophilic bacteria *Methanococcus jannaschii* and *Pyrococcus furiosus*.[14] For information on the presence and discovery of ncRNAs in *E. coli* and other bacteria several recent reviews are available, see refs.[13,15-17]

Noncoding RNAs: Molecular Biology and Molecular Medicine, edited by Jan Barciszewski and Volker A. Erdmann.

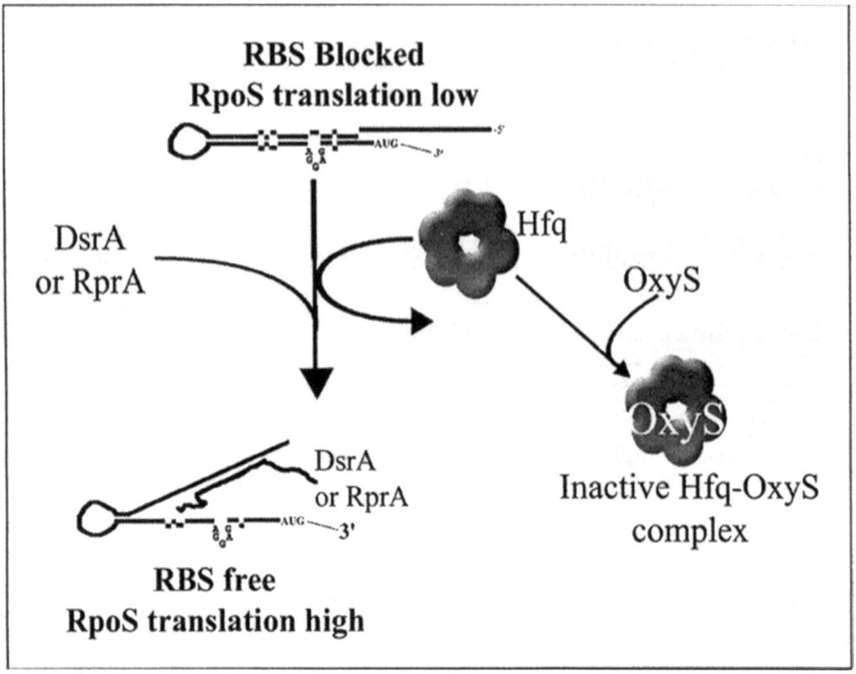

Figure 1. Model of ncRNA and Hfq regulation of *rpoS* translation. In this model, Hfq is necessary for the base pairing of the *rpoS* mRNA 5' leader region with DsrA or RprA ncRNAs. The pairing between either ncRNA stabilizes an alternative conformer of *rpoS* mRNA that leads to increased translation by exposing the ribosome binding site.[20,21] Although Hfq is necessary for the DsrA-mediated regulation of *rpoS* mRNA,[59] it is not known whether Hfq is necessary for regulation by RprA. During oxidative stress high levels of OxyS ncRNA are synthesized and bind to Hfq.[22] This could prevent Hfq from binding to DsrA or RpoS mRNA. Alternatively, OxyS might inhibit translation directly by binding to the *rpoS* mRNA ribosome binding site.

Functions of Noncoding RNAs in Gram Negatives

The precise functions of ncRNAs are varied, but of those characterized, many are involved in the regulation of translation. An interesting example is the regulation of the *E. coli* alternate sigma factor *rpoS*. The *rpoS* mRNA contains a long 5′-untranslated region (UTR) that forms a secondary structure that inhibits translation by blocking the ribosome binding site (RBS)[18] (Fig. 1). Mutations that abolish or alter the formation of this secondary structure increase the translation of *rpoS*, while mutations that stabilize this structure decrease its translation.[18] Two ncRNAs, DsrA and RprA have been identified that enhance the translation of *rpoS* by base pairing with the *rpoS* mRNA UTR, altering the secondary structure and freeing the RBS to interact with the ribosome[19-21] (Fig. 1). A third ncRNA, OxyS, represses *rpoS* translation via interactions with the protein Hfq[22] possibly by blocking binding of DsrA and RprA (Fig. 1) or by directly base pairing to the ribosome binding site. Thus, the translation of 1 gene is regulated by at least 3 ncRNAs!

Although DsrA binds to the *rpoS* mRNA and activates translation, DsrA also contains a separate domain that is complementary to a region near the RBS of another gene, *hns*. In this case when DsrA binds to the *hns* mRNA it inhibits translation and decreases the stability of the mRNA.[19] The OxyS ncRNA also has a second regulatory activity. OxyS contains a domain that is complementary to a region that overlaps the RBS of *fhlA* mRNA.[23] Similar to the DsrA and *hns* mRNA system, this leads to an inhibition of *fhlA* translation and decreases mRNA stability.[23] This type of antisense regulation has also been postulated for many other RNAs[4] including two recently identified ncRNAs, RhyB[24] and GcvB.[25] Thus, depending on the target mRNA, ncRNAs can have both positive and negative regulatory effects.

A different type of ncRNA regulation is by sequestration of a regulatory protein. Two examples of this are the regulation of carbon storage metabolism in *E. coli* by *csrA* and B[26] and 6S RNA inhibition of RNA polymerase mediated promoter recognition and transcription.[27]

The genes necessary for glycogen storage are encoded by the *glgCAP* operon. The RNA binding protein CsrA binds to and destabilizes the *glgCAP* mRNA, leading to decreased accumulation of intracellular glycogen.[9,28] The CsrA protein also binds to the ncRNA CsrB.[9] CsrB contains 18 CsrA binding motifs and antagonizes CsrA action by binding and sequestration into a large ribonucleoprotein complex.[26] This leads to increased stability and translation of *glgCAP* mRNA and increased glycogen storage. Since CsrB ncRNA stability is unaffected by CsrA[29] it is not known if the formation of the ribonucleoprotein complex is reversible. Whether these ncRNA:mRNA or ncRNA:protein interactions are reversible is a key unexplored aspect of ncRNA mediated regulation.

A similar mechanism of action has been proposed for 6S RNA.[27] 6S RNA was originally identified as a small stable *E. coli* RNA in polyacrylamide gels,[30] but its function was unknown for many years. Recently, 6S RNA was shown to specifically bind to *E. coli* RNA polymerase but only when complexed with the major vegetative sigma factor, sigma 70. 6S RNA did not bind to RNA polymerase complexed with alternative sigma factors.[27] Since sigma factors are involved in promoter recognition, this sequestration of the sigma 70 RNA polymerase complex by 6S RNA resulted in decreased transcription of sigma 70 promoters relative to other types of promoters, and broad changes in global gene expression.[27] Most of these changes led to increased expression of stationary phase genes during exponential growth.

Noncoding RNAs in Gram-Positive Bacteria

Gram-positive bacteria also use ncRNAs to regulate gene expression. The most extensively studied Gram-positive ncRNA is the Agr RNAIII of the human pathogen *Staphylococcus aureus*. RNAIII is an unusual ncRNA because it also encodes a small delta lysin protein.[31] However, regulation by RNA III is not dependent on delta lysin production since mutants that no longer synthesize the protein still regulated gene expression.[31] In fact, alleles of RNAIII exist that regulate gene expression but do not encode a functional delta lysin.[32] RNAIII could be a case of an ncRNA evolving from a more common mRNA. The molecular mechanisms of RNAIII-mediated gene regulation are unclear. Recent evidence suggests that RNAIII works via sequestration by binding to the *S. aureus* global regulator SarA.[33] SarA and related *sar* genes are transcriptional repressors but, when SarA was bound to RNAIII it could no longer repress transcription.[33] Consequently increased expression of RNAIII leads to increased transcription of Sar regulated genes.[33] Thus, RNAIII shares regulatory similarities to both CsrB and 6S RNAs.

Another ncRNA, FasX, has been identified in another Gram-positive pathogenic bacteria, *Streptococcus pyogenes*. Although its mechanism of action is not known, FasX clearly functions as a noncoding RNA to alter expression of several extra cellular virulence proteins.[34]

Regulating the Regulators. How Are ncRNAs Regulated?

The regulation of expression of a number of ncRNAs occurs at the level of transcription. DsrA RNA expression is controlled by two independent mechanisms, temperature[35,36] and the protein LeuO.[37,38] The accumulation of DsrA was 25-fold higher in cells grown at 20°C compared to cells grown at 37°C.[35,36] This increased accumulation was a result of both increased transcription of *dsrA*[35] and increased stability of the DsrA RNA.[36] The expression of *rpoS* (a stationary phase sigma factor) during exponential growth at 20°C, as assayed by Western blots and Rpos:LacZ fusions, was comparable to stationary phase expression at 37°C.[35] In a *dsrA* mutant, *rpoS* expression was not regulated by temperature.[35] This DsrA-dependent accumulation of Rpos was sufficient to cause increased expression of Rpos-dependent genes, which are normally expressed only at stationary phase at 37°C. The regulatory signals that control the temperature regulation are intrinsic to the DsrA promoter and reside in the –10 box.[36] This has led to the hypothesis that DsrA acts as a cellular thermometer to control the expression of *rpoS*.[36]

Temperature independent regulation of *dsrA* occurs via transcriptional repression by LeuO.[38] Although expression of *leuO* appears to be regulated by entry into stationary phase and the response of cells to amino acid starvation its role in DsrA-mediated regulation is not known.[38,39] Mutational analysis of the *dsrA* promoter suggests that it contains two other transcription activating regions that respond to unknown regulators/signals.[36] Although the functional significance is not known, DsrA might also be regulated by RNA processing or stability since two different forms of DsrA have been observed.[20,36]

The transcription of two other ncRNA regulators of *rpoS*, OxyS and RprA, are regulated by oxidative stress[8] and cell surface stress[40] respectively. These ncRNAs allow the precise regulation of *rpoS* translation by a multitude of environmental conditions. This is similar to the MicF-mediated regulation of *ompF* translation where the *micF* promoter was regulated by at least 4 different transcription factors.[41-43]

Although the ncRNAs studied to date are regulated at the level of transcription or RNA stability the possibility exists that they could respond directly to environmental signals or even bind to small molecules. It has been shown in vivo that, mRNA structures and translational efficiency can be directly affected by growth temperature.[44,45] Even more interesting is the possibility that small molecules can bind to mRNAs and alter their translational efficiency. So far this has only been observed in genes involved in vitamin biosynthesis,[46-49] but the probability that a small molecule could bind to an ncRNA and alter its regulatory activity is high.

The RNA-Binding Protein Hfq and ncRNAs

Hfq (also known as host factor I, HF-I) was originally identified as an *E. coli* RNA-binding protein that was essential for the in vitro replication of the RNA bacteriophage, Q β.[50,51] More than 20 years later, the in vivo requirement for Hfq in Q β replication was demonstrated when an *E. coli* strain with an inactivated *hfq* gene showed greatly reduced Q β replication.[52]

Hfq is a small (11.2 kD), abundant heat-stable protein that functions as a hexamer in vivo and in vitro.[53,54] The expression of *hfq* peaks in exponential phase at an estimated 55,000 molecules per cell, so it is not surprising that Hfq is involved in a variety of cellular processes.[55,56] Analysis of the phenotypes of an *hfq* null mutation suggested a possible role in stationary phase gene regulation. But, the pleiotropic phenotypes of an *hfq* mutant (including osmosensitivity, ultraviolet light sensitivity, cell growth defects and increased cell size) make it difficult to discern the role of Hfq in normal cells.[57,58]

What Is the Function of Hfq?

It is now clear that Hfq has a broad impact on the normal physiology of the cell through its interactions with various cellular RNAs.[22,52,59-69] These regulatory effects have also been seen in other Gram- bacteria including *Brucella*,[65] and *Azorhizobium*.[68] In *E. coli*, Hfq increases the polyadenylation of mRNAs[63] and interferes with ribosome binding.[62] Hfq binds to a short poly A tract on *ompA* mRNA and facilitates its lengthening by polyadenylate polymerase I (PAP I).[63] The increased length of the poly A tail leads to decreased stability of the *ompA* mRNA. Hfq also binds to the 5′-end of the *ompA* mRNA and occludes the ribosome binding site, decreasing translational efficiency.[62,63] This explains why an *hfq* mutant exhibited increased translation and amount of *OmpA* mRNA.[66] Hfq also decreases the amount and stability of its own mRNA, *miaA* (tRNA modification) and *mutS* (DNA repair).[69]

Besides binding to and altering the translation of mRNAs, Hfq was shown to alter intramolecular interactions in the Q β RNA phage. In this case, Hfq action resembles that of an RNA chaperone.[70] Hfq is thought to bind the 3' end of the (+) strand and melt-out a secondary structure, allowing Q β replicase to synthesize the (-) strand.[70] Secondary structure predictions of Q β adaptive mutants, which do not require Hfq for replication, demonstrate that the 3' end has lost long-range intramolecular base-pairing and may be more accessible to the Q β replicase.[70] Another mutant of Q β that extends the 3' end of the (+) strand by two residues also leads to Hfq independence, supporting Hfq increasing access of the 3' end of Q β (+) strand.[71]

While Hfq seems to be acting as an RNA chaperone in Q β replication, other functions exist. Although Hfq is necessary for the DsrA-mediated translational regulation of *hns* and *rpoS*, the secondary structure of DsrA is not altered.[7,19] In fact, the requirement of Hfq can be overcome by high-level expression of DsrA. Since DsrA seems to directly bind to *rpoS* and *hns* mRNAs, an additional role of Hfq might be to facilitate intermolecular interactions of RNAs.

The ability of Hfq to enhance intermolecular RNA interactions was shown in vitro and in vivo between RNAs that shared complementarity with Spot 42 and OxyS ncRNAs.[53,54] This was contrary to Hfq's destabilizing effect on numerous transcripts.[63,69] Hfq was shown to increase the stability of Spot 42 ncRNA, a translational repressor of galactose operon genes.[54] Spot 42 was shown to interact with Hfq in vitro via gel shift analysis and in vivo by coimmunoprecipitation with Hfq antisera. The Hfq binding site, identified by hydroxyl radical footprinting of the Hfq-Spot 42 complex, showed protection of three U-rich regions of Spot 42. Gel shift analysis and hydroxyl radical footprinting demonstrated enhanced in vitro binding of Spot 42 ncRNA to its target, *galK* mRNA (*galK* translational initiation region), in the presence of Hfq.

Hfq also facilitates the binding of OxyS ncRNA, a global regulator inducible under oxidative stress, to its target mRNAs.[53] Hfq interacts with OxyS in vitro and in vivo[22] and does not destabilize OxyS RNA.[53] The Hfq binding site on OxyS was determined by minimal binding analysis of the OxyS-Hfq complex by limited alkaline hydrolysis. Hfq was shown to interact with a U-rich linker on OxyS. Unlike Hfq effects on Spot 42 and DsrA ncRNAs, Hfq binding was shown to slightly open the OxyS secondary structure. Single-strand specific nuclease cleavage of OxyS showed enhanced cleavage in two stems in the presence of Hfq, indicating diminished secondary structure. Similar to the effect on Spot 42, Hfq facilitates in vitro binding of OxyS to its targets, *fhlA* mRNA and *rpoS* mRNA as demonstrated by gel shift assays.[53] It is unclear how Hfq mediates these RNA:RNA interactions but see Figure 3 for one possibility.

What Structural Feature or Sequence of RNA Does Hfq Recognize?

Since the discovery of Hfq binding to the RNA bacteriophage Q β, the RNA recognition sequence for Hfq has been investigated.[72,73] Early studies demonstrated that Hfq binds to polyA and polyU RNAs with a higher affinity than polyG or polyC RNA.[72,73] Hfq also seems to bind nonspecifically to DNA, but the importance of this binding is not known.[61,74] In the case of polyA RNA, length makes a difference in the binding affinity. Ten-fold tighter binding was noted with a polyA 18 mer compared to a polyA 15 mer.[73] It is not known whether this same length dependence is true for polyU and polyG RNAs. Recent equilibrium dissociation binding studies measured the affinity of Hfq for a series of short oligos: AUUUUUG (RNA-U), AAAAAAG (RNA-A), ACCCCCG(RNA-C), or AGGGGGG(RNA-G).[75] Again supporting previous Hfq-Q β studies, there was no measurable binding of Hfq to RNA-C and RNA-G bound weakly; whereas, Hfq was shown to bind RNA-A and RNA-U with high affinity. Recently, the interactions of Hfq with Spot 42,[54] OxyS[53] or DsrA ncRNAs[76] have been studied independently. Nuclease probing of each RNA:Hfq complex has given a similar answer; Hfq

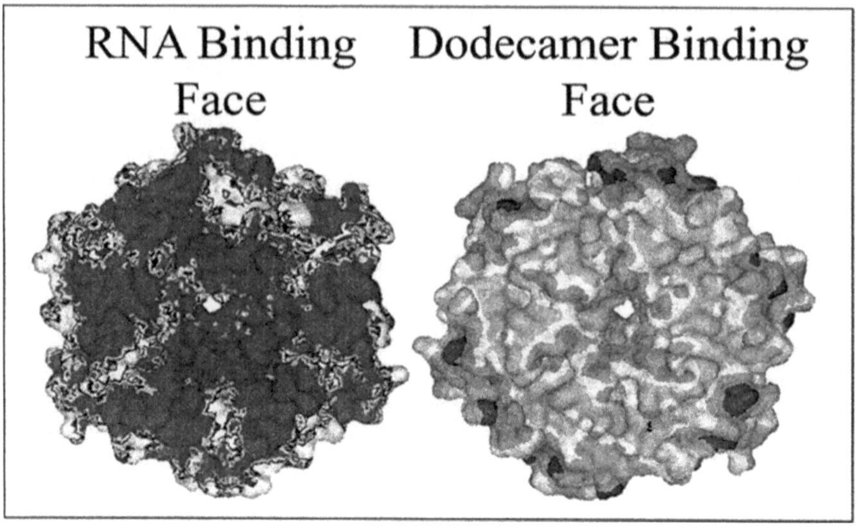

Figure 2. The Two Faces of Hfq. The structure and coordinates are from ref.[75] The RNA binding face of Hfq is highly charged (dark gray). The opposite face of Hfq is much more hydrophobic (white and light gray). An interaction between the hydrophobic faces of two Hfq hexamers allows formation of the Hfq dodecamer.

binds to single-stranded A/U rich regions that are flanked by stem-loop structures. The DsrA RNA contains three structural and functional domains.[20] Domain I (*rpoS* regulation stem-loop), Domain II (single stranded poly A/U rich region, Hfq binding region and *hns* regulation region) and Domain III (transcription terminator stem-loop). Hfq protected the single-stranded A/U rich sequence present in Domain II of DsrA from nuclease cleavage that was expected to be the binding site of Hfq. However, upon mutation of the presumed binding site, there was no measurable change in binding affinity, indicating that a simple A/U rich sequence was not essential for Hfq binding.[76] This was further supported by a minimal binding assay that demonstrated that nearly all of Domain I and Domain II as well as part of Domain III were required for tight binding of DsrA to Hfq. It is our belief that Hfq initially binds to A/U rich regions, however Hfq requires interactions with secondary or tertiary RNA elements of RNA for strong binding.

How Is Hfq Interacting with RNAs?

The conserved Hfq homologs, found in the genomes of most the sequenced Gram-negative bacteria and some of the Gram-positive bacteria[77] are a useful tool in understanding the interactions of Hfq with RNAs. Recently, it was hypothesized that Hfq is related to a family of eucaryotic and archeabacterial RNA binding proteins called Sm and Sm-like proteins.[53,75] This was supported by crystallographic studies showing the *S. aureus* Hfq monomer contained the characteristic Sm fold.[75] The Sm fold is comprised of an α-helix at the N-terminal end, followed by a variable region and a bent five-stranded β-sheet.[75] Members of this class of proteins (including Hfq) form ring-shaped multimers, bind poly U RNA and are necessary for intermolecular RNA interactions.[53,54] Cocrystallization of *S. aureus* Hfq with a small single-stranded RNA (RNA-U described above), demonstrated RNA binding around the central pore in a region of the Hfq hexamer that is electropositive, which was specific to one face of the hexamer (Fig. 2).[75] This region corresponded the Hfq Sm1 and Sm2 motifs as well as at the highly

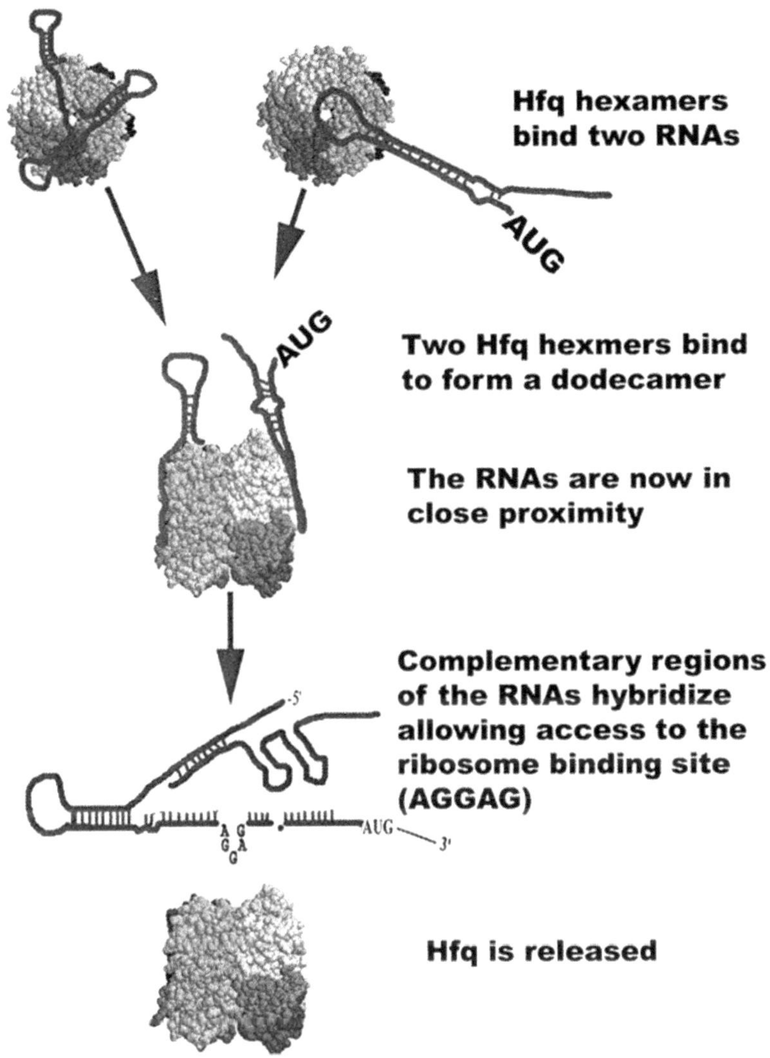

Figure 3. A model of how Hfq enhances RNA:RNA interactions. In the absence of an ncRNA the 5′-leader region of the *rpoS* mRNA (denoted by the AUG) forms a secondary structure that blocks translation.[18] In this model one Hfq hexamer binds to an ncRNA (DsrA) while the other Hfq hexamer binds to the target mRNA (*rpoS*). The two Hfq hexamers form a dodecamer, via the hydrophobic face, bringing the two RNAs into close proximity and promoting complementary base pairing. The intermolecular base pairing alters the secondary structures and Hfq binding of both RNAs ncRNA:mRNA complex releasing the from Hfq. When in the RNA complex, the ribosome binding site of the *rpoS* mRNA (AGGAG) is now freed to interact with the ribosome, increasing the translation of the gene.[19,20] It is also possible that Hfq functions as an RNA chaperone to assist in the unfolding of one or both RNAs. A similar mechanism could also explain negative regulation by ncRNAs. The model of the DsrA:Hfq complex is from Brescia et al.[76] The model of *rpoS* mRNA:Hfq complex is speculative.

conserved Gln8 residue present in the α1-helix.[75] Sm proteins were also recently reported to contain a second RNA-binding site beyond the poly U binding pore[78] and it has been proposed that Hfq contains two RNA-binding sites.[76] In cases where Hfq enhances intermolecular RNA interactions (OxyS /*fhlA*. mRNA, *rpoS* mRNA, Spot 42 /*galK* mRNA or DsrA/*rpoS* mRNA), two Hfq hexamers with single-stranded RNAs that are bound to the cationic face may come together, in forming a dodecamer, thus bringing the RNAs in close proximity to interact (Fig. 3). Another possibility is that Hfq can overcome inhibitory cis-acting secondary structures within the RNAs (RNA chaperone activity) by binding to the RNAs and holding them in an unfolded state; however, it is unlikely that this is the action of Hfq on DsrA ncRNA, as Hfq does not affect the secondary structure of DsrA.[76]

Conclusion

The conservation of Hfq among the three major branches of life along with its abundance in *E. coli* and its ability to enhance RNA interactions suggests that Hfq and Hfq-like proteins play a critical role in the physiology of cells. This is also supported by the recent discovery that Hfq binds to a sizable number of *E. coli* ncRNAs in vivo that are predicted to be involved in regulation.[11] Hfq is clearly multifunctional since it can destabilize some RNAs, yet also increase the translational efficiency and intra- and intermolecular interactions of others. Although Hfq is a conserved protein little is known about how Hfq functions in vivo. Understanding the functions of Hfq and its interaction with the legions of ncRNAs found in bacteria will give us a more complete understanding of these understudied and hidden regulatory networks.

Acknowledgements

We wish to thank National Institutes of Health (RO1-GM56448) for support.

References

1. Tomizawa J. Control of ColE1 Plasmid replication: The process of binding of RNA I to the primer transcript. Cell 1984; 38:861-870.
2. Tomizawa J. Control of Col E1 plasmid replication: Initial interaction of RNAI and the primer transcript is reversible. Cell 1988; 40:527-535.
3. Tomizawa J-i, Itoh T. The importance of RNA secondary structure in colE1 primer formation. Cell 1982; 31:575-583.
4. Eguchi Y, Itoh T, Tomizawa J. Antisense RNA. Annu Rev Biochem 1991; 60:631-652.
5. Eguchi Y, Tomizawa J. Complexes formed by complementary RNA stem-loops. Their formations, structures and interaction with ColE1 Rom protein. J Mol Biol 1991; 220:831-842.
6. Aiba H, Matsuyama S, Mizuno T et al. Function of *micF* as an antisense RNA in osmoregulatory expression of the *ompF* gene in *Escherichia coli*. J Bacteriol 1987; 169:3007-3012.
7. Sledjeski D, Gottesman S. A small RNA acts as an antisilencer of the H-NS-silenced *rcsA* gene of *Escherichia coli*. Proc Nat Acad Sci USA 1995; 92:2003-2007.
8. Altuvia S, Weinstein-Fischer D, Zhang A et al. A Small, stable RNA induced by oxidative stress: Role as a pleiotropic regulator and antimutator. Cell 1997; 89:1-20.
9. Liu MY, Gui G, Wei B et al. The RNA molecule CsrB binds to the global regulatory protein CsrA and antagonizes its activity in *Escherichia coli*. J Biol Chem 1997; 272:17502-17510.
10. Zheng M, Wang X, Templeton LJ et al. DNA microarray-mediated transcriptional profiling of the *Escherichia coli* response to hydrogen peroxide. J Bacteriol 2001; 183:4562-4570.
11. Wassarman KM, Repoila F, Rosenow C et al. Identification of novel small RNAs using comparative genomics and micro arrays. Genes Dev 2001; 15:1637-1651.
12. Argaman L, Hershberg R, Vogel J et al. Novel small RNA-encoding genes in the intergenic regions of *Escherichia coli*. Curr Biol 2001; 11:941-950.
13. Storz G. An expanding universe of noncoding RNAs. Science 2002; 296:1260-1263.
14. Klein RJ, Misulovin Z, Eddy SR. Noncoding RNA genes identified in AT-rich hyperthermophiles. Proc Natl Acad Sci USA 2002; 99:7542-7547.

15. Eddy SR. Computational genomics of noncoding RNA genes. Cell 2002; 109:137-140.
16. Wassarman KM. Small RNAs in bacteria: diverse regulators of gene expression in response to environmental changes. Cell 2002; 109:141-144.
17. Gottesman S. Stealth regulation:Biological circuits with small RNA switches. Genes Dev 2002; 16:in press.
18. Brown L, Elliott T. Mutations that increase expression of the *rpoS* gene and decrease its dependence on *hfq* function in *Salmonella typhimurium*. J. Bacteriol 1997; 179:656-662.
19. Lease RA, Cusick ME, Belfort M. Riboregulation in *Escherichia coli*: DsrA RNA acts by RNA:RNA interactions at multiple loci. Proc Natl Acad Sci USA 1998; 95:12456-12461.
20. Majdalani N, Cunning C, Sledjeski D et al. DsrA RNA regulates translation of RpoS message by an anti-antisense mechanism, independent of its action as an antisilencer of transcription. Proc Natl Acad Sci USA 1998; 95:12462-12467.
21. Majdalani N, Chen S, Murrow J et al. Regulation of RpoS by a novel small RNA: the characterization of RprA. Mol Microbiol 2001; 39:1382-1394.
22. Zhang A, Altuvia S, Tiwari A et al. The OxyS regulatory RNA represses *rpoS* translation and binds the Hfq (HF-I) protein. EMBO J 1998; 17:6061-6068.
23. Altuvia S, Zhang A, Argaman L et al. The *Escherichia coli* OxyS regulatory RNA represses *fhlA*. translation by blocking ribosome binding. EMBO J 1998; 17:6069-6075.
24. Masse E, Gottesman S. A small RNA regulates the expression of genes involved in iron metabolism in *Escherichia coli*. Proc Natl Acad Sci USA 2002; 99:4620-4625.
25. Urbanowski ML, Stauffer LT, Stauffer GV. The *gcvB* gene encodes a small untranslated RNA involved in expression of the dipeptide and oligopeptide transport systems in *Escherichia coli*. Mol Microbiol 2000; 37:856-868.
26. Romeo T. Global regulation by the small RNA-binding protein CsrA and the non coding RNA molecule CsrB. Mol Microbiol 1998; 29:1321-1330.
27. Wassarman KM, Storz G. 6S RNA regulates *E. coli* RNA polymerase activity. Cell 2000; 101:613-623.
28. Liu MY, Romeo T. The global regulator CsrA of *Escherichia coli* is a specific mRNA- binding protein. J Bacteriol 1997; 179:4639-4642.
29. Gudapaty S, Suzuki K, Wang X et al. Regulatory interactions of Csr components: The RNA binding protein CsrA activates *csrB* transcription in *Escherichia coli*. J Bacteriol 2001; 183:6017-6027.
30. Hindley J. Fractionation of 32P-labelled ribonucleic acids on polyacrylamide gels and their characterization by fingerprinting. J Mol Biol 1967; 30:125-136.
31. Janzon L, Arvidson S. The role of the delta-lysine gene (*hld*) in the regulation of virulence genes by the accessory gene regulator (*agr*) in *Staphylococcus aureus*. EMBO J 1990; 9:1391-1399.
32. Li S, Arvidson S, Mollby R. Variation in the *agr*-dependent expression of alpha-toxin and protein A among clinical isolates of *Staphylococcus aureus* from patients with septicemia. FEMS Microbiol Lett 1997; 152:155-161.
33. Arvidson S, Tegmark K. Regulation of virulence determinants in *Staphylococcus aureus*. Int J Med Microbiol 2001; 291:159-170.
34. Kreikemeyer B, Boyle MD, Buttaro BA et al. Group A streptococcal growth phase-associated virulence factor regulation by a novel operon (Fas) with homologies to two-component-type regulators requires a small RNA molecule. Mol Microbiol 2001; 39:392-406.
35. Sledjeski DD, Gupta A, Gottesman S. The small RNA, DsrA, is essential for the low temperature expression of RpoS during exponential growth in *Escherichia coli*. EMBO J 1996; 15:3993-4000.
36. Repoila F, Gottesman S. Signal transduction cascade for regulation of RpoS: temperature regulation of DsrA. J Bacteriol 2001; 183:4012-4023.
37. Majumder A, Fang M, Tsai KJ et al. LeuO expression in response to starvation for branched-chain amino acids. J Biol Chem 2001; 276:19046-19051.
38. Klauck E, Böhringer J, Hengge-Aronis R. The LysR-like regulator LeuO in *Escherichia coli* is involved in the translational regulation of *rpoS* by affecting the expression of the small regulatory DsrA-RNA. Mol Microbiol 1997; 25:559-569.
39. Fang M, Majumder A, Tsai KJ et al. ppGpp-dependent *leuO* expression in bacteria under stress. Biochem Biophys Res Commun 2000; 276:64-70.

40. Majdalani N, Hernandez D, Gottesman S. Regulation and mode of action of the second small RNA activator of RpoS translation, RprA. Mol Microbiol 2002; 46:813-826.
41. Delihas N, Rokita SE, Zheng P. Natural antisense RNA/target RNA interactions: possible models for antisense oligonucleotide drug design. Nat Biotechnol 1997; 15:751-753.
42. Pratt LA, Hsing W, Gibson KE et al. From acids to *osmZ*: multiple factors influence synthesis of the OmpF and OmpC porins in *Escherichia coli*. Mol Microbiol 1996; 20:911-917.
43. Delihas N, Forst S. MicF: an antisense RNA gene involved in response of *Escherichia coli* to global stress factors. J Mol Biol 2001; 313:1-12.
44. Morita MT, Tanaka Y, Kodama TS et al. Translational induction of heat shock transcription factor sigma32: evidence for a built-in RNA thermo sensor. Genes Dev 1999; 13:655-665.
45. Johansson J, Mandin P, Renzoni A et al. An RNA Thermo sensor controls expression of virulence genes in *Listeria monocytogenes*. Cell 2002; 110:551.
46. Stormo GD, Ji Y. Do mRNAs act as direct sensors of small molecules to control their expression? Proc Natl Acad Sci USA 2001; 98:9465-9467.
47. Miranda-Rios J, Navarro M, Soberon M. A conserved RNA structure (*thi* box) is involved in regulation of thiamin biosynthetic gene expression in bacteria. Proc Natl Acad Sci USA 2001; 98:9736-9741.
48. Nou X, Kadner RJ. Adenosylcobalamin inhibits ribosome binding to *btuB* RNA. Proc Natl Acad Sci USA 2000; 97:7190-7195.
49. Mironov AS, Gusarov I, Rafikov R et al. Sensing small molecules by nascent RNA: A mechanism to control transcription in bacteria. Cell 2002.
50. Franze de Fernandez MT, Eoyang L, August JT. Factor fraction required for the synthesis of bacteriophage Qbeta-RNA. Nature 1968; 219:588-590.
51. Franze de Fernandez MT, Hayward WS, August JT. Bacterial proteins required for replication of phage Q β ribonucleic acid. Purification and properties of host factor I, a ribonucleic acid- binding protein. J Biol Chem 1972; 247:824-831.
52. Su Q, Schuppli D, Tsui Hc T et al. Strongly reduced phage Qbeta replication, but normal phage MS2 replication in an *Escherichia coli* K12 mutant with inactivated Qbeta host factor (*hfq*) gene. Virology 1997; 227:211-214.
53. Zhang A, Wassarman KM, Ortega J et al. The Sm-like Hfq protein increases OxyS RNA interaction with target mRNAs. Mol Cell 2002; 9:11-22.
54. Møller T, Franch T, Højrup P et al. Hfq: A bacterial Sm-like protein that mediates RNA-RNA interaction. Mol Cell 2002; 9:23-30.
55. Kajitani M, Kato A, Wada A et al. Regulation of the *Escherichia coli hfq* gene encoding the host factor for phage Q beta. J Bacteriol 1994; 176:531-534.
56. Azam TA, Ishihama A. Twelve species of the nucleoid-associated protein from *Escherichia coli*. Sequence recognition specificity and DNA binding affinity. J Biol Chem 1999; 274:33105-33113.
57. Muffler A, Traulsen DD, Fischer D et al. The RNA-binding protein HF-I plays a global regulatory role which is largely, but not exclusively, due to its role in expression of the sigmaS subunit of RNA polymerase in *Escherichia coli*. J Bacteriol 1997; 179:297-300.
58. Tsui HC, Leung HC, Winkler ME. Characterization of broadly pleiotropic phenotypes caused by an *hfq* insertion mutation in *Escherichia coli* K-12. Mol Microbiol 1994; 13:35-49.
59. Sledjeski DD, Whitman C, Zhang A. Hfq is necessary for regulation by the untranslated RNA DsrA. J Bacteriol 2001; 183:1997-2005.
60. Brown L, Elliott T. Efficient translation of the RpoS sigma factor in *Salmonella typhimurium* requires host factor I, an RNA-binding protein encoded by the *hfq* gene. J Bacteriol 1996; 178:3763-3770.
61. Shi X, Bennett GN. Plasmids bearing hfq and the *hns*-like gene *stpA* complement *hns* mutants in modulating arginine decarboxylase gene expression in *Escherichia coli*. J Bacteriol 1994; 176:6769-6775.
62. Vytvytska O, Moll I, Kaberdin VR et al. Hfq (HF1) stimulates *ompA* mRNA decay by interfering with ribosome binding. Genes Dev 2000; 14:1109-1118.
63. Hajnsdorf E, Regnier P. Host factor Hfq of *Escherichia coli* stimulates elongation of poly(A) tails by poly(A) polymerase I. Proc Natl Acad Sci USA 2000; 97:1501-1515.

64. Takada A, Wachi M, Nagai K. Negative regulatory role of the *Escherichia coli hfq* gene in cell division. Biochem Biophys Res Commun 1999; 266:579-583.
65. Robertson GT, Roop Jr RM. The *Brucella* abortus host factor I (HF-I) protein contributes to stress resistance during stationary phase and is a major determinant of virulence in mice. Mol Microbiol 1999; 34:690-700.
66. Vytvytska O, Jakobsen JS, Balcunaite G et al. Host factor I, Hfq, binds to *Escherichia coli ompA* mRNA in a growth rate-dependent fashion and regulates its stability. Proc Natl Acad Sci USA 1998; 95:14118-14123.
67. Wachi M, Takada A, Nagai K. Overproduction of the outer-membrane proteins FepA and FhuE responsible for iron transport in *Escherichia coli hfq::cat* mutant. Biochem Biophys Res Commun 1999; 264:525-529.
68. Kaminski PA, Elmerich C. The control of *Azorhizobium caulinodans nifA* expression by oxygen, ammonia and by the HF-I-like protein, NrfA. Mol Microbiol 1998; 28:603-613.
69. Tsui HC, Feng G, Winkler ME. Negative regulation of *mutS* and *mutH* repair gene expression by the Hfq and RpoS global regulators of *Escherichia coli* K-12. J Bacteriol 1997; 179:7476-7487.
70. Schuppli D, Miranda G, Tsui HC et al. Altered 3'-terminal RNA structure in phage Qbeta adapted to host factor-less *Escherichia coli*. Proc Natl Acad Sci USA 1997; 94:10239-10242.
71. Schuppli D, Georgijevic J, Weber H. Synergism of mutations in bacteriophage Qbeta RNA affecting host factor dependence of Qbeta replicase. J Mol Biol 2000; 295:149-154.
72. Carmichael GG, Weber K, Niveleau A et al. The host factor required for RNA phage Qbeta RNA replication in vitro. Intracellular location, quantitation, and purification by polyadenylate- cellulose chromatography. J Biol Chem 1975; 250:3607-3612.
73. de Haseth PL, Uhlenbeck OC. Interaction of *Escherichia coli* host factor protein with Q beta ribonucleic acid. Biochem 1980; 19:6146-6151.
74. Takada A, Wachi M, Kaidow A et al. DNA binding properties of the *hfq* gene product of *Escherichia coli*. Biochem Biophys Res Commun 1997; 236:576-579.
75. Schumacher MA, Pearson RF, Moller T et al. Structures of the pleiotropic translational regulator Hfq and an Hfq- RNA complex: a bacterial Sm-like protein. EMBO J 2002; 21:3546-3556.
76. Brescia CC, Mikulecky PJ, Feig AL et al. Identification of the Hfq binding site on DsrA RNA: Hfq binds without altering DsrA secondary structure. RNA 2002; in press.
77. Sun X, Zhulin I, Wartell RM. Predicted structure and phyletic distribution of the RNA binding protein Hfq. Nucl Acids Res 2002; 30:In Press.
78. Thore S, Mayer C, Sauter C et al. Crystal structures of the pyrococcus abyssi Sm core and its complex with RNA: common features of RNA-binding in Archaea and Eukarya. J Biol Chem 2002.

Index

F

G

H

I

J

K

L

M

N

O

P

Q

R

S

T

U

V

W

X

Z

MIX
Papier aus verantwortungsvollen Quellen
Paper from responsible sources
FSC® C105338

If you have any concerns about our products,
you can contact us on
ProductSafety@springernature.com

In case Publisher is established outside the EU,
the EU authorized representative is:
Springer Nature Customer Service Center GmbH
Europaplatz 3, 69115 Heidelberg, Germany

Printed by Libri Plureos GmbH
in Hamburg, Germany